Springer Series in Reliability Engineering

For further volumes:
http://www.springer.com/series/6917

Toshio Nakagawa

Stochastic Processes

with Applications to Reliability Theory

Springer

Dr. Toshio Nakagawa
Department of Business Administration
Aichi Institute of Technology
1247 Yachigusa, Yakusa-cho
470-0392 Toyota
Japan
e-mail: toshi-nakagawa@aitech.ac.jp

ISSN 1614-7839

ISBN 978-0-85729-273-5 e-ISBN 978-0-85729-274-2

DOI 10.1007/978-0-85729-274-2

Springer London Dordrecht Heidelberg New York

British Library Cataloguing in Publication Data
A catalogue record for this book is available from the British Library

Cover design: eStudio Calamar, Berlin/Figueres

Printed on acid-free paper

Springer is part of Springer Science+Business Media (www.springer.com)

Preface

I have learned first reliability theory from the book *Mathematical Theory of Reliability* [1] written by Barlow in 1965 that is the most famous and excellent one theoretically up to now. I was a marvelously lucky reader, unfortunately, I could not understand throughly this book at that time because my poor mathematical tools and abysmal ignorance about stochastic processes. I have now already published three monographs *Maintenance Theory of Reliability* [2], *Shock and Damage Models in Reliability Theory* [3], and *Advanced Reliability Models and Maintenance Policies* [4] in which I have surveyed mainly maintenance policies in reliability theory on the research results of the author and my colleagues. Most of the three books have been written based on basic theory of stochastic processes and their mathematical tools. A number of graduate students, researchers, and engineers demand of us some book written in an easy style on stochastic processes to be able to understand readily reliability theory.

Recently, plants, satellites, computer and information systems have become more large-scale and complex, and most products are distributed all over the world. If they or some of them would fail and have trouble, it might incur many serious and catastrophic situations and heavy damage. More and better maintenance is required constantly from the economical and environmental points as public infrastructures and operating plants become old in advanced nations. Reliability and maintenance theory is more useful for protecting such severe matters and environmental considerations, and moreover, for making good and safe living.

Stochastic processes can be described by probabilistic phenomena in some space at each point of time t. The knowledge of stochastic processes and mathematical tools is indispensable for engineers, managers and researchers in reliability and maintenance. This book introduces basic stochastic processes from the viewpoint of elementary mathematics, mainly based on the books *Stochastic Processes* [5], *Applied Stochastic System Modeling* [6], and *Introduction to Stochastic Processes* [7]. Many reliability examples are cited in this book to explain concretely stochastic processes. They are quoted mainly from the books [1–4]. Furthermore, several interesting reliability examples appeared in two famous and classical books [8, 9]. However, they have become old and their important contributions are going to be forgotten. This book is aimed at quoting possible examples from the two books.

This book is composed of eight chapters: Chapter 1 is devoted to the introduction to stochastic processes and reliability theory. Chapters 2 –5 are devoted to standard and fundamental stochastic processes such as Poisson processes, renewal processes, Markov chains, Markov processes, and Markov renewal processes. These four chapters are written mainly based on the books [5–7] and are introduced from the viewpoint of elementary mathematics without using hard proof. Chapter 6 is devoted to cumulative processes that are related greatly to shock and damage models in one field of reliability. Chapter 7 introduces simply Brownian motion and Lévy processes, that might be omitted at the first reading. To understand and explain stochastic processes easily and actually, a lot of examples are quoted from reliability models through each chapter. As final examples of reliability models, Chapter 8 takes up redundant systems and shows systematically how to use well the tools of stochastic processes to analyze them. The reader could learn both stochastic process and reliability theory from this book at the same time.

I wish to thank most kindly Professor E. Çinlar, Professor S. M. Ross, and Professor S. Osaki for referring to the good books written by them and the other books in references. I wish to express my special thanks to Professor Fumio Ohi and Professor Mingchih Chen for their careful reviews of this book, and to Dr. Kodo Ito, Dr. Satoshi Mizutani, Dr. Sayori Maeji, and Mr. Xufeng Zhao for their support in writing and typing this book. Finally, I would like to express my sincere appreciation to Professor Hoang Pham, Rutgers University, and Editor Anthony Doyle, Springer-Verlag, London, for providing the opportunity for me to write this book.

Toyota, April 2010 Toshio Nakagawa

References

1. Barlow RE, Proschan F (1965) Mathematical theory of reliability. Wiley, NewYork
2. Nakagawa T (2005) Maintenance theory of reliability. Springer, London
3. Nakagawa T (2007) Shock and damage models in reliability theory. Springer, London
4. Nakagawa T (2008) Advanced reliability models and maintenance policies. Springer, London
5. Ross SM (1983) Stochastic processes. Wiley, New York
6. Osaki S (1992) Applied stochastic system modeling. Springer, Berlin
7. Çinlar E (1975) Introduction to stochastic processes. Prentice-Hall, Englewood Cliffs, NJ
8. Takács L (1960) Stochastic processes. Wiley, New York
9. Cox DR (1962) Renewal theory. Methuen, London

Contents

Chapter 1
Introduction

All things from near oneself to space form stochastic processes whether large or small because their real lives are limited. Of course, each lifetime of human being forms a stochastic process [1], however, such process could not be analyzed exactly because given conditions and constrains would be too many and much complex at the present, as shown in a toy of life games. Most reliability models are much simpler and more clear than human being, and much more, they have to be analyzed practically and be of great use in actual fields.

The importance of reliability will be greatly enhanced by environmental considerations, and moreover, for the protection of natural resources and the earth. Reliability techniques have to be developed and expanded as objective models become more complex and large-scale. They also will be applied not only to daily life, but also to a variety of other fields because consumers, workers, and managers must make, buy, sell, and use, and handle articles and products with a sense of safety and security. A wide knowledge of probabilities, statistics, and stochastic processes are needed for learning reliability theory mathematically.

Stochastic processes have been applied in many fields such as engineering, physics, biology, electronics, management science, economics, and psychology, and in operations research, they have spread widely over queueing, finance, and insurance. Especially, in reliability theory, stochastic processes are the most powerful mathematical tools for analyzing reliability models.

We aim to write this book about stochastic processes from viewpoints of reliability theory and applications: This book introduces basic stochastic processes such as Poisson processes, renewal processes, Markov chains and processes, and furthermore, introduces cumulative processes, Brownian motion, and Lévy processes. A lot of examples cited from reliability models demonstrate how to apply stochastic processes correctly.

T. Nakagawa, *Stochastic Processes*, Springer Series in Reliability Engineering,
DOI: 10.1007/978-0-85729-274-2_1,

1.1 Reliability Models

1.1.1 Redundancy

High system reliability can be achieved by redundancy and maintenance. A classical problem is to determine how reliability can be improved by using redundant units. Various expressions of reliability quantities of many redundant systems were summarized [2, 3]. A two-unit system is the most fundamental redundant model, and its reliability measures and optimum maintenance policies were surveyed deeply [4]. For systems with redundant or spare units, we have to determine how many number of units should be provided initially.

It is well-known as *fault tolerance* [5] in which systems continue to function correctly in the presence of hardware failures and software errors. Redundancy techniques of systems for improving reliability and achieving fault tolerance are classified commonly in the following forms [5, 6]: hardware redundancy, software redundancy, information redundancy, and time redundancy.

1.1.2 Maintenance

Maintenance of units after failure may be costly, and sometimes requires a long time. Thus, the most important problem is to determine when and how to maintain preventively units before failure. However, it is not wise to maintain units with unnecessary frequency. From such viewpoints, the commonly considered maintenance policies are preventive replacement for units without repair, and preventive maintenance for units with repair on a specific schedule. Classifying into three large groups, planned maintenance of units are done at a certain age, a periodic time, or a specified number of occurrences. Consequently, the object of optimization problems is to determine the frequency and timing of several kinds of maintenance and replacement policies in accordance with maintenance costs and effects, and system structures and circumstances.

1.2 Stochastic Processes

All reliability models consisting of random time factors form stochastic processes. We briefly explain stochastic processes as examples of reliability systems with maintenance: Consider the simplest one-unit system with repair or replacement whose time is negligible, i.e., the unit is operating and is repaired or replaced at failure, where the time required for repair and replacement is negligible. When the repair or replacement is completed, the unit becomes as good as new and begins to operate.

If failures occur exponentially, i.e., the unit fails constantly during the time interval $(t, t + \mathrm{d}t]$ irrespective of time t, the system forms a Poisson process. Roughly speaking, failures occur randomly in $[t, t + \mathrm{d}t]$ with probability $\lambda \mathrm{d}t$ for constant $\lambda > 0$, and interarrival times between failures have an exponential distribution $(1 - \mathrm{e}^{-\lambda t})$. Then, it is said that failures occur in a Poisson process with rate λ. A Poisson process is the simplest and the most fundamental stochastic process. In Chap. 2, we define a Poisson process theoretically and explain various properties of an exponential distribution with memoryless property. As examples of a Poisson process, we consider a parallel system, a spare part problem, an inspection policy, a backward time problem, and so on. Furthermore, we introduce a nonhomogeneous Poisson process in which failures occur in $[t, t + \mathrm{d}t]$ with probability $\lambda(t)\mathrm{d}t$, and a compound Poisson process in which an amount of some quantities such as customers, damage, costs and profits occurs in a Poisson process. To apply two processes to reliability models, we take up periodic replacements, and shock and damage models.

If failures occur generously as an extended process of a Poisson one, the system forms a renewal process, i.e., a renewal theory arises from the study of self-renewing aggregates. The process plays a major role in the analysis of probability models with sums of independent nonnegative random variables, and is more useful tool in reliability theory because many reliability models have a renewal property by replacement and maintenance. In Chap. 3, we define a renewal process theoretically, and investigate and give a renewal function in theorem forms. The most important matters in a renewal process are how to compute renewal functions and to form renewal equations. Optimum age and periodic replacement policies are shown as practical examples. Furthermore, we introduce an alternating renewal process where the repair or replacement time needs a nonnegligible time for a one-unit system, and a discrete renewal process with only discrete times. As examples of two processes, we give a one-unit system with repair and a discrete age replacement, and obtain their reliability quantities using a renewal theory.

We have considered a Poisson process with only one state in which the unit is operating. When we consider a one-unit system whose repair time is nonnegligible, the process forms an alternating renewal process with two states that repeats up and down alternately. In addition, when we consider two types of repair maintenance according to minor or major failures of the unit, the process forms a stochastic process with three states. In such ways, if the process transits among many states and has a Markov property in which the future behavior depends only on the present state and is independent of its past history, it forms a Markov process. If the duration times of states are multiples of time unit such as day, week, month, year, and a certain number of occurrences, then the process is called a discrete-time Markov chain, and if the duration time are distributed exponentially, it is called a continuous-time Markov chain. In Chap. 4, we show how to compute transition probabilities of a Markov chain and to obtain its several useful quantities from them. As examples of reliability models, we take up a one-unit system with repair and a two-unit parallel system, and derive their reliability measures by using the techniques of a Markov chain.

We consider a one-unit system with two types of repair maintenance. If neither up and down times are exponential, the process does not have the Markov property at any time except only transition times which are called regeneration points. The process forms a Markov renewal process or a semi-Markov process that is a marriage of renewal processes and Markov chains. If we are interested mainly in the number of visits to states, the process is called a Markov renewal process, and if we are interested in what state the process is at time t, it is called a semi-Markov process. However, there is not so much difference between two given names. In Chap. 5, we derive first-passage distributions, transition probabilities, and renewal functions by forming renewal equations. As an example of a standby system with repair, we obtain a mean time to failure, transition probabilities, and expected numbers of visits to states.

Markov renewal processes give powerful plays for analyzing reliability systems because they include renewal and Markov processes. Learning the techniques of Markov renewal processes is positively necessary for analyzing reliability models. Naturally, such processes are not all-round. For example, we consider a two-unit parallel system with one repair person when both failure and repair times are not exponential that seems to be a simpler system. However, we cannot analyze the system because there is no generation point, i.e., the process has no Markov property at any time. Next, consider a two-unit standby system with one repair person. Then, the process has two regeneration points only at times when one unit fails and another is in standby, and the repair of one unit is completed and another waits for repair [4]. From the above viewpoints, we introduce two types of Markov renewal processes with nonregeneration points and analyze an n-unit parallel and a two-unit standby systems with repair.

Suppose that shocks occur in a stochastic process and each amount of damage due to shocks is additive. This process forms a cumulative process that is also called a jump process [7] or doubly stochastic process [8]. The analysis of such processes are more difficult than the already defined stochastic processes because the process is made up of two dependent stochastic processes. In Chap. 6, we define a cumulative process and investigate its properties. There are many examples of shock models, cumulative damage models, and independent damage models, and furthermore, their optimum maintenance policies.

State space of the above stochastic processes may be finite because they are usually defined as the number of operating or failed units and the degree of wear for most reliability models. As far as reliability applications are concerned, we consider the processes with only a finite number of states. A continuous-time Markov process with continuous state space forms a Brownian motion, a Wiener process, and a Lévy process. This is now one of the most useful stochastic processes in applied sciences such as physics, economics, communication theory, biology, management science, and mathematical statistics. However, such processes cannot be much applicable to reliability models because of continuous state space. In Chap. 7, we make only the definitions of a Brownian motion and a Lévy process. As examples, we give continuous wear models in which wear increases continuously with time and consider their replacement policies.

Finally, in Chap. 8, as examples of some redundant systems, we show how to analyze them by ready and handy tools of the stochastic processes. This would be greatly helpful for understanding the stochastic processes and learning reliability theory.

1.3 Further Studies

This book is written mainly based on the books [9–11] for learning basic stochastic processes. In similar contents of such books, we recommend to read the books [12, 13] for a deeper understanding. There were contained many examples of stochastic processes in various fields of applications [14, 15] and in reliability systems [16, 17]. The names of stochastic processes such as stationary process, Gaussian process, martingales, branching process, and so on, appeared [18]. For more applications of stochastic processes, there were [19–21] for queuing theory, [22] for insurance and finance, and [20] for computer applications. It would be impossible to form stochastic models and to analyze them for more complex systems with too many constrains. Then, simulation would be a useful tool for analyzing such systems. We do not touch on the simulation at all in this book and recommend to master its techniques by [21, 23, 24]. We introduce simply the Brownian motion and Lévy process because these processes are not very applicable to reliability models and are more difficult to understand over our mathematical level. For further studies of such processes, refer to [25–27].

This book would be suited well to learn elementary stochastic processes for students with a major of reliability, and researchers and workers engaged in reliability engineering and management, with reference to books [6, 28–30]. As other text books, refer to recent published books [31–33]. For more further and recent studies of reliability, edited books and hand books [34–36] would be advisable to look through. In Appendix A and B, we present Laplace and Laplace-Stieltjes transforms and make answers to selected problems.

1.4 Problems 1

1.1 Search for examples of stochastic processes on your daily life, or read the books [14–17] and give some examples.

1.2 Examine why there is no generation point for a two-unit parallel system with one repair person where the failure and repair times are not exponential [4]. What about the same system with two repair persons?

References

1. Klein JP, Moeschberger ML (2003) Survival analysis. Springer, New York
2. Ushakov IA (1994) Handbook of reliability engineering. Wiley, New York
3. Birolini A (1999) Reliability engineering theory and practice. Springer, New York
4. Nakagawa T (2002) Two-unit redundant models. In: Osaki S (eds) Stochastic models in reliability and maintenance. Springer, Berlin, pp 165–185
5. Lala PK (1985) Fault tolerant and fault testable hardware design. Prentice-Hall, London
6. Nakagawa T (2008) Advanced reliability models and maintenance policies. Springer, London
7. Abdel-Hameed M (1984) Life distribution properties of devices subject to a pure jump damage process. J Appl Probab 21:816–825
8. Grandell J (1976) Doubly stochastic poisson process. Lecture notes in mathematics 529. Springer, New York
9. Ross SM (1983) Stochastic processes. Wiley, New York
10. Osaki S (1992) Applied stochastic system modeling. Springer, Berlin
11. Çinlar E (1975) Introduction to stochastic processes. Prentice-Hall, Englewood Cliffs
12. Karlin S, Taylor HM (1975) A first course in stochastic processes. Academic Press, New York
13. Tijms HC (2003) A first course in stochastic models. Wiley, Chichester
14. Ross SM (2000) Introduction to probability models. Academic Press, San Diego
15. Bhat UN, Miller GK (2002) Elements of applied stochastic processes. Wiley, Hoboken, NJ
16. Beichelt FE, Fatti LP (2002) Stochastic processes and their applications. CRC Press, Boca Raton
17. Beichelt FE (2006) Stochastic processes in science, engineering and finance. Chapman & Hall, Boca Raton
18. Prabhu NU (2007) Stochastic processes. World Scientific, Singapore
19. Bhat UN, Basawa IV (1992) Queeing and related models. Oxford University Press, Oxford
20. Trivedi KS (2002) Probabilty and statistics with reliability, queueing and computer science applications. Wiley, New York
21. Stewart WJ (2009) Probability, Markov chains, queues, and simulation. Princeton University Press, Princeton
22. Rolski T, Schmidli H, Schmidt V, Teugels J (1999) Stochastic processes for insurance and finance. Wiley, Chichester
23. Ross SM (1997) Simulation. Academic Press, New York
24. Papadopoulos C, Limnious N (2002) Quick Monte Carlo methods in stochastic systems and reliability. In: Osaki S (eds) Stochastic models in reliability and maintenance. Springer, Berlin, pp 307–333
25. Karatzas I, Shreve S (1991) Brownian motion and stochastic calculus. Springer, New York
26. Bertoin J (1996) Lévy processes. Cambridge University Press, Cambridge
27. Barndorff-Nielsen OE, Mikosch T, Resnick SI (eds) (2001) Lévy processes: theory and applications. Birkhävser, Boston
28. Barlow RE, Proschan F (1965) Mathematical theory of reliability. Wiley, New York
29. Nakagawa T (2005) Maintenance theory of reliability. Springer, London
30. Nakagawa T (2007) Shock and damage models in reliability theory. Springer, London
31. Dhillon BS (2002) Applied reliability and quality. Springer, London
32. O'Connor PDT (2002) Practical reliability engineering. Wiley, Chichester
33. Rausand M, Hϕyland A (2004) System reliability theory. Wiley, Hoboken
34. Osaki S (eds) (2002) Stochastic models in reliability and maintenance. Springer, Berlin
35. Pham H (eds) (2003) Handbook of reliability engineering. Springer, London
36. Misra KB (2008) Handbook of performability engineering. Springer, London

Chapter 2
Poisson Processes

It is well-known that most units operating in a useful life period, and complex systems that consist of many kinds of components, fail normally due to random causes independently over the time interval. Then, it is said in technical terms of stochastic processes, that failures occur in a *Poisson process* that counts the number of failures through time. This is a natural modeling tool in reliability problems. Some reliability measures such as MTTF (Mean Time To Failure), availability, and failure rate are estimated statistically from life data and are in practical use under such modelings without much theoretical arguments. Furthermore, because a Poisson process has stationary and independent properties, it is much convenient for formulating stochastic models in mathematical reliability theory.

A Poisson process is the simplest stochastic process that arises in many applications for arrival processes. It is of great importance to know the properties of a Poisson process and to learn how to apply the process to reliability models in the beginning: In Sect. 2.1, we investigate properties of an exponential distribution, because most units are assumed practically to have its distribution as their failure times. Then, it is shown that the number of failures has a Poisson distribution when the failure time has an exponential distribution. As applied examples, we discuss about the number of spare units for constructing parallel and standby systems that should be previously prepared.

In Sect. 2.2, a Poisson process is defined theoretically, based on such above facts, and its interarrival times are shown to have an exponential distribution. Furthermore, when the failure was detected at time t, its failure occurs uniformly over $[0, t]$. As such applied examples, we take up an inspection and two backup models, and derive their optimum policies.

It has been assumed often in maintenance models that a unit undergoes minimal repair at failures. This means that the failure rate is not changed by any maintenance, and forms a *nonhomogeneous Poisson process*. Underlying the process, failures occur successively, depending on the operating time t. In Sect. 2.3, we summarize the properties of the process. In Sect. 2.4, we apply these results to the replacement model where the unit is replaced at the Nth failure and the software

T. Nakagawa, *Stochastic Processes*, Springer Series in Reliability Engineering,
DOI: 10.1007/978-0-85729-274-2_2, © Springer-Verlag London Limited 2011

reliability model, and investigate its properties. Finally, in Sect. 2.5, we consider a cumulative damage model where shocks occur in a Poisson process and the total damage due to shocks are additive. This is called a *compound Poisson process*. We survey probabilistic properties of the process and apply damage models with a failure level K where the unit fails when the total damage exceeds K. For a more deep study of Poisson processes, refer to the book [1].

2.1 Exponential Distribution

Suppose that a unit begins to operate at time 0 and is replaced with a new identical one immediately at each failure, where the time required for replacement is negligible. The successive operating units have failure times X_k $(k = 1, 2, \dots)$ from the beginning of their operation to failure and $S_n \equiv \sum_{k=1}^{n} X_k$ in Fig. 2.1, where the horizontal axis represents the time-axis t from $t = 0$ to the right direction in all figures of this book. It is assumed that X_k are independent and identically distributed random variables, and have an identical exponential distribution

$$F(t) \equiv \Pr\{X_k \le t\} = 1 - \mathrm{e}^{-\lambda t}, \tag{2.1}$$

and its density function

$$f(t) \equiv \frac{\mathrm{d}F(t)}{\mathrm{d}t} = F'(t) = \lambda \mathrm{e}^{-\lambda t}$$

for $\lambda > 0$ and any k, that decreases strictly from λ to 0.

The mean and variance are

$$\begin{aligned} E\{X_k\} &= \int_0^\infty t\lambda \mathrm{e}^{-\lambda t}\mathrm{dt} = \frac{1}{\lambda}, \\ V\{X_k\} &= \int_0^\infty t^2\lambda \mathrm{e}^{-\lambda t}\mathrm{d}t - (E\{X_k\})^2 = \frac{1}{\lambda^2}. \end{aligned} \tag{2.2}$$

Then, it is said that the unit has a failure time X, omitting subscript k, and fails according to an exponential distribution $(1 - \mathrm{e}^{-\lambda t})$ with parameter λ or MTTF $1/\lambda$. Furthermore, from Appendix A, the LS (Laplace-Stieltjes) transform of an exponential distribution is

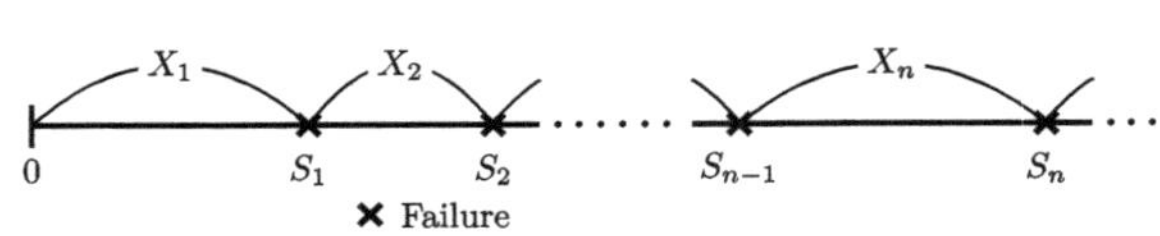

Fig. 2.1 Process of failures

$$F^*(s) \equiv E\{e^{-sX}\} = \int_0^\infty e^{-st}dF(t) = \int_0^\infty e^{-st}\lambda e^{-\lambda t}dt = \frac{\lambda}{s+\lambda} \tag{2.3}$$

for any $Re(s) > 0$, where note that we omit $Re(s) > 0$ throughout this book. From the assumptions that random variables are independent and identical,

$$E\{X_1 + X_2 + \cdots + X_n\} = nE\{X\} = \frac{n}{\lambda},$$
$$V\{X_1 + X_2 + \cdots + X_n\} = nV\{X\} = \frac{n}{\lambda^2},$$
$$E\{e^{-s(X_1+X_2+\cdots+X_n)}\} = (E\{e^{-sX}\})^n = \left(\frac{\lambda}{s+\lambda}\right)^n. \tag{2.4}$$

If two random variables X_1 and X_2 are independent and have the respective exponential distributions with mean $1/\lambda_1$ and $1/\lambda_2$, then

$$\Pr\{X_1 \le X_2\} = \int_0^\infty e^{-\lambda_2 t}\lambda_1 e^{-\lambda_1 t}dt = \frac{\lambda_1}{\lambda_1 + \lambda_2}.$$

In general, if k random variables X_k are independent and have an exponential distribution with mean $1/\lambda_k$ $(k = 1, 2, \ldots)$, and $X_m = \min\{X_1, X_2, \ldots, X_n\}$, then

$$\Pr\{X_1 = X_m\} = \int_0^\infty e^{-(\lambda_2+\lambda_3+\cdots+\lambda_n)t}\lambda_1 e^{-\lambda_1 t}dt = \frac{\lambda_1}{\lambda_1 + \lambda_2 + \cdots + \lambda_n}.$$

2.1.1 Properties of Exponential Distribution

We investigate and summarize various properties of an exponential distribution $(1 - e^{-\lambda t})$ appearing frequently in reliability theory:

(1) From the MTTF or the mean life time of the unit $1/\lambda$,

$$F(1/\lambda) = 1 - e^{-1} \approx 0.632,$$

i.e., 63.2% of units have failed at the time of MTTF (Example 2.13).

(2) The reliability function $R(t) \equiv \overline{F}(t) = 1 - F(t)$ that is the probability that the unit is operating at time t without failure is

$$R(t) = e^{-\lambda t}.$$

Thus, the failure rate or hazard rate is [2, p. 5, 3, p. 22]

$$h(t) \equiv \frac{f(t)}{R(t)} = \frac{\lambda e^{-\lambda t}}{e^{-\lambda t}} = \lambda,$$

irrespective of time t. Thus, if the failure rate would be estimated from actual life data, then the MTTF would be obtained by taking the reciprocal of λ. For example, when $\lambda = 10^{-4}$ (1/h), the MTTF is 10^4 h and vice versa.

Therefore, because $h(t)\Delta t \approx \Pr\{t < X \le t + \Delta t | X > t\}$ means the probability that the unit with age t will fail in $(t, t + \Delta t]$ for small Δt, the probability that a failure occurs in a small interval $(t, t + \Delta t]$ is constant for any time t. That is, the failure rate of an exponential distribution has the property of random failures occurring constantly in a small interval time t, irrespective of ages or elapsed operating times. In general, the failure rate curve, i.e., the bathtub curve [4, p. 188] (Fig. 2.5) is divided into three periods; (1) the initial failure period in which $h(t)$ is DFR (Decreasing Failure Rate), (2) the random failure rate period in which $h(t)$ is CFR (Constant Failure Rate), and (3) the wearout failure period in which $h(t)$ is IFR (Increasing Failure Rate). In this case, an exponential distribution has the properties of DFR and IFR, and shows the failure phenomenon observed in a random failure period. Thus, when units have an exponential failure distribution, there is no necessity for making any maintenance of them theoretically.

For such reasons, the exponential distribution might be of limited applications. However, the exponential distribution plays a fitting role for complex systems that consist of many kinds of components, each of which has a variety pattern of malfunction and replacement and makes the failure pattern of the system as a whole [2, p. 18]. It is well-known that the exponential distribution is fitted well to life data of electric and electronic components.

(3) Using the Taylor expansion of an exponential function,

$$1 - e^{-\lambda h} = \lambda h - \frac{(\lambda h)^2}{2!} + \frac{(\lambda h)^3}{3!} - \cdots = \lambda h + o(h),$$

where $o(h)$ is an infinite decimal higher than h, i.e., $o(h)/h \to 0$ as $h \to 0$. Thus, we have the inequalities

$$\lambda h - \frac{(\lambda h)^2}{2} < 1 - e^{-\lambda h} < \lambda h$$

for a small $\lambda h > 0$. These inequalities are very useful for computing equations numerically, involving exponential functions because λt is very small in actual reliability models, as shown in Table 2.1.

Table 2.1 Comparative table of $1 - e^{-\lambda h}$ and approximation $\lambda h - (\lambda h)^2/2$ for λh

λh	$1 - e^{-\lambda h}$	$\lambda h - (\lambda h)^2/2$
1	0.6321205	0.5
10^{-1}	0.951625×10^{-1}	0.95×10^{-1}
10^{-2}	0.99501×10^{-2}	0.995×10^{-2}
10^{-3}	0.99949×10^{-3}	0.9995×10^{-3}
10^{-4}	0.9999×10^{-4}	0.9999×10^{-4}

Example 2.1 (MTTF of exponential distribution) When the failure distribution is an exponential distribution $(1 - e^{-\lambda t})$, we ask the following questions:

(a) How much is the failure rate λ and the MTTF $1/\lambda$ when $R(t) = 0.99$ at $t = 10^3$ h?
(b) What fraction time t of the MTTF are to satisfy that $F(t) = 10^{-1}, 10^{-2}, 10^{-3}$, and 10^{-4}?

The solution of (a) is easily given as

$$e^{-\lambda \times 10^3} = 0.99,$$

i.e.,

$$\lambda = -10^{-3} \log 0.99 \approx 1.005 \times 10^{-5} \text{ (1/h)},$$
$$\text{MTTF} = \frac{1}{\lambda} = 99{,}499 \text{ (h)}.$$

We can compute approximately, $\lambda \times 10^3 = 1 - 0.99 = 0.01$, i.e.,

$$\widetilde{\lambda} = 10^{-5} \text{(1/h)}, \quad \widetilde{\text{MTTF}} = 10^5 \text{ (h)}.$$

In this case, the difference between the exact and approximate values are very small.

The solution of (b) is given by computing

$$\lambda t = \frac{t}{1/\lambda} = \frac{t}{\text{MTTF}}.$$

For example, $1 - e^{-\lambda t} = 0.1, \lambda t = -\log 0.9 = 0.10536$. Approximately, from (3), $\widetilde{\lambda t} = 0.1 = 1/10$. The time t that satisfies $F(t) = 0.1$ is also called a B_{10} *life*. Table 2.2 presents λt and their approximate values $\widetilde{\lambda t}$ for $F(t) = 10^{-1}, 10^{-2}, 10^{-3}$, and 10^{-4}. This table indicates that when $F(t) = 10^{-n}$ $(n = 1, 2, \ldots)$, the fraction time t of the MTTF is approximately $10^{-n} = 1/10^n$, and its value becomes very good as n becomes large. From such results, we can easily estimate the fraction time of the MTTF for a small $F(t)$ by setting $t/\text{MTTF} = F(t)$.

Example 2.2 (Preventive maintenance) [3, p. 8] Consider operating units such as scales, computers, and production systems where its preventive maintenance is

Table 2.2 Comparative table of λt and approximation $\widetilde{\lambda t}$ for $F(t)$

$F(t)$	λt	$\widetilde{\lambda t}$
10^{-1}	1.0536×10^{-1}	10^{-1}
10^{-2}	1.0050×10^{-2}	10^{-2}
10^{-3}	1.0005×10^{-3}	10^{-3}
10^{-4}	1.0000×10^{-4}	10^{-4}

done only at time T $(0<T\leq\infty)$ and its failure is detected. It is supposed that the unit has a failure distribution $F(t)$. We get some earning per unit of time and does not have any earning during the interval if it fails before time T. The average time during $[0,T]$ in which we get some earning is

$$l(T)=0\times F(T)+T\overline{F}(T)=T\overline{F}(T), \tag{2.5}$$

and $l(0)=l(\infty)=0$. Differentiating $l(T)$ with respect to T and setting it equal to zero,

$$\overline{F}(T)-Tf(T)=0, \quad \text{i.e.,} \quad h(T)=\frac{f(T)}{\overline{F}(T)}=\frac{1}{T}. \tag{2.6}$$

Thus, an optimum time T^* that maximizes $l(T)$ in (2.5) is given by a unique solution of (2.6), when $F(t)$ is IFR. In particular, when $F(t)=1-\mathrm{e}^{-\lambda t}, T=1/\lambda$, i.e., we should do the preventive maintenance at the MTTF.

Example 2.3 (Discount cost) [5, p. 82] When we consider the total expected cost of an operating unit for an infinite time span, we should evaluate the present values of all replacement costs by using an appropriate discount rate [3, p. 78]. Suppose that a continuous discount with rate α $(0<\alpha<\infty)$ is used for the cost incurred at replacement time. That is, the present value of replacement cost c at time t is $c\mathrm{e}^{-\alpha t}$ at time 0. Then, because the total replacement cost is $c\sum_{n=1}^{\infty}\exp[-\alpha(X_1+X_2+\cdots+X_n)]$, the expected cost is

$$C\equiv c\sum_{n=1}^{\infty}E\{\exp[-\alpha(X_1+X_2+\cdots+X_n)]\}=c\sum_{n=1}^{\infty}\left(E\{\mathrm{e}^{-\alpha X}\}\right)^n.$$

Thus, when X is distributed exponentially, from (2.4),

$$C=c\sum_{n=1}^{\infty}\left(\int_0^{\infty}\mathrm{e}^{-\alpha t}\lambda\mathrm{e}^{-\lambda t}\mathrm{d}t\right)^n=c\sum_{n=1}^{\infty}\left(\frac{\lambda}{\alpha+\lambda}\right)^n=\frac{c\lambda}{\alpha}.$$

For example, if the mean failure time is 1,000 h, the replacement cost is 20 dollars, and the interest rate is 5 % per year, i.e., $\lambda=1/1{,}000, c=20$, and $\alpha=0.05/(365\times 24)$, then $C=20\times 365\times 24/(1000\times 0.05)=3{,}504$ dollars. In general, for any negative function $g(t)$ for $t\geq 0$ (Problem 2.1),

$$E\left\{\sum_{n=1}^{\infty}g(S_n)\right\}=\lambda\int_0^{\infty}g(t)\mathrm{d}t.$$

In this example, $g(t)=c\mathrm{e}^{-\alpha t}$.

(4) From the assumption that X is a random variable with an exponential distribution $(1-\mathrm{e}^{-\lambda t})$, the conditional survival probability that the unit does not fail in a finite interval time t, given that it is operating at time u $(u\geq 0)$ is [4, p. 74]

$$\Pr\{X > u + t | X > u\} = \frac{\Pr\{X > u + t\}}{\Pr\{X > u\}} = \frac{e^{-\lambda(u+t)}}{e^{-\lambda u}}$$
$$= e^{-\lambda t} = \Pr\{X > t\},$$

that is independent of any time u, i.e., for all $t, u \geq 0$,

$$\Pr\{X > u + t\} = \Pr\{X > u\} \Pr\{X > t\}. \tag{2.7}$$

The condition that the unit is operating at time u, i.e., its age is u, has no effect on the next operating time t, i.e., its residual or remaining time. This is called the *memoryless property* or *lack of memory property*. Conversely, the only continuous and positive function that satisfies (2.7) is an exponential function $\Pr\{X > t\} = e^{-\lambda t}$ (Problem 2.2). That is, only the exponential distribution has the memoryless property. For such reasons, as one of failure distributions, an exponential distribution comes to be frequently used for forming simpler and analyzing easier stochastic models.

Example 2.4 (Parallel system) Consider a parallel system that consists of n identical units and fails when all units have failed, i.e., at least one of n units is operating, the system is also operating [6, p. 8]. Each unit has an independent and identical failure distribution $(1 - e^{-\lambda t})$. Because the system with n units has a failure distribution $(1 - e^{-\lambda t})^n$, its MTTF is (Problems 2.3–2.8)

$$\int_0^\infty \left[1 - (1 - e^{-\lambda t})^n\right] dt = \frac{1}{\lambda} \int_0^1 \frac{1 - x^n}{1 - x} dx = \frac{1}{\lambda} \sum_{k=1}^{n} \frac{1}{k}. \tag{2.8}$$

On the other hand, let $X_{(k)}$ $(k = 1, 2, \ldots, n)$ be the failure time from the $(k-1)$th to kth unit of n units. The failure time of the system is given by $X_{(1)} + X_{(2)} + \cdots X_{(n)}$. Then, because $X_{(k)}$ has an exponential distribution $\left[1 - e^{-\lambda(n-k+1)t}\right]$ with mean $1/[(n - k + 1)\lambda]$ from the memoryless property, the MTTF is

$$E\{X_{(1)} + X_{(2)} + \cdots + X_{(n)}\} = \frac{1}{\lambda} \sum_{k=1}^{n} \frac{1}{n - k + 1} = \frac{1}{\lambda} \sum_{k=1}^{n} \frac{1}{k},$$

that agrees with (2.8), and it is called a *harmonic series* (Problem 2.8).

We have the inequalities [7]

$$\log n + \frac{1}{n} < \sum_{k=1}^{n} \frac{1}{k} < \log n + 1 \quad (n = 2, 3, \ldots), \tag{2.9}$$

and

$$\sum_{k=1}^{n} \frac{1}{k} \approx C + \log n,$$

where C is Euler's constant and $C \equiv 0.577215\ldots$. From such results, the MTTF of a parallel system increases with n very slowly and logarithmically to infinity, and is given approximately by (2.9) for large n (Problem 2.5).

Next, because the reliability function is

$$R_n(t) = 1 - \left(1 - e^{-\lambda t}\right)^n, \tag{2.10}$$

its failure density function is

$$f_n(t) = -R_n'(t) = n\lambda e^{-\lambda t}\left(1 - e^{-\lambda t}\right)^{n-1}.$$

Hence,

$$\frac{R_n(t)}{f_n(t)} = \frac{1}{n\lambda}\frac{1-\left(1-e^{-\lambda t}\right)^n}{e^{-\lambda t}\left(1-e^{-\lambda t}\right)^{n-1}} = \frac{1}{n\lambda}\sum_{k=0}^{n-1}\left(1-e^{-\lambda t}\right)^{-k},$$

that decreases with t. Thus, the failure rate

$$h_n(t) = \frac{f_n(t)}{R_n(t)} = \frac{n\lambda}{\sum_{k=0}^{n-1}\left(1-e^{-\lambda t}\right)^{-k}} \tag{2.11}$$

increases strictly with t from 0 to λ $(n \geq 2)$ that is the failure rate of each unit. Furthermore, because

$$\frac{1}{h_{n+1}(t)} - \frac{1}{h_n(t)} = \frac{1}{n(n+1)\lambda}\sum_{k=0}^{n-1}\left[\left(1-e^{-\lambda t}\right)^{-n} - \left(1-e^{-\lambda t}\right)^{-k}\right] > 0,$$

the failure rate $h_n(t)$ decreases strictly with n $(n \geq 1)$ from λ to 0 for a given $t > 0$.

2.1.2 Poisson and Gamma Distributions

We are often interested in the number of failures and replacements of units that have occurred in the interval $[0, t]$. For this purpose, we introduce the total sum of random variables $S_n \equiv X_1 + X_2 + \cdots + X_n$ $(n = 1, 2, \ldots)$ that represents the time of the nth failure (Fig. 2.1) when $F(t) \equiv \Pr\{X_k \leq t\} = 1 - e^{-\lambda t}$ for all $k \geq 1$. Then, we derive the distribution

$$\Pr\{S_n \leq t\} = \int_0^t \frac{\lambda(\lambda u)^{n-1}}{(n-1)!}e^{-\lambda u}du = \sum_{k=n}^{\infty}\frac{(\lambda t)^k}{k!}e^{-\lambda t}$$

$$= 1 - \sum_{k=0}^{n-1}\frac{(\lambda t)^k}{k!}e^{-\lambda t} \quad (n = 1, 2, \ldots), \tag{2.12}$$

that is called an *Erlang distribution* or a *gamma distribution* with order n.

We prove (2.12) by the mathematical induction as follows: Using the total sum of independent random variables with an exponential distribution $\left(1 - e^{-\lambda t}\right)$,

$$\Pr\{S_2 \le t\} = \Pr\{X_1 + X_2 \le t\} = \int_0^t \left[1 - e^{-\lambda(t-u)}\right] \lambda e^{-\lambda u} du$$
$$= 1 - (1 + \lambda t)e^{-\lambda t}.$$

Assuming that when $n = k$,

$$\Pr\{S_k \le t\} = \int_0^t \frac{\lambda(\lambda u)^{k-1}}{(k-1)!} e^{-\lambda u} du,$$

we easily have

$$\Pr\{S_{k+1} \le t\} = \Pr\{S_k + X_{k+1} \le t\} = \int_0^t \left[1 - e^{-\lambda(t-u)}\right] \frac{\lambda(\lambda u)^{k-1}}{(k-1)!} e^{-\lambda u} du$$
$$= \sum_{j=k}^{\infty} \frac{(\lambda t)^j}{j!} e^{-\lambda t} - \frac{(\lambda t)^k}{k!} e^{-\lambda t} = \sum_{j=k+1}^{\infty} \frac{(\lambda t)^j}{j!} e^{-\lambda t},$$

that concludes that (2.12) holds for all $n \ge 1$.

Equation (2.12) is alternatively proved by the technique of the LS transform in Appendix A: Recalling that the LS transform of a random variable X_k with an exponential distribution $\left(1 - e^{-\lambda t}\right)$ is given in (2.3), the LS transform of S_n is

$$E\{e^{-sS_n}\} = E\{e^{-s(X_1 + X_2 + \cdots + X_n)}\} = \left(\frac{\lambda}{s + \lambda}\right)^n.$$

On the other hand, the LS transform of (2.12) is

$$\int_0^\infty e^{-st} \frac{\lambda(\lambda t)^{n-1}}{(n-1)!} e^{-\lambda t} dt = \left(\frac{\lambda}{s + \lambda}\right)^n.$$

Therefore, because an original function is in one-to-one correspondence with its LS transform, it is shown that S_n has a gamma distribution in (2.12), and its mean and variance are also given in (2.4). A gamma distribution with order n represents the distribution of the nth failure of a unit with an exponential distribution. If a unit fails every at the nth minor fault, each of which occurs exponentially, its failure time has a gamma distribution. In other words, a standby redundant system (Example 2.6) with n units fail according to a gamma distribution with order n.

A density function of S_n given in (2.12) is generalized as

$$g(t) \equiv \frac{\lambda(\lambda t)^{\alpha-1}}{\Gamma(\alpha)} e^{-\lambda t} \quad \lambda, \alpha > 0, \tag{2.13}$$

where $\Gamma(\alpha) \equiv \int_0^\infty x^{\alpha-1}e^{-x}dx$ for $\alpha > 0$, and is called a gamma function with order α. The mean, variance, and LS transform of a random variable X with a gamma density function in (2.13) are, respectively,

$$E\{X\} = \frac{\alpha}{\lambda}, \quad V\{X\} = \frac{\alpha}{\lambda^2}, \quad E\{e^{-sX}\} = \left(\frac{\lambda}{s+\lambda}\right)^\alpha, \tag{2.14}$$

and the failure rate is

$$h(t) = \frac{(\lambda t)^{\alpha-1}e^{-\lambda t}}{\int_t^\infty (\lambda u)^{\alpha-1}e^{-\lambda u}du}. \tag{2.15}$$

Thus,

$$\frac{1}{h(t)} = \int_t^\infty \left(\frac{u}{t}\right)^{\alpha-1} e^{-\lambda(u-t)}du = \int_0^\infty \left(1+\frac{x}{t}\right)^{\alpha-1} e^{-\lambda x}dx,$$

that implies that $h(t)$ decreases from ∞ to λ for $0<\alpha<1$, is constant λ for $\alpha = 1$, and increases from 0 to λ for $\alpha > 1$.

We define $N(t) \equiv \max_n\{S_n \le t\}$ that represents the number of failures in $[0, t]$, where $\{N(t)\}$ is called a *counting process*, that is one of stochastic processes. If $S_1 = X_1 > t$, then $N(t) = 0$, i.e., no failure occurs in $[0, t]$. Then, we have the relation

$$\{S_n \le t\} \Leftrightarrow \{N(t) \ge n\}, \tag{2.16}$$

because if the nth failure occurs up to time t, then the number of failures in time t is greater than n, and conversely, if the number of failures in time t is greater than n, then the nth failure occurs up to time t (Fig. 2.1).

Thus, from (2.12) and (2.16),

$$\Pr\{N(t) \ge n\} = \Pr\{S_n \le t\} = \int_0^t \frac{\lambda(\lambda u)^{n-1}}{(n-1)!}e^{-\lambda u}du = \sum_{k=n}^\infty \frac{(\lambda t)^k}{k!}e^{-\lambda t}.$$

Hence,

$$\begin{aligned}\Pr\{N(t) = n\} &= \Pr\{S_n \le t\} - \Pr\{S_{n+1} \le t\} \\ &= \frac{(\lambda t)^n}{n!}e^{-\lambda t} \quad (n = 0, 1, 2, \ldots),\end{aligned} \tag{2.17}$$

and its mean and variance are

$$\begin{aligned}E\{N(t)\} &= \sum_{n=1}^\infty n\frac{(\lambda t)^n}{n!}e^{-\lambda t} = \lambda t, \\ V\{N(t)\} &= \sum_{n=1}^\infty n^2\frac{(\lambda t)^n}{n!}e^{-\lambda t} - (\lambda t)^2 = \lambda t.\end{aligned} \tag{2.18}$$

Equation (2.17) is called a Poisson distribution with mean λt that was originally derived as a limiting distribution of a binomial distribution by S.O. Poisson (1873) (Problem 2.9). The number of defects due to random causes normally in some time, area, volume, and so on, has a Poisson distribution.

Example 2.5 (Poisson distribution) A unit fails exponentially at a failure rate $\lambda = 10^{-3}$(1/h). Then, calculate the probabilities that k $(k = 0, 1, 2)$ failures occur in 500 h. In this case, because $\lambda t = 10^{-3} \times 500 = 0.5$,

$$\begin{aligned}
\Pr\{N(500) = 0\} &= \mathrm{e}^{-0.5} \approx 0.607,\\
\Pr\{N(500) = 1\} &= 0.5\mathrm{e}^{-0.5} \approx 0.303,\\
\Pr\{N(500) = 2\} &= \frac{(0.5)^2}{2!}\mathrm{e}^{-0.5} \approx 0.076.
\end{aligned}$$

Furthermore, the probability that the number of failures is equal to or more than 3 in 500 h is

$$\begin{aligned}
\Pr\{N(500) \geq 3\} &= 1 - \Pr\{N(500) \leq 2\}\\
&= 1 - \left(1 + 0.5 + \frac{0.25}{2}\right)\mathrm{e}^{-0.5} \approx 0.0144.
\end{aligned}$$

This probability is very small because the mean failure time $1/\lambda = 10^3$ h is two times of $t = 500$.

Example 2.6 (Scheduling problem) Consider a standby system where an operating unit is replaced with one of identical spares at each failure and each unit has an exponential failure distribution $\left(1 - \mathrm{e}^{-\lambda t}\right)$. We are interested in how many number of spare units should be provided initially in order to assure with probability α $(0 < \alpha < 1)$ that the system will remain operating in time t [2, p. 49]. An optimum number n^* for fixed α and t is given by a unique minimum that satisfies

$$\Pr\{N(t) \leq n\} = \sum_{k=0}^{n} \frac{(\lambda t)^k}{k!}\mathrm{e}^{-\lambda t} \geq \alpha.$$

Next, consider the scheduling problem where the system has to work for a job with working time S and its distribution $W(t) \equiv \Pr\{S \leq t\}$ [6, p. 84] and $\overline{W}(t) \equiv 1 - W(t)$. It is assumed that if the work of a job is accomplished when at least one unit is operating, it needs cost c_2, if the work is not accomplished when all units have failed, it needs cost c_1 with $c_1 > c_2$, and the cost of a standby system with n units is $c_0 n$. Then, the expected cost is

$$C(n) = c_2 + (c_1 - c_2)\int_0^\infty \overline{W}(t)\mathrm{d}F^{(n)}(t) + c_0 n \quad (n = 0, 1, 2, \ldots). \tag{2.19}$$

Table 2.3 Optimum numbers n^* of units for a standby system

$c_0/(c_1 - c_2)$	λ/w		
	1.0	2.0	5.0
0.5	0	0	0
0.3	1	1	0
0.1	3	3	3
0.05	4	5	7
0.01	6	9	16

In particular, when $W(t) = 1 - \mathrm{e}^{-wt}$,

$$C(n) = c_2 + (c_1 - c_2)[F^*(w)]^n + c_0 n. \tag{2.20}$$

To find an optimum number n^* that minimizes $C(n)$, from the inequality $C(n+1) - C(n) \geq 0$,

$$[F^*(w)]^n[1 - F^*(w)] \leq \frac{c_0}{c_1 - c_2} \quad (n = 0, 1, 2, \ldots), \tag{2.21}$$

whose left-hand decreases strictly from $1 - F^*(w)$ to 0. Thus, if $F^*(w) < (c_1 - c_2 - c_0)/(c_1 - c_2)$, then there exists a finite and unique minimum n^* $(1 \leq n^* < \infty)$ that satisfies (2.21). If $F^*(w) \geq (c_1 - c_2 - c_0)/(c_1 - c_2)$, then $n^* = 0$, i.e., we should not provide any units for a job.

In addition, when $F(t) = 1 - \mathrm{e}^{-\lambda t}$, (2.21) is

$$\frac{w\lambda^n}{(w + \lambda)^{n+1}} \leq \frac{c_0}{c_1 - c_2} \quad (n = 0, 1, 2, \ldots). \tag{2.22}$$

If $w/(w + \lambda) > c_0/(c_1 - c_2)$, then a positive n^* exists. Table 2.3. presents the optimum n^* that decreases with $c_0/(c_1 - c_2)$ (Problem 2.10).

2.2 Poisson Process

We consider a stochastic process $\{X(t), t \in \boldsymbol{T}\}$ in which $X(t)$ changes randomly with time $t \in \boldsymbol{T}$ in a space by the probabilistic law, i.e., $X(t)$ is a random variable for each t in time set and $\{X(t), t \in \boldsymbol{T}\}$ is its collection. We restrict ourselves to a nonnegative time parameter t for the time set $\boldsymbol{T}$. If the time set is countable, i.e., $\boldsymbol{T} = \{0, 1, 2, \ldots\}$, the process $\{X(k), k = 0, 1, 2, \ldots\}$ is called a *discrete-time stochastic process*, and if the time set is continuous, i.e., $\boldsymbol{T}$ = Set of {a nonnegative real number}, the process $\{X(t), t \geq 0\}$ is called a *continuous-time stochastic process*. If a random variable $X(t)$ presents the working condition of a unit at every time and at every day or week, the time set is continuous and discrete, respectively. We omit $\{k = 0, 1, 2, \ldots\}$ and $\{t \geq 0\}$ for simplicity unless otherwise stated.

A set of all possible values of $X(t)$ for the process $\{X(t)\}$ is called a *state space*. We restrict ourselves to a *discrete state space*, where $X(t) = 0, 1, 2, \ldots$, except for the cumulative processes in Chap. 6, and Brownian motion and Lévy process in

Chap. 7. When the state space in such stochastic processes is continuous, we need more mathematical tools for their analysis. If the process $\{X(t)\}$ is observed over time t, such realization of $X(t)$ is called a *sample path* or *sample function*.

Furthermore, we introduce the following two properties of a stochastic process: It is said that the difference $X(t_k) - X(t_{k-1})$ is called the *increment*, and $X(t)$ has *independent increments* if

$$X(t_1) - X(t_0), X(t_2) - X(t_1), \ldots, X(t_n) - X(t_{n-1})$$

are independent for all $0 \le t_0 < t_1 < \cdots < t_{n-1} < t_n$, where it is called that t_k is an *arrival time* of the kth event and $t_k - t_{k-1}$ is an *interarrival time*. In addition, the process $\{X(t)\}$ is said to have *stationary* or *homogeneous increments* if $X(t+u) - X(t)$ has the same distribution for all $t \ge 0$. If $X(t)$ has independent and stationary increments, then $X(t)$ and $X(t+u) - X(t)$ are independent, and $X(t+u) - X(t)$ has the same distribution as $X(u)$. Thus, we have the relation

$$E\{X(t+u)\} = E\{X(t+u) - X(t) + X(t)\} = E\{X(u)\} + E\{X(t)\}.$$

The properties of stationary and independent increments in reliability mean intuitively that any failures occur independently and stationarity in $[t, t+u]$, being dependent only on time interval u irrespective of the beginning time t and other intervals. For example, when we consider two intervals $[t_1, t_1 + u]$ and $[t_2, t_2 + u]$ for $t_1 + u < t_2$, failures occur independently in two intervals and the probability of failures in two intervals is equal to that in $[0, u]$. In general, the property of independent increment is more important than that of stationary increment for most reliability models, because they might be easy to fail with their age.

We consider the stochastic process $\{N(t)\}$, where $N(t) = 0, 1, 2, \ldots$, that is called a *counting process* $\{N(t)\}$, because $N(t)$ counts the total number of events occurred in time t.

Definition 2.1 If

(i) $N(0) = 0$,
(ii) the process $\{N(t)\}$ has stationary and independent increments,
(iii) $\Pr\{N(h) = 1\} = \lambda h + o(h)$,
(iv) $\Pr\{N(h) \ge 2\} = o(h)$,

then a counting process $\{N(t)\}$ is said to be a *Poisson process* with rate $\lambda > 0$.

Using this definition, we derive the probability

$$P_k(t) \equiv \Pr\{N(t) = k | N(0) = 0\} = \Pr\{N(t) = k\} \quad (k = 0, 1, 2, \ldots), \tag{2.23}$$

that represents the probability that k events occur in $[0, t]$, and $\sum_{k=0}^{\infty} P_k(t) = 1$. From (ii), (iii), and (iv),

$$\Pr\{N(t+h) = k \,|\, N(t) = j\} = \begin{cases} 1 - \lambda h + o(h) & \text{for } k = j, \\ \lambda h + o(h) & \text{for } k = j+1, \\ o(h) & \text{for } k > j+1. \end{cases} \tag{2.24}$$

Thus, from (ii) and (2.24),

$$\begin{aligned}P_0(t+h) &= \Pr\{N(t+h)=0\} = \Pr\{N(t)=0, N(t+h)-N(t)=0\}\\ &= \Pr\{N(t)=0\}\Pr\{N(t+h)-N(t)=0\}\\ &= \Pr\{N(t)=0\}\Pr\{N(h)=0\} = P_0(t)[1-\lambda h+o(h)].\end{aligned}$$

Hence,

$$\frac{P_0(t+h)-P_0(t)}{h} = -\lambda P_0(t) + \frac{o(h)}{h}.$$

Letting $h \to 0$, we have the differential equation

$$\frac{\mathrm{d}P_0(t)}{\mathrm{d}t} = -\lambda P_0(t), \quad \text{i.e.,} \quad \frac{\mathrm{d}P_0(t)/\mathrm{d}t}{P_0(t)} = -\lambda.$$

By integration and setting $P_0(0)=1$ from (i),

$$P_0(t) = \mathrm{e}^{-\lambda t}. \tag{2.25}$$

Similarly, from (2.24),

$$\begin{aligned}P_k(t+h) &= \Pr\{N(t+h)=k\}\\ &= \Pr\{N(t)=k-1, N(t+h)-N(t)=1\}\\ &\quad + \Pr\{N(t)=k, N(t+h)-N(t)=0\}\\ &\quad + \sum_{j=2}^{k}\Pr\{N(t)=k-j, N(t+h)-N(t)=j\}\\ &= P_{k-1}(t)[\lambda h+o(h)] + P_k(t)[1-\lambda h+o(h)] + o(h).\end{aligned}$$

Hence,

$$\frac{P_k(t+h)-P_k(t)}{h} = \lambda P_{k-1}(t) - \lambda P_k(t) + \frac{o(h)}{h}.$$

Letting $h \to 0$,

$$\frac{\mathrm{d}P_k(t)}{\mathrm{d}t} = \lambda P_{k-1}(t) - \lambda P_k(t),$$

i.e.,

$$\mathrm{e}^{\lambda t}\left[P_k'(t) + \lambda P_k(t)\right] = \frac{\mathrm{d}\left[\mathrm{e}^{\lambda t}P_k(t)\right]}{\mathrm{d}t} = \lambda \mathrm{e}^{\lambda t}P_{k-1}(t). \tag{2.26}$$

Recalling that $P_0(t) = \mathrm{e}^{-\lambda t}$ in (2.25), when $k=1$,

$$\frac{\mathrm{d}[\mathrm{e}^{\lambda t}P_1(t)]}{\mathrm{d}t} = \lambda.$$

By integration under the initial condition of $P_1(0) = 0$,

$$P_1(t) = \lambda t \mathrm{e}^{-\lambda t}.$$

Using the mathematical induction by setting $P_{k-1}(t) = \left[(\lambda t)^{k-1}/(k-1)!\right]\mathrm{e}^{-\lambda t}$ in (2.26),

$$\frac{\mathrm{d}\left[\mathrm{e}^{\lambda t}P_k(t)\right]}{\mathrm{d}t} = \frac{\lambda(\lambda t)^{k-1}}{(k-1)!}.$$

Similarly, by integration under $P_k(0) = 0$,

$$P_k(t) = \frac{(\lambda t)^k}{k!}\mathrm{e}^{-\lambda t} \quad (k = 0, 1, 2, \ldots), \tag{2.27}$$

that corresponds to a Poisson distribution in (2.17).

Conversely, when a counting process $N(t)$ has a Poisson distribution in (2.27), the interarrival times X_k $(k = 1, 2, \ldots)$ are independent and identically distributed according to an exponential distribution $\left(1 - \mathrm{e}^{-\lambda t}\right)$ [4, p. 69]: It is easily shown that from (2.16),

$$\Pr\{X_1 \le t\} = \Pr\{S_1 \le t\} = \Pr\{N(t) \ge 1\} = 1 - \Pr\{N(t) = 0\} = 1 - \mathrm{e}^{-\lambda t},$$

i.e., X_1 has an exponential distribution $\left(1 - \mathrm{e}^{-\lambda t}\right)$. Similarly,

$$\begin{aligned}\Pr\{X_2 \le t | X_1 = u\} &= \Pr\{N(t+u) - N(u) \ge 1 | X_1 = u\} \\ &= 1 - \Pr\{N(t+u) - N(u) = 0\} = 1 - \Pr\{N(t) = 0\} = 1 - \mathrm{e}^{-\lambda t}.\end{aligned}$$

Generally,

$$\begin{aligned}\Pr\{X_{n+1} \le t | S_n = u\} &= \Pr\{N(t+u) - N(u) \ge 1 | S_n = u\} \\ &= 1 - \Pr\{N(t+u) - N(u) = 0\} = 1 - \mathrm{e}^{-\lambda t}.\end{aligned}$$

Therefore, each interarrival time X_k is independent and identically distributed exponentially in (2.1). Thus, a Poisson process has memoryless property. In addition, because an exponential density $\lambda \mathrm{e}^{-\lambda t}$ decreases strictly, a Poisson process has shorter interarrival times than longer ones. From (2.16), the arrival time S_n of the nth event has a gamma distribution

$$\Pr\{S_n \le t\} = \sum_{k=n}^{\infty} \frac{(\lambda t)^k}{k!}\mathrm{e}^{-\lambda t}.$$

Applying the stationary property of (ii) to (2.27),

$$\Pr\{N(t+u) - N(u) = k | N(0) = 0\} = \Pr\{N(t) = k\} = P_k(t)$$

for any $t, u \ge 0$. Using this result, we can make an alternative definition of a Poisson process [4, p. 67, 8, p. 31]:

Definition 2.2 If

(i) $N(0) = 0$,
(ii) the process $\{N(t)\}$ has independent increments,
(iii) the probability that k events occur in any interval t is

$$\Pr\{N(t+u) - N(u) = k\} = \frac{(\lambda t)^k}{k!} \mathrm{e}^{-\lambda t} \quad (k = 0, 1, 2, \ldots)$$

for any $t, u \geq 0$, then a counting process $\{N(t)\}$ is a Poisson process.

It can be easily seen that two definitions are equivalent: From the above results, Definition 2.1 implies Definition 2.2. It can be clearly verified that the process has stationary increments from (iii) of Definition 2.2. Furthermore, from (iii) and (3) of Sect. 2.1.1,

$$\Pr\{N(h) = 1\} = \lambda h \mathrm{e}^{-\lambda h} = \lambda h + o(h),$$
$$\Pr\{N(h) \geq 2\} = \sum_{k=2}^{\infty} \frac{(\lambda h)^k}{k!} \mathrm{e}^{-\lambda h} = o(h),$$

that follows that Definition 2.2 implies Definition 2.1. A Poisson process involves all properties shown in Definitions 2.1 and 2.2.

Example 2.7 (Poisson distribution) It has been estimated from the past life data that failures of some unit occur in a Poisson process and its expected number is twice the average per year, i.e., $365 \times 24 = 8,760$ h. Then, from $\lambda t = 2$, the MTTF is

$$\frac{1}{\lambda} = \frac{t}{2} = \frac{8,760}{2} = 4,380\,\mathrm{h}.$$

Thus, the unit has a failure exponential distribution $\left(1 - \mathrm{e}^{-t/4380}\right)$ and failures occur in a Poisson distribution $\left\{[t/4380]^k / k!\right\} \mathrm{e}^{-t/4380}$ $(k = 0, 1, 2, \ldots)$.

The number of units that should be provided to assure with probability 0.90 to operate continuously during one year without replenishment is given by the minimum number that satisfies, from Example 2.6,

$$\sum_{k=0}^{n} \frac{2^k}{k!} \mathrm{e}^{-2} \geq 0.9,$$

i.e., its number is 5, including the first operating unit.

Example 2.8 (Superposition of Poisson process) [5, p. 87] Let $N_1(t)$ be the number of failed units from one factory arriving at a repair shop in $[0, t]$ and $N_2(t)$ be the number of failed units from the other. If the arrival times from two factories are independent and have the respective Poisson processes with rates λ_1 and λ_2, the total number $N(t)$ of failed units arriving at a repair shop has the probability

$$\Pr\{N(t)=n\}=\Pr\{N_1(t)+N_2(t)=n\}=\sum_{k=0}^{n}\Pr\{N_1(t)=k,N_2(t)=n-k\}$$
$$=\sum_{k=0}^{n}\frac{(\lambda_1 t)^k}{k!}e^{-\lambda_1 t}\frac{(\lambda_2 t)^{n-k}}{(n-k)!}e^{-\lambda_2 t}$$
$$=\frac{[(\lambda_1+\lambda_2)t]^n}{n!}e^{-(\lambda_1+\lambda_2)t}\sum_{k=0}^{n}\frac{n!}{k!(n-k)!}\left(\frac{\lambda_1}{\lambda_1+\lambda_2}\right)^k\left(\frac{\lambda_2}{\lambda_1+\lambda_2}\right)^{n-k}$$
$$=\frac{[(\lambda_1+\lambda_2)t]^n}{n!}e^{-(\lambda_1+\lambda_2)t}.$$

Thus, the process $\{N(t)\}$ is also a Poisson process with rate $\lambda_1+\lambda_2$ (Problem 2.11).

Example 2.9 (Decomposition of Poisson process) [5, p. 88] Let $N(t)$ be the number of failed units occurring at a factory in $[0,t]$ and be a Poisson process with rate λ. Classifying into two large groups of failed units, the number $N_1(t)$ of minor ones occurs with probability p and the number $N_2(t)$ of major ones occurs with probability $q\equiv 1-p$, independent of $N_1(t)$. Then, because the number $N_1(t)$ has a binomial distribution with p, given that n failures have occurred, the joint probability is

$$\Pr\{N_1(t)=k,N_2(t)=n-k\}$$
$$=\Pr\{N_1(t)=k,N_2(t)=n-k|N(t)=n\}\Pr\{N(t)=n\}$$
$$=\binom{n}{k}p^k q^{n-k}\frac{(\lambda t)^n}{n!}e^{-\lambda t}=\frac{(p\lambda t)^k}{k!}e^{-p\lambda t}\frac{(q\lambda t)^{n-k}}{(n-k)!}e^{-q\lambda t}.$$

Thus, the two Poisson processes $\{N_1(t)\}$ and $\{N_2(t)\}$ are independent and have the Poisson processes with rates $p\lambda$ and $q\lambda$, respectively (Problem 2.12).

In general, when the total number $N(t)$ of failed units with a Poisson process with rate λ are classified into k groups of $N_j(t)$ $(j=1,2,\ldots,k)$ with probability p_j where $\sum_{j=1}^{k}N_j(t)=N(t)$ and $\sum_{j=1}^{k}p_j=1$, the joint probability is

$$\Pr\{N_1(t)=n_1,N_2(t)=n_2,\ldots,N_k(t)=n_k\}$$
$$=\frac{(p_1\lambda t)^{n_1}}{n_1!}e^{-p_1\lambda t}\frac{(p_2\lambda t)^{n_2}}{n_2!}e^{-p_2\lambda t}\cdots\frac{(p_k\lambda t)^{n_k}}{n_k!}e^{-p_k\lambda t}, \tag{2.28}$$

that is called a *multi-Poisson process*.

Next, when events occur in a Poisson process, we obtain the distribution of the interarrival time X, given that there was an event in $[0,t]$ [4, p.71, 8, p.36]. This probability is given by, for $u\le t$,

$$\Pr\{X \leq u | N(t) = 1\} = \frac{\Pr\{N(u) = 1, N(t) - N(u) = 0\}}{\Pr\{N(t) = 1\}}$$
$$= \frac{\lambda u \mathrm{e}^{-\lambda u} \mathrm{e}^{-\lambda(t-u)}}{\lambda t \mathrm{e}^{-\lambda t}} = \frac{u}{t}, \tag{2.29}$$

that is a uniform distribution over $[0, t]$. That is, when the event was detected at time t, it occurs constantly over $[0, t]$.

Example 2.10 (Binomial distribution) [9, p. 76] Let $N(t)$ be a Poisson process with rate λ. Then, from Definition 2.2, for $0 \leq u \leq t$,

$$\Pr\{N(u) = k | N(t) = n\} = \frac{\Pr\{N(u) = k, N(t) = n\}}{\Pr\{N(t) = n\}}$$
$$= \frac{\Pr\{N(u) = k\} \Pr\{N(t) - N(u) = n - k\}}{\Pr\{N(t) = n\}}$$
$$= \frac{[(\lambda u)^k / k!] \mathrm{e}^{-\lambda u} \left\{ [\lambda(t-u)]^{n-k} / (n-k)! \right\} \mathrm{e}^{-\lambda(t-u)}}{[(\lambda t)^n / n!] \mathrm{e}^{-\lambda t}}$$
$$= \binom{n}{k} \left(\frac{u}{t}\right)^k \left(1 - \frac{u}{t}\right)^{n-k} \quad (k = 0, 1, 2, \ldots, n), \tag{2.30}$$

that is a binomial distribution with parameters n and u/t.

A uniform distribution sometimes appears in the cases where events occur only in the finite interval and when an event was detected at time t, without any information of its arrival time, as shown in (2.29).

Example 2.11 (Inspection policy) [2, p. 113] Consider an inspection policy where the failure time of a unit is uniformly distributed over $[0, S]$ that is called a working interval [6, p. 96]. To detect a failure, the unit is checked at planned times T_k $(k = 1, 2, \ldots, N)$, where $T_N = S$ (Fig. 2.2).

Let c_1 be the cost of one check and c_2 be the cost per unit of time for the time elapsed between a failure and its detection at the next checking time. Then, the expected cost until failure detection is

$$\boldsymbol{C}_1(N) = \sum_{k=0}^{N-1} \int_{T_k}^{T_{k+1}} [(k+1)c_1 + c_2(T_{k+1} - t)] \frac{\mathrm{d}t}{S}. \tag{2.31}$$

Differentiating $\boldsymbol{C}_1(N)$ with respect to T_k and setting it equal to zero,

$$T_{k+1} - T_k = T_k - T_{k-1} - \frac{c_1}{c_2}.$$

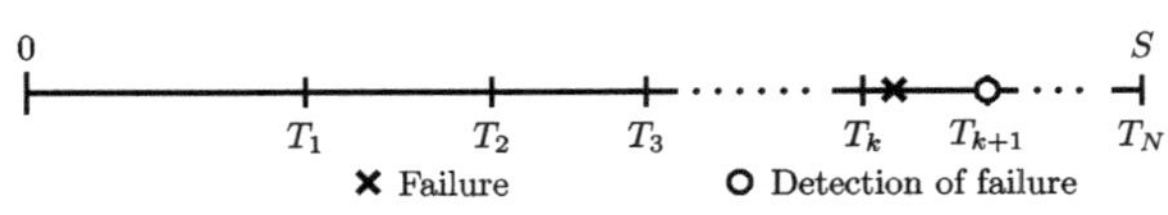

Fig. 2.2 Process of an inspection policy

Solving for T_k,

$$T_k = kT_1 - \frac{k(k-1)}{2}\frac{c_1}{c_2}.$$

Setting $T_N = S$,

$$T_k = \frac{kS}{N} + k(N-k)\frac{c_1}{2c_2} \quad (k = 0, 1, 2, \ldots, N).$$

From $T_{k+1} - T_k > 0$,

$$\frac{S}{N} + \frac{c_1}{2c_2}(N - 2k - 1) > 0.$$

When $k = N - 1$,

$$\frac{N(N-1)}{2} < \frac{c_2 S}{c_1}, \quad \text{i.e.,} \quad \frac{N(N+1)}{2} \geq \frac{c_2 S}{c_1}. \tag{2.32}$$

Thus, N must be a unique minimum that satisfies $\sum_{k=1}^{N} k \geq c_2 S/c_1$. Note that the summation of integers from 1 to N plays an important rule in obtaining optimum policies for some partition models [6, p. 39] (Problem 2.13). For example, when $S = 100$ and $c_1/c_2 = 7$, the checking number is $N = 5$ and checking times are 34, 61, 81, 94, 100.

Next, consider the backup policy with checkpoint times in which we execute the rollback operation to the latest checkpoint [6, p. 96] and reconstruct the consistent state. Let c_2 be the cost per unit of time for the backup operation to the latest checking time T_k when the failure was detected between T_k and T_{k+1} (Fig. 2.3). The other assumptions are the same as those in the inspection policy. Then, the expected cost until the backup operation is done to the latest checking time when a unit has failed at time t is

$$\boldsymbol{C}_2(N) = \sum_{k=0}^{N-1} \int_{T_k}^{T_{k+1}} [kc_1 + c_2(t - T_k)] \frac{\mathrm{d}t}{S}. \tag{2.33}$$

Differentiating $\boldsymbol{C}_2(N)$ with respect to T_k and setting it equal to zero,

$$T_{k+1} - T_k = T_k - T_{k-1} - \frac{c_1}{c_2}.$$

Therefore, the problem of optimizing checkpoint intervals of the backup operation corresponds to that of the inspection policy when the failure time has a uniform distribution. This is the reason, as failures occur uniformly over any interval.

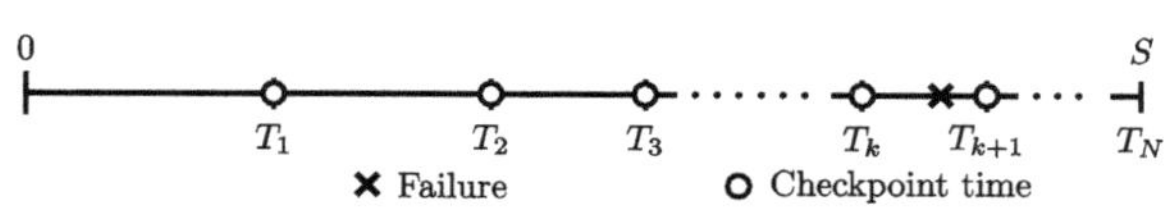

Fig. 2.3 Process of backup policy

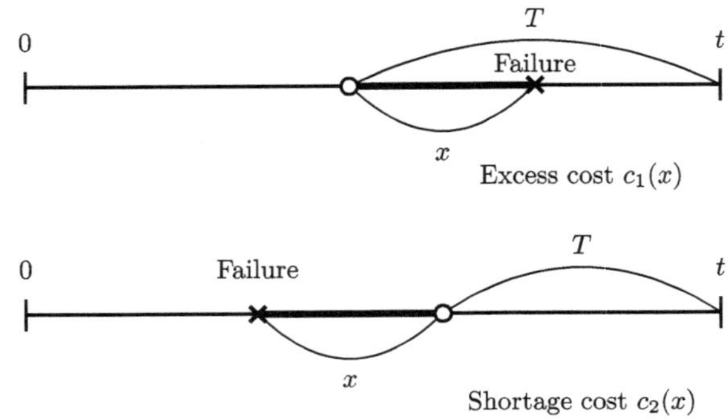

Fig. 2.4 Excess cost and shortage cost of backward time T

Example 2.12 (Backward time problem) [6, p. 88] Suppose that when a unit has failed, and its failure time is unknown, we go back to time T $(0 \le T \le t)$ from time t to detect its failure and call T a *planned backward time*. Then, we introduce the following costs: Cost $c_1(x)$ is the excess cost for the time x from a failure to the backward time, $c_2(x)$ is the shortage cost for the time x from the backward time to a failure, and $c_0(T)$ is the cost required for the backward time with $c_0(0) \equiv 0$ (Fig. 2.4). It is assumed that a failure is distributed uniformly over $[0, t]$ based on the result obtained in (2.29). Then, the expected cost for the backward time T is

$$C(T|t) = \frac{1}{t}\left[\int_0^T c_1(x)\mathrm{d}x + \int_0^{t-T} c_2(x)\mathrm{d}x\right] + c_0(T). \tag{2.34}$$

In particular, when $c_i(t) = c_i t$ $(i = 0, 1, 2)$, the expected cost is

$$C(T|t) = \frac{1}{2t}\left[c_1 T^2 + c_2(t-T)^2\right] + c_0 T. \tag{2.35}$$

Differentiating $C(T|t)$ with respect to T and setting it equal to zero,

$$T = \frac{c_2 - c_0}{c_2 + c_1} t. \tag{2.36}$$

Therefore, an optimum $T^*(0 < T^* < t)$ that minimizes (2.35) is given by (2.36) for $c_2 > c_0$. If $c_2 \le c_0$, then $T^* = 0$, i.e., we should not go back at all because the backward cost is high (Problem 2.14).

This result in (2.29) is generalized as follows [4, p. 71, 8, p. 37]: The kth arrival times S_k $(k = 1, 2, \ldots, n)$, given that $N(t) = n$, has the probability

$$\begin{aligned}\Pr\{t_k < S_k < t_k + h_k, k = 1, 2, \ldots, n | N(t) = n\} &= \frac{\lambda h_1 \mathrm{e}^{-\lambda h_1} \ldots \lambda h_n \mathrm{e}^{-\lambda h_n} \mathrm{e}^{-\lambda(t - h_1 - h_2 \cdots - h_n)}}{[(\lambda t)^n / n!]\mathrm{e}^{-\lambda t}} \\ &= \frac{n!}{t^n} h_1 h_2 \ldots h_n\end{aligned}$$

for any $0<t_1<t_2<\cdots<t_n<t_{n+1}=t$ and $t_k+h_k<t_{k+1}$ for small h_k, and hence, its density function is

$$f(t_1,t_2,\ldots,t_n|N(t)=n)=\frac{n!}{t^n}. \tag{2.37}$$

That is, n arrival times $S_1,S_2,\ldots,S_n$ given that $N(t)=n$ are independent and identically distributed uniformly over the interval $[0,t]$. In addition [10, p. 15], from Example 2.10,

$$\begin{aligned}\Pr\{S_k\le u|N(t)=n\} &= \frac{\Pr\{S_k\le u, N(t)=n\}}{\Pr\{N(t)=n\}}\\ &= \frac{\Pr\{N(u)\ge k, N(t)=n\}}{\Pr\{N(t)=n\}}\\ &= \frac{\sum_{j=k}^{n}\Pr\{N(u)=j\}\Pr\{N(t)-N(u)=n-j\}}{\Pr\{N(t)=n\}}\\ &= \sum_{j=k}^{n}\binom{n}{j}\left(\frac{u}{t}\right)^j\left(1-\frac{u}{t}\right)^{n-j}.\end{aligned} \tag{2.38}$$

Clearly (Problem 2.15),

$$E\{S_k|N(t)=n\}=\frac{kt}{n+1},\quad E\{S_k-S_{k-1}|N(t)=n\}=\frac{t}{n+1}. \tag{2.39}$$

2.3 Nonhomogeneous Poisson Process

Suppose that a unit begins to operate at time 0. If the unit fails, then it undergoes minimal repair and begins to operate again. It is assumed that the time for repair is negligible. Let us denote $0\equiv S_0\le S_1\le\cdots\le S_{n-1}\le S_n\le\ldots$ be the successive failure times of the unit, and $X_n\equiv S_n-S_{n-1}$ $(n=1,2,\ldots)$ be the times between failures (Fig. 2.1) with $F(t)\equiv\Pr\{X_1\le t\}$ and $\overline{F}(t)\equiv 1-F(t)$. We define to make minimal repair at failure as follows [3, p. 96]:

Definition 2.3 The unit undergoes minimal repair at failures if and only if

$$\Pr\{X_n\le x|S_{n-1}=t\}=\frac{F(t+x)-F(t)}{\overline{F}(t)}\quad (n=2,3,\ldots) \tag{2.40}$$

for $x>0$ such that $F(t)<1$.

The function $[F(t+x)-F(t)]/\overline{F}(t)$ is called the *failure rate* and represents the probability that a unit with age t fails in the interval $(t,t+x]$. The definition in reliability models means that the failure rate remains undisturbed by any minimal repair of failures, i.e., a unit after each minimal repair has the same failure rate as previous failure. Imperfect PM (Preventive Maintenance) models in which the failure rate after PM reduces were considered and their optimum policies were discussed analytically [3, p. 171]. A variety of failure rate modelling was extensively collected [12].

Assume that $F(t)$ has a density function $f(t)$ and $h(t) \equiv f(t)/\overline{F}(t)$ is continuous. The function $h(t)$ is also called the instantaneous failure rate or simply the *failure rate* and has the same monotone property as $[F(t+x)-F(t)]/\overline{F}(t)$ [2, p. 23]. Moreover, $H(t) \equiv \int_0^t h(u)\mathrm{d}u$ is called the *cumulative hazard function* and satisfies the relation $\overline{F}(t) = \mathrm{e}^{-H(t)}$ that decreases with t. Because, from the definition of $h(t)$,

$$\frac{-\left[\overline{F}(t)\right]'}{\overline{F}(t)} = -\frac{\mathrm{d}\left[\log \overline{F}(t)\right]}{\mathrm{d}t} = h(t).$$

By integration under the initial condition $\overline{F}(0) = 1$,

$$\log \overline{F}(t) = -\int_0^t h(u)\mathrm{d}u = -H(t), \quad \text{i.e.,} \quad \overline{F}(t) = \mathrm{e}^{-H(t)}. \tag{2.41}$$

We derive the distribution of the nth failure time S_n;

$$G_n(t) \equiv \Pr\{S_n \le t\} = \sum_{k=n}^{\infty} \frac{[H(t)]^k}{k!}\mathrm{e}^{-H(t)} = 1 - \sum_{k=0}^{n-1} \frac{[H(t)]^k}{k!}\mathrm{e}^{-H(t)} \quad (n = 1, 2, \ldots). \tag{2.42}$$

By the mathematical induction,

$$G_1(t) = \Pr\{X_1 \le t\} = F(t) = 1 - \mathrm{e}^{-H(t)},$$

$$\begin{aligned} G_{n+1}(t) &= \int_0^\infty \Pr\{X_{n+1} \le t - u | S_n = u\}\mathrm{d}G_n(u) \\ &= \int_0^t \frac{F(t) - F(u)}{\overline{F}(u)} \frac{[H(u)]^{n-1}}{(n-1)!}\mathrm{d}F(u) \\ &= 1 - \sum_{k=0}^{n-1} \frac{[H(t)]^k}{k!}\mathrm{e}^{-H(t)} - \mathrm{e}^{-H(t)} \int_0^t \frac{[H(u)]^{n-1}}{(n-1)!} h(u)\mathrm{d}u \\ &= 1 - \sum_{k=0}^{n} \frac{[H(t)]^k}{k!}\mathrm{e}^{-H(t)} \quad (n = 1, 2, \ldots). \end{aligned}$$

The density function of $G_n(t)$ is

$$g_n(t) \equiv G_n'(t) = \frac{[H(t)]^{n-1}}{(n-1)!}\mathrm{e}^{-H(t)}h(t) = \frac{[H(t)]^{n-1}}{(n-1)!}f(t),$$

and hence, we have the relation

$$G_n(t) = \int_0^t \frac{[H(u)]^{n-1}}{(n-1)!} \mathrm{d}F(u) = \sum_{k=n}^{\infty} \frac{[H(t)]^k}{k!} \mathrm{e}^{-H(t)}. \tag{2.43}$$

From the above results,

$$\begin{aligned} E\{S_n\} &\equiv \int_0^\infty \overline{G}_n(t)\mathrm{d}t = \sum_{k=0}^{n-1} \int_0^\infty \frac{[H(t)]^k}{k!} \mathrm{e}^{-H(t)}\mathrm{d}t, \\ E\{X_n\} &= E\{S_n\} - E\{S_{n-1}\} = \int_0^\infty \frac{[H(t)]^{n-1}}{(n-1)!} \mathrm{e}^{-H(t)}\mathrm{d}t. \end{aligned} \tag{2.44}$$

Next, we prove that when $h(t)$ increases (decreases), $E\{X_n\}$ decreases (increases) with n and converge to $1/h(\infty)$ as $n \to \infty$ [3, p. 98]. Using the relation (Problem 2.16)

$$\frac{[H(t)]^n}{n!} = \int_0^t \frac{[H(u)]^{n-1}}{(n-1)!} h(u)\mathrm{d}u,$$

and from the assumption that $h(t)$ increases, it follows that

$$\begin{aligned} \int_0^\infty \frac{[H(t)]^n}{n!} \mathrm{e}^{-H(t)}\mathrm{d}t &= \int_0^\infty \left\{ \int_0^t \frac{[H(u)]^{n-1}}{(n-1)!} h(u)\mathrm{d}u \right\} \mathrm{e}^{-H(t)}\mathrm{d}t \\ &\le \int_0^\infty \left\{ \int_0^t \frac{[H(u)]^{n-1}}{(n-1)!} \mathrm{d}u \right\} h(t)\mathrm{e}^{-H(t)}\mathrm{d}t \\ &= \int_0^\infty \frac{[H(t)]^{n-1}}{(n-1)!} \mathrm{e}^{-H(t)}\mathrm{d}t, \end{aligned}$$

and hence, $E\{X_n\}$ decreases with n. Furthermore, from (2.43),

$$\begin{aligned} \int_0^\infty \frac{[H(t)]^n}{n!} \mathrm{e}^{-H(t)}\mathrm{d}t &= \int_0^\infty \frac{[H(t)]^n}{n!} \frac{h(t)}{h(t)} \mathrm{e}^{-H(t)}\mathrm{d}t \\ &\ge \frac{1}{h(\infty)} \int_0^\infty \frac{[H(t)]^n}{n!} \mathrm{d}F(t) = \frac{1}{h(\infty)}. \end{aligned}$$

On the other hand, for any $0<T<\infty$,

$$\int_0^\infty \frac{[H(t)]^n}{n!}\mathrm{e}^{-H(t)}\mathrm{d}t = \int_0^T \frac{[H(t)]^n}{n!}\mathrm{e}^{-H(t)}\mathrm{d}t + \int_T^\infty \frac{[H(t)]^n}{n!}\mathrm{e}^{-H(t)}\mathrm{d}t$$
$$\leq \int_0^T \frac{[H(t)]^n}{n!}\mathrm{e}^{-H(t)}\mathrm{d}t + \frac{1}{h(T)}.$$

Thus, because

$$\lim_{n\to\infty}\int_0^\infty \frac{[H(t)]^n}{n!}\mathrm{e}^{-H(t)}\mathrm{d}t \leq \frac{1}{h(T)},$$

and T is arbitrary,

$$\lim_{n\to\infty}\int_0^\infty \frac{[H(t)]^n}{n!}\mathrm{e}^{-H(t)}\mathrm{d}t = \frac{1}{h(\infty)}.$$

Similarly, when $h(t)$ decreases, we can prove that $E\{X_n\}$ increases with n and converge to $1/h(\infty)$ as $n \to \infty$, where we set that $1/h(\infty) \equiv 0$ for $h(\infty) = \infty$ and $1/h(\infty) \equiv \infty$ for $h(\infty) = 0$. Note that when $h(t)$ increases strictly to $\infty, E\{X_n\}$ decreases strictly to 0. The failure rate $h(t)$ and the cumulative hazard function $H(t) \equiv \int_0^t h(u)\mathrm{d}u$ in reliability theory are called an *intensity function* and a *mean value function* in stochastic processes, respectively.

Furthermore, let $N(t)$ be the number of failures of a unit in $[0, t]$, i.e., $N(t) \equiv \max_n\{S_n \leq t\}$. Then, using the relation with $N(t)$ and S_n in (2.16),

$$P_n(t) \equiv \Pr\{N(t) = n\} = \Pr\{S_n \leq t < S_{n+1}\} = G_n(t) - G_{n+1}(t)$$
$$= \frac{[H(t)]^n}{n!}\mathrm{e}^{-H(t)} \quad (n = 0, 1, 2, \ldots), \tag{2.45}$$
$$E\{N(t)\} = V\{N(t)\} = H(t).$$

In particular, when $F(t) = 1 - \mathrm{e}^{-\lambda t}$, i.e., $H(t) = \lambda t$,

$$\Pr\{S_n \leq t\} = G_n(t) = 1 - \sum_{k=0}^{n-1}\frac{(\lambda t)^k}{k!}\mathrm{e}^{-\lambda t}, \quad E(S_n) = \frac{n}{\lambda},$$
$$\Pr\{X_n \leq t\} = 1 - \mathrm{e}^{-\lambda t}, \quad E\{X_n\} = \frac{1}{\lambda}, \quad E\{N(t)\} = V\{N(t)\} = \lambda t,$$

that agree with those of the Poisson process with rate λ.

Referring to the above results and expanding Definition 2.1, we can generalize the Poisson process with the parameter of time t [4, p. 77, 8, p. 46]:

Definition 2.4 If

(i) $N(0) = 0$,
(ii) the process $\{N(t)\}$ has independent increments,
(iii) $\Pr\{N(t+h) - N(t) = 1\} = \lambda(t)h + o(h)$,
(iv) $\Pr\{N(t+h) - N(t) \geq 2\} = o(h)$,

then a counting process $\{N(t)\}$ is said to be a *nonhomogeneous* or *nonstationary Poisson process* that eliminates the stationary from a Poisson process.

Using this definition, we derive the probability

$$P_n(u) \equiv \Pr\{N(t+u) - N(t) = n\} = \frac{[H(t+u) - H(t)]^n}{n!} \mathrm{e}^{-[H(t+u)-H(t)]} \quad (n = 0, 1, 2, \ldots) \tag{2.46}$$

for a fixed t. From (ii), (iii), and (iv),

$$\begin{aligned} P_0(u+h) &= \Pr\{N(t+u+h) - N(t) = 0\} \\ &= \Pr\{N(t+u) - N(t) = 0, N(t+u+h) - N(t+u) = 0\} \\ &= \Pr\{N(t+u) - N(t) = 0\} \Pr\{N(t+u+h) - N(t+u) = 0\} \\ &= P_0(u)[1 - \lambda(t+u)h + o(h)]. \end{aligned}$$

Hence,

$$\frac{P_0(u+h) - P_0(u)}{h} = -\lambda(t+u)P_0(u) + \frac{o(h)}{h}.$$

Letting $h \to 0$, we have the differential equation

$$P_0'(u) = -\lambda(t+u)P_0(u), \quad \text{i.e.,} \quad \frac{\mathrm{d}P_0(u)/\mathrm{d}u}{P_0(u)} = -\lambda(t+u).$$

By integration and setting $P_0(0) = 1$,

$$\log P_0(u) = -\int_0^u \lambda(t+x)\mathrm{d}x = -[H(t+u) - H(t)],$$

that implies

$$P_0(u) = \mathrm{e}^{-[H(t+u)-H(t)]}.$$

Similarly, by the same method used in the Poisson process, we can prove (2.46). Clearly,

$$\begin{aligned} E\{N(t+u) - N(t)\} &= \sum_{n=0}^{\infty} n \frac{[H(t+u) - H(t)]^n}{n!} \mathrm{e}^{-H(t+u)-H(t)} \\ &= H(t+u) - H(t). \end{aligned}$$

Finally, the distribution of the interarrival time $X_1 = S_1$, given that there was an event in $[0, t]$, is [4, p. 78], for $u \leq t$,

$$\Pr\{S_1 \leq u | N(t) = 1\} = \frac{\Pr\{N(u) = 1, N(t) - N(u) = 0\}}{\Pr\{N(t) = 1\}}$$
$$= \frac{\Pr\{N(u) = 1\} \Pr\{N(t) - N(u) = 0\}}{\Pr\{N(t) = 1\}}$$
$$= \frac{H(u)\mathrm{e}^{-H(u)}\mathrm{e}^{-[H(t)-H(u)]}}{H(t)\mathrm{e}^{-H(t)}} = \frac{H(u)}{H(t)}. \tag{2.47}$$

Example 2.13 (Expected number of failures) A random variable $Y \equiv H(X)$, that represents the expected number of failures until the failure time, has the following distribution

$$\Pr\{Y \leq t\} = \Pr\{H(X) \leq t\} = \Pr\{X \leq H^{-1}(t)\} = 1 - \mathrm{e}^{-t}, \tag{2.48}$$

where H^{-1} is the inverse function of H. Thus, Y has an exponential distribution with mean 1, and $E\{H(X)\} = V\{H(X)\} = 1$. In other words, by transforming the failure times $S_1, S_2, \ldots$ to $H(S_1), H(S_2), \ldots$, a nonhomogeneous Poisson process corresponds to a Poisson process with rate 1.

Moreover, x_1 that satisfies $H(x_1) = 1$ is called a *characteristic life* in the probability paper of a Weibull distribution. This represents the mean lifetime that about 63.2 % of units have failed until time x_1, as shown in (1) of Sect. 2.1.1.

In general, x_n that satisfies $H(x_n) = n$ $(n = 1, 2, \ldots)$ represents the time that the expected number of failures is n, when failures occur in a nonhomogeneous Poisson with a mean value function $H(t)$. From $H(x_n) = n$,

$$\int_{x_{n-1}}^{x_n} h(t)\mathrm{d}t = 1 \quad \text{or} \quad \int_0^{x_n} h(t)\mathrm{d}t = n \quad (n = 1, 2, \ldots). \tag{2.49}$$

In particular, when $F(t)$ is IFR,

$$x_n - x_{n-1} \leq \frac{1}{h(x_{n-1})}, \quad \text{i.e.,} \quad x_n \leq \frac{1}{h(x_{n-1})} + x_{n-1}.$$

Thus, denoting that

$$\overline{x}_n \equiv \frac{1}{h(x_{n-1})} + x_{n-1} \quad (n = 2, 3, \ldots),$$

$x_n \leq \overline{x}_n$. Similarly,

$$x_n - \frac{1}{h(x_n)} \geq x_{n-1}.$$

Thus, denoting $\underline{x}_n$ that satisfies

$$\underline{x}_n - \frac{1}{h(\underline{x}_n)} = x_{n-1} \quad (n = 2, 3, \ldots),$$

$x_n \geq \underline{x}_n$. The function $1/h(t)$ is called an *instantaneous MTBF* [11, p. 71].

Furthermore, from Theorem 4.4 of [2, p. 27] when F is IFR (DFR) and $\mu \equiv E\{X\}, \overline{F}(\mu) \geq (\leq) \mathrm{e}^{-1}$. Thus, $H(\mu) \leq (\geq) H(x_1)$, i.e., $\mu \leq (\geq) x_1$.

Example 2.14 (Weibull distribution) Suppose that $H(t) = \lambda t^\alpha$ for $\alpha > 0$, i.e., the failure time has a Weibull distribution $F(t) = 1 - \exp(-\lambda t^\alpha)$ with a shape parameter α. Then, the failure rate is $h(t) = \lambda \alpha t^{\alpha-1}$, and decreases for $0 < \alpha < 1$, constant for $\alpha = 1$, and increases for $1 < \alpha < \infty$. So that, the Weibull distribution represents a typical failure distribution that has three patterns in the bath-tub curve (Fig. 2.5). In this case, the process is sometimes called a *Weibull process*. Furthermore, from (2.49), $x_n = (n/\lambda)^{1/\alpha}$ $(n = 1, 2, \ldots)$ and $\mu = (1/\lambda)^{1/\alpha}\Gamma(1 + 1/\alpha)$. Therefore, (i) $x_1 < \mu$ for $0 < \alpha < 1$, (ii) $x_1 = \mu$ for $\alpha = 1$, and $x_1 > \mu$ for $\alpha > 1$. Table 2.4. presents $x_n, \overline{x}_n$ and $\underline{x}_n$ when $\overline{x}_1 = \underline{x}_1 = x_1$, and $E(S_n)$ from (2.44) for $\alpha = 2$ and $\lambda = 0.01$. This indicates that $x_n > E(S_n)$, however, $\underline{x}_n$ are almost the same as $E(S_n)$ for $n \geq 2$. Many Weibull models were summarized [13].

Let $0 \equiv S_0 < S_1 < S_2 < \ldots$ be the successive times of a nonhomogeneous Poisson process with a mean value function $H(t)$. Then, we have the relation, from Example 2.13,

$$\{S_{n+1} - S_n > t\} \Leftrightarrow \{H(S_{n+1}) > H(S_n + t)\}.$$

In addition, because $H(S_{n+1}) - H(S_n)$ is independent of $S_1, S_2, \ldots, S_n$ and has an exponential distribution with mean 1,

$$\begin{aligned}
\Pr\{S_{n+1} - S_n > t | S_1, S_2, \ldots, S_n\} &= \Pr\{H(S_{n+1}) > H(S_n + t) | S_1, S_2, \ldots, S_n\} \\
&= \Pr\{H(S_{n+1}) - H(S_n) > H(S_n + t) - H(S_n)\} \\
&= \mathrm{e}^{-[H(S_n+t) - H(S_n)]}.
\end{aligned}$$

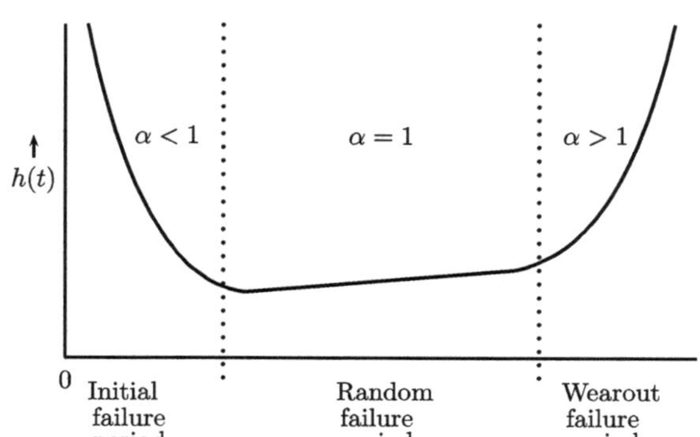

Fig. 2.5 Bath-tub curve of the failure rate

Table 2.4 Comparisons of times of expected number when $\alpha = 2$ and $\lambda = 0.01$

n	x_n	$\overline{x}_n(\overline{x}_1 = x_1)$	$\underline{x}_n$	$E(S_n)$
1	10.00	10.00	10.00	8.86
2	14.14	15.00	13.66	13.29
3	17.32	18.33	17.07	16.62
4	20.00	21.06	19.84	19.39
5	22.36	23.43	22.25	21.81
6	24.49	25.56	24.41	23.99
7	26.46	27.52	26.39	25.99
8	28.28	29.34	28.23	27.85
9	30.00	31.04	29.95	29.59
10	31.62	32.65	31.59	31.23

The interarrival times for a nonhomogeneous Poisson are not independent, however, still have conditionally exponential distribution [5, p. 97].

2.4 Applications to Reliability Models

2.4.1 Replacement at the Nth *Failure*

A new unit begins to operate at time 0, and when it fails, only minimal repair is made. It is assumed that the repair and replacement times are negligible. The failure times of the unit have a density function $f(t)$ and a distribution $F(t)$ with finite mean $\mu \equiv \int_0^\infty \overline{F}(t)\mathrm{d}t$, failure rate $h(t) \equiv f(t)/\overline{F}(t)$, and cumulative hazard rate $H(t) \equiv \int_0^t h(u)\mathrm{d}u$.

Suppose that the unit is replaced only at the Nth failure. Then, the mean time to the Nth failure is, from (2.44),

$$E\{S_N\} = \sum_{k=0}^{N-1} \int_0^\infty P_k(t)\mathrm{d}t,$$

where $P_k(t) \equiv \{[H(t)]^k/k!\}\mathrm{e}^{-H(t)}$ $(k = 0, 1, 2, \ldots)$.

Consider one cycle from the beginning of operation to the replacement. Let c_1 be the cost of minimal repair and c_2 be the replacement cost at the Nth failure. The expected cost rate is [3, p. 106]

$$C(N) \equiv \frac{\text{Expected cost until replacement}}{\text{Mean time to replacement}} = \frac{c_1(N-1) + c_2}{\sum_{k=0}^{N-1} \int_0^\infty P_k(t)\mathrm{d}t} \quad (N = 1, 2, \ldots). \tag{2.50}$$

When the failure rate $h(t)$ increases strictly to $h(\infty)$, we derive an optimum number N^* that minimizes $C(N)$ in (2.50). From the inequality $C(N+1)-C(N)\geq 0$,

$$\frac{\sum_{k=0}^{N-1}\int_0^\infty P_k(t)\mathrm{d}t}{\int_0^\infty P_N(t)\mathrm{d}t}-(N-1)\geq\frac{c_2}{c_1}\quad(N=1,2,\ldots).\tag{2.51}$$

Letting $L(N)$ be the left-hand side of (2.51),

$$L(N+1)-L(N)=\sum_{k=0}^{N}\int_0^\infty P_k(t)\mathrm{d}t\left[\frac{1}{\int_0^\infty P_{N+1}(t)\mathrm{d}t}-\frac{1}{\int_0^\infty P_N(t)\mathrm{d}t}\right]>0,$$

because $\int_0^\infty P_k(t)\mathrm{d}t$ decreases strictly from (2.44) to $1/h(\infty)$ when $h(t)$ increases strictly. Thus, if a solution to (2.51) exists, then it is unique, and the resulting cost rate is

$$\frac{c_1}{\int_0^\infty P_{N^*-1}(t)\mathrm{d}t}<C(N^*)\leq\frac{c_1}{\int_0^\infty P_{N^*}(t)\mathrm{d}t}.\tag{2.52}$$

Furthermore, we have the inequality (Problem 2.17)

$$L(N)>\frac{\mu}{\int_0^\infty P_N(t)\mathrm{d}t}\quad(N=2,3,\ldots).$$

Therefore, if

$$\lim_{N\to\infty}\frac{\mu}{\int_0^\infty P_N(t)\mathrm{d}t}\geq\frac{c_2}{c_1},$$

i.e., if $h(\infty)\geq c_2/(\mu c_1)$, then a solution to (2.51) exists, and it is unique. In addition, if $\overline{N}$ is a unique minimum that satisfies

$$\int_0^\infty P_N(t)\mathrm{d}t\leq\frac{\mu c_1}{c_2},$$

then $N^*\leq\overline{N}$.

Example 2.15 (Optimum number) Suppose that the failure time of a unit has a Weibull distribution, i.e., $\overline{F}(t)=\exp(-\lambda t^\alpha)$ for $\alpha>1$. Then, $h(t)$ increases strictly from 0 to infinity, and (Problem 2.18)

$$\int_0^\infty P_N(t)\mathrm{d}t=\frac{1}{\alpha}\frac{\Gamma(N+1/\alpha)}{\lambda^{1/\alpha}\Gamma(N+1)},\quad\sum_{k=0}^{N-1}\int_0^\infty P_k(t)\mathrm{d}t=\frac{\Gamma(N+1/\alpha)}{\lambda^{1/\alpha}\Gamma(N)}.$$

Thus, there exists a finite and unique minimum N^* that satisfies (2.51) which is given by

$$N^* = \left[\frac{c_2 - c_1}{(\alpha - 1)c_1}\right]^{1/\alpha} + 1, \tag{2.53}$$

where $[x]$ denotes the greatest integer contained in x.

2.4.2 *Software Reliability Model*

In software reliability modeling [14], it is assumed that $H(t)$ is the expected number of faults detected by time t and the number of faults detected at any time is proportional to the current number of faults. Let $a > 0$ be the total number of faults, and $b(t)$ be the error detection rate per fault at time t. Then, because the number of faults at time t is $a - H(t)$, we have the differential equation

$$\frac{\mathrm{d}H(t)}{\mathrm{d}t} = b(t)[a - H(t)].$$

Solving this equation under the initial condition $H(0) \equiv 0$ (Problem 2.19),

$$H(t) = a\left\{1 - \exp\left[-\int_0^t b(u)\mathrm{d}u\right]\right\}. \tag{2.54}$$

This is called a software reliability growth model. In particular, when $b(t) = b > 0$,

$$H(t) = a\left(1 - \mathrm{e}^{-bt}\right),$$

and $H(\infty) = a$. A variety of generalized software reliability growth models were summarized extensively [11, 15], and several interesting new topics were cited [16].

2.5 Compound Poisson Process

Consider a standard cumulative damage model [17, p. 16, 18]: Shocks occur in a Poisson Process with rate λ. A unit is subjected to shocks and suffers some damage due to shocks. Let random variables X_k $(k = 1, 2, \ldots)$ denote a sequence of interarrival times between successive shocks, and random variables W_k $(k = 1, 2, \ldots)$ denote the damage produced by the kth shock, where $W_0 \equiv 0$ (Fig. 2.6). It is assumed that the sequence of $\{W_k\}$ is nonnegative, independently, and identically distributed according to $\Pr\{W_k \le x\} \equiv G(x)$ with finite mean, and furthermore, W_k is independent of X_j $(j \ne k)$.

Fig. 2.6 Process for a standard cumulative damage model

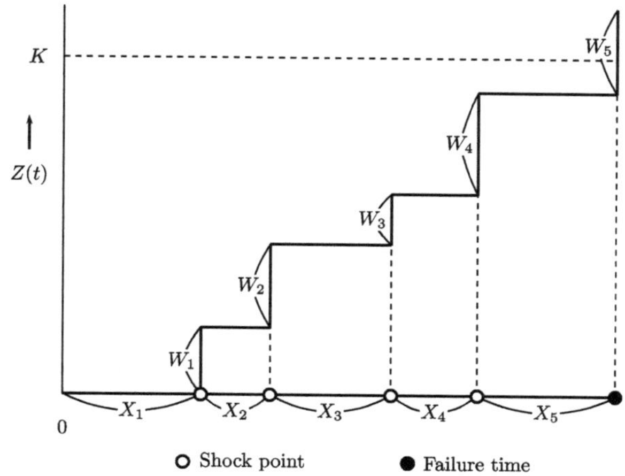

Let $N(t)$ denote the total number of shocks up to time t. Then, define a random variable

$$Z(t) \equiv \sum_{k=0}^{N(t)} W_k \quad (N(t) = 0, 1, 2, \ldots), \tag{2.55}$$

where $Z(t)$ represents the total damage at time t. This is called a *compound Poisson Process* and will be discussed in detail in Chap. 7.

The distribution of $Z(t)$ defined in (2.55) is

$$\begin{aligned}
\Pr\{Z(t) \le x\} &= \Pr\left\{\sum_{j=0}^{N(t)} W_j \le x\right\} \\
&= \sum_{k=0}^{\infty} \Pr\left\{\sum_{j=0}^{k} W_j \le x \,\middle|\, N(t) = k\right\} \Pr\{N(t) = k\} \\
&= \sum_{k=0}^{\infty} \Pr\{W_0 + W_1 + \cdots + W_k \le x\} \frac{(\lambda t)^k}{k!} \mathrm{e}^{-\lambda t} \\
&= \sum_{k=0}^{\infty} G^{(k)}(x) \frac{(\lambda t)^k}{k!} \mathrm{e}^{-\lambda t}.
\end{aligned} \tag{2.56}$$

Letting $G^*(s)$ be the LS transform of $G(x)$, the LS transform of (2.56) is

$$\begin{aligned}
\int_0^{\infty} \mathrm{e}^{-sx} \mathrm{d}\Pr\{Z(t) \le x\} &= \sum_{k=0}^{\infty} [G^*(s)]^k \frac{(\lambda t)^k}{k!} \mathrm{e}^{-\lambda t} \\
&= \exp\{-\lambda t [1 - G^*(s)]\}.
\end{aligned} \tag{2.57}$$

Because $E\{N(t)\} = V\{N(t)\} = \lambda t$ in (2.18),

$$E\{Z(t)\} = E\left\{E\left\{\sum_{k=1}^{N(t)} W_k \middle| N(t)\right\}\right\} = E\{N(t)\}E\{W_k\} = \lambda t E\{W\},$$
$$V\{Z(t)\} = E\{Z^2(t)\} - [E\{Z(t)\}]^2$$
$$= E\left\{E\left\{\sum_{k=1}^{N(t)} W_k \sum_{j=1}^{N(t)} W_j \middle| N(t)\right\}\right\} - [E\{Z(t)\}]^2$$
$$= V\{N(t)\}[E\{W\}]^2 + E\{N(t)\}V\{W\} = \lambda t E\{W^2\}. \tag{2.58}$$

When a random variable X has the LS transform function $F^*(s)$, i.e., $F^*(s) \equiv \int_0^\infty \mathrm{e}^{-sx}\mathrm{d}\Pr\{X \le x\}$, the above results are derived by (Appendix A)

$$E\{X\} = -\left.\frac{\mathrm{d}F^*(s)}{\mathrm{d}s}\right|_{s=0}, \quad E\{X^2\} = \left.\frac{\mathrm{d}^2F^*(s)}{\mathrm{d}s^2}\right|_{s=0}.$$

Next, it is assumed that the unit fails when the total damage has exceeded a failure level K $(0 < K < \infty)$. Usually, a failure level K is statistically estimated and is already known. We are interested in a random variable defined by $Y \equiv \min_t\{Z(t) > K\}$, i.e., $\Pr\{Y \le t\}$ represents the distribution of the failure time and $E\{Y\}$ is the MTTF of the unit. In stochastic processes, $\Pr\{Y \le t\}$ is called the *first-passage distribution* to some event.

Because the events of $\{Y \le t\}$ and $\{Z(t) > K\}$ are equivalent, from (2.56),

$$\Phi(t) \equiv \Pr\{Y \le t\} = \Pr\{Z(t) > K\}$$
$$= \sum_{k=0}^{\infty}\left[G^{(k)}(K) - G^{(k+1)}(K)\right]\sum_{j=k+1}^{\infty}\frac{(\lambda t)^j}{j!}\mathrm{e}^{-\lambda t}, \tag{2.59}$$

and its LS transform is

$$\Phi^*(s) \equiv \int_0^\infty \mathrm{e}^{-st}\mathrm{d}\Phi(t) = \sum_{k=0}^{\infty}\left[G^{(k)}(K) - G^{(k+1)}(K)\right]\left(\frac{\lambda}{s+\lambda}\right)^{k+1}.$$

Thus, the MTTF and variance are

$$E\{Y\} = -\left.\frac{\mathrm{d}\Phi^*(s)}{\mathrm{d}s}\right|_{s=0} = \frac{1}{\lambda}\sum_{k=0}^{\infty}G^{(k)}(K) = \frac{1}{\lambda}[1 + M_G(K)],$$
$$V\{Y\} = \left.\frac{\mathrm{d}^2\Phi^*(s)}{\mathrm{d}s^2}\right|_{s=0} - (E\{Y\})^2 = \frac{2}{\lambda^2}\sum_{k=0}^{\infty}(k+1)G^{(k)}(K) - (E\{Y\})^2, \tag{2.60}$$

where $M_G(K) \equiv \sum_{k=1}^{\infty} G^{(k)}(K)$ represents the expected number of shocks before the total damage exceeds a failure level K.

Furthermore, the failure rate of the unit with a failure level K is

$$r(t) \equiv \frac{\Phi'(t)}{1-\Phi(t)} = \frac{\sum_{k=0}^{\infty}\left[G^{(k)}(K) - G^{(k+1)}(K)\right]\left[\lambda(\lambda t)^k/k!\right]}{\sum_{k=0}^{\infty} G^{(k)}(K)\left[(\lambda t)^k/k!\right]}. \tag{2.61}$$

Example 2.16 (MTTF) Suppose that shocks occur in a Poisson process with rate λ and each damage due to shocks is exponential with mean $1/\theta$, i.e., $F(t) = 1 - e^{-\lambda t}$ and $G(x) = 1 - e^{-\theta x}$. Then, from (2.57),

$$\int_0^\infty e^{-sx} d\Pr\{Z(t) \le x\} = e^{-\lambda t[s/(s+\theta)]}.$$

By inversion [2, p. 80, 17, p. 23],

$$\Pr\{Z(t) \le x\} = e^{-\lambda t}\left[1 + \sqrt{\lambda\theta t}\int_0^x e^{-\theta u} u^{-1/2} I_1(2\sqrt{\lambda\theta t u})du\right], \tag{2.62}$$

where $I_i(x)$ is the Bessel function of order i for the imaginary argument defined by

$$I_i(x) \equiv \sum_{k=0}^{\infty}\left(\frac{x}{2}\right)^{2k+i}\frac{1}{k!(k+i)!}.$$

Thus, from (2.62),

$$\Pr\{Y \le t\} = 1 - e^{-\lambda t}\left[1 + \sqrt{\lambda\theta t}\int_0^K e^{-\theta u} u^{-1/2} I_1\left(2\sqrt{\lambda\theta t u}\right)du\right]. \tag{2.63}$$

Furthermore, from (2.58) and (2.60) (Problem 2.20),

$$\begin{aligned} E\{Z(t)\} &= \frac{\lambda t}{\theta}, \quad E\{Y\} = \frac{\theta K + 1}{\lambda}, \\ V\{Z(t)\} &= \frac{2\lambda t}{\theta^2}, \quad V\{Y\} = \frac{2\theta K + 1}{\lambda^2}, \end{aligned} \tag{2.64}$$

where note that $E\{Z(t)\}$ increases linearly with time t. In addition, we have the relation

$$\frac{E\{Z(t)\}}{K + 1/\theta} = \frac{t}{E\{Y\}}.$$

Finally, we give two modified damage models in reliability theory:

(1) Independent Damage Model [17, p. 21, 19]. Suppose that the total damage is not additive, i.e., any shock does no damage unless its amount has not exceeded a failure level K. If the damage due to some shock has exceeded for the first time a failure level K, then the unit fails. The same assumptions as those of the previous model are made except that the total damage is additive. A typical example of this model is the fracture of brittle materials such as glasses, and semiconductor parts that have failed by some over current of fault voltage.

In this model, the probability that the unit fails exactly at the $(k+1)$th shock $(k = 0, 1, 2, \ldots)$ is $[G(K)]^k - [G(K)]^{k+1}$. Thus, the distribution of time to failure is

$$\Pr\{Y \leq t\} = 1 - \exp\{-\lambda t[1 - G(K)]\}, \tag{2.65}$$

that is an exponential distribution with parameter $\lambda[1 - G(K)]$, and hence, the MTTF and the failure rate are

$$\begin{aligned} E\{Y\} &= \frac{1}{\lambda[1 - G(K)]}, \quad V\{Y\} = \frac{1}{\lambda^2[1 - G(K)]^2}, \\ r(t) &= \lambda[1 - G(K)] = \frac{1}{E\{Y\}}, \end{aligned} \tag{2.66}$$

that is constant for any $t \geq 0$.

(2) Random Failure Level [17, p. 29, 19]. Suppose that a failure level K is not constant and is a random variable with a distribution $L(x) \equiv \Pr\{K \leq x\}$, where $L(0) \equiv 0$. Then, for the cumulative damage model, the distribution of time to failure is, from (2.59)

$$\Pr\{Y \leq t\} = \sum_{k=0}^{\infty} \int_0^{\infty} \left[G^{(k)}(x) - G^{(k+1)}(x)\right] \mathrm{d}L(x) \sum_{j=k+1}^{\infty} \frac{(\lambda t)^j}{j!} \mathrm{e}^{-\lambda t}, \tag{2.67}$$

and the MTTF and the failure rate are, respectively,

$$\begin{aligned} E\{Y\} &= \frac{1}{\lambda} \sum_{k=0}^{\infty} \int_0^{\infty} G^{(k)}(x) \mathrm{d}L(x), \\ r(t) &= \frac{\sum_{k=0}^{\infty} \int_0^{\infty} \left[G^{(k)}(x) - G^{(k+1)}(x)\right] \mathrm{d}L(x) \left[\lambda(\lambda t)^k / k!\right]}{\sum_{k=0}^{\infty} \int_0^{\infty} G^{(k)}(x) \mathrm{d}L(x) \left[(\lambda t)^k / k!\right]}. \end{aligned} \tag{2.68}$$

For the independent damage model,

$$\Pr\{Y \le t\} = \sum_{k=0}^{\infty} \int_0^{\infty} \left\{[G(x)]^k - [G(x)]^{k+1}\right\} \mathrm{d}L(x) \sum_{j=k+1}^{\infty} \frac{(\lambda t)^j}{j!} \mathrm{e}^{-\lambda t},$$

$$E\{Y\} = \frac{1}{\lambda} \sum_{k=0}^{\infty} \int_0^{\infty} [G(x)]^k \mathrm{d}L(x). \tag{2.69}$$

Example 2.17 (Exponential distribution) (Problem 2.21) Suppose that $F(t) = 1 - \mathrm{e}^{-\lambda t}, G(x) = 1 - \mathrm{e}^{-\theta x}$ and $L(x) = 1 - \mathrm{e}^{-\beta x}$. Then, for the independent damage model,

$$\Pr\{Y \le t\} = 1 - \exp(-\lambda t \mathrm{e}^{-\theta K}), \quad E\{Y\} = \frac{1}{r(t)} = \frac{1}{\lambda} \mathrm{e}^{\theta K}.$$

For the random failure level,

$$\int_0^{\infty} \left[G^{(k)}(x) - G^{(k+1)}(x)\right] \mathrm{d}L(x) = \frac{\beta \theta^k}{(\theta + \beta)^{k+1}},$$

$$\int_0^{\infty} \mathrm{e}^{-st} \mathrm{d} \Pr\{Y \le t\} = \sum_{k=0}^{\infty} \left(\frac{\lambda}{s + \lambda}\right)^{k+1} \frac{\beta \theta^k}{(\theta + \beta)^{k+1}} = \frac{\lambda \beta}{s(\theta + \beta) + \lambda \beta}.$$

By inversion,

$$\Pr\{Y \le t\} = 1 - \exp\left(-\frac{\lambda \beta t}{\theta + \beta}\right), \tag{2.70}$$

$$E\{Y\} = \frac{1}{r(t)} = \frac{\theta + \beta}{\lambda \beta}, \quad V\{Y\} = \left(\frac{\theta + \beta}{\lambda \beta}\right)^2. \tag{2.71}$$

In addition, for the independent damage model,

$$\Pr\{Y > t\} = \int_0^{\infty} \exp(-\lambda t \mathrm{e}^{-\theta x}) \beta \mathrm{e}^{-\beta x} \mathrm{d}x$$

$$= \sum_{k=0}^{\infty} \frac{(-\lambda t)^k}{k!} \int_0^{\infty} \beta \mathrm{e}^{-(\beta + k\theta)x} \mathrm{d}x = \sum_{k=0}^{\infty} \frac{(-\lambda t)^k}{k!} \frac{\beta}{\beta + k\theta},$$

$$E\{Y\} = \frac{1}{r(t)} = \frac{1}{\lambda} \int_0^{\infty} \mathrm{e}^{\theta x} \beta \mathrm{e}^{-\beta x} \mathrm{d}x = \begin{cases} \frac{\beta}{\lambda(\beta - \theta)} & (\beta > \theta), \\ \infty & (\beta \le \theta). \end{cases}$$

(3) Nonhomogeneous Poisson process. Suppose that shocks occur in a nonhomogeneous Poisson process with an intensity function $h(t)$ and a mean value function $H(t)$. Then, because the probability of $N(t)$ is given in (2.45),

$$\Pr\{Z(t) \le x\} = \sum_{k=0}^{\infty} G^{(k)}(x) \frac{[H(t)]^k}{k!} \mathrm{e}^{-H(t)}, \quad E\{Z(t)\} = E\{W\}H(t),$$

$$E\{Y\} = \sum_{k=0}^{\infty} G^{(k)}(K) \int_0^{\infty} \frac{[H(t)]^k}{k!} \mathrm{e}^{-H(t)} \mathrm{d}t.$$

For the independent damage model,

$$\Pr\{Y \le t\} = 1 - \mathrm{e}^{-[1-G(K)]H(t)}, \quad E\{Y\} = \int_0^{\infty} \mathrm{e}^{-[1-G(K)]H(t)} \mathrm{d}t.$$

The failure rate is

$$r(t) = [1 - G(K)]h(t),$$

that has the same property as that of an intensity function $h(t)$.

2.6 Problems 2

2.1 For any negative function $g(t)$ in Example 2.3 [5, p. 84], show that

$$E\left\{\sum_{n=1}^{\infty} g(S_n)\right\} = \lambda \int_0^{\infty} g(t)\mathrm{d}t.$$

2.2 Show that a continuous failure distribution $F(t)$ that satisfies the memoryless properly, i.e., $\overline{F}(t+u) = \overline{F}(t)\overline{F}(u)$ for any $t, u \ge 0$, is only an exponential distribution [8, p. 24].

2.3 Let $X_{(k)} (k = 1, 2, \ldots, n)$ be the kth failure time of n units in Example 2.4. Then, noting that $X_{(1)}$ be the failure time of a series system in which when one of n units fails, the system fails, and $X_{(n)}$ be the failure time of a parallel system when all of n units have failed, the system fails. Then, show that the failure distributions of the two systems are, respectively,

$$\Pr\{X_{(1)} \le t\} = \int_0^t n\lambda \mathrm{e}^{-n\lambda u} \mathrm{d}u = 1 - \mathrm{e}^{-n\lambda t},$$

$$\Pr\{X_{(n)} \le t\} = \int_0^t n(1 - \mathrm{e}^{-\lambda t})^{n-1} \lambda \mathrm{e}^{-\lambda t} \mathrm{d}t = (1 - \mathrm{e}^{-\lambda t})^n,$$

$$E\{X_{(1)}\} = \frac{1}{n\lambda}, \quad E\{X_{(n)}\} = \frac{1}{\lambda} \sum_{k=1}^{n} \frac{1}{k}.$$

2.4 When $F(t)$ is IFR in Example 2.4, show that the MTTF μ_n of a parallel system with n units has the inequality [2, p. 65, 6, p. 9]

$$\mu_1 \leq \mu_n \leq \mu_1 \sum_{k=1}^{n} \frac{1}{k}.$$

2.5 Check numerically in Example 2.4 that when the failure time has a Weibull distribution $F(t) = 1 - \mathrm{e}^{-t^\alpha}\,(\alpha \geq 1)$, the MTTF of a parallel system is given by approximately [20],

$$\left(\sum_{k=1}^{n} \frac{1}{k}\right)^{1/\alpha}, \quad (C + \log n)^{1/\alpha}.$$

2.6 When the acquisition cost of one unit is c_1 and the replacement cost of a failed system is c_2 in Example 2.4, show that the expected cost rate is [6, p. 8]

$$C(n) = \frac{nc_1 + c_2}{(1/\lambda)\sum_{k=1}^{n}(1/k)} \quad (n = 1, 2, \ldots),$$

and derive analytically an optimum n^* that minimizes $C(n)$.

2.7 When the failure time is exponential, obtain the MTTF of a k-out-of-n system.

2.8 A harmonic series in (2.8) appears in a record time [8, p. 47]: Let $X_1, X_2, \ldots$ be independent and identically distributed random variables. It is said that a record occurs at time n if $X_n > \max\{X_1, X_2, \ldots, X_{n-1}\}$. In addition, let N_n denote the total number of records until time n. Then, show [8, p. 286] that

$$E\{N_n\} = \sum_{k=1}^{n} \frac{1}{k}, \quad V\{N_n\} = \sum_{k=1}^{n} \frac{1}{k}\left(1 - \frac{1}{k}\right).$$

This is applied to a breaking test [7]: Suppose that there are n wooden beams. To know the strength of which beams can bear a minimum burden, we make the following experiment: We record a burden B_1 on which the first beam breaks. Next, we bear B_1 on the second beam. If the beam does not break, then $B_2 > B_1$. Conversely, if it breaks, we record a burden B_2. Similarly, we bear $\min\{B_1, B_2\}$ on the third beam. If the beam does not break, then $B_3 > \min\{B_1, B_2\}$, and conversely, if it breaks, we record B_3. Then, in the same method of a record time, show that the expected number of broken beams by such experiment is $\sum_{k=1}^{n}(1/k)$. Compute the expected number of broken beams when n ($n = 10, 100, 1000$) experiments are made.

2.9 Show that the limiting distribution of binomial distribution is a Poisson distribution [4, p. 40].

2.10 For a n units parallel system in Example 2.6, where each unit has an identical failure distribution $F(t)$, the expected cost rate is [6, p. 84]

$$C(n) = c_2 + (c_1 - c_2) \int_0^\infty \overline{W}(t) \mathrm{d}[F(t)]^n + c_0 n \quad (n = 0, 1, 2, \ldots).$$

An optimum number n^* is given by a unique minimum that satisfies

$$\int_0^\infty [F(t)]^n \overline{F}(t) \mathrm{d}W(t) \leq \frac{c_0}{c_1 - c_2}.$$

In particular, when $W(t) = 1 - \mathrm{e}^{-\omega t}$ and $F(t) = 1 - \mathrm{e}^{-\lambda t}$, the above equation is

$$\sum_{k=0}^{n} \binom{n}{k} (-1)^k \frac{\omega}{\omega + (k+1)\lambda} \leq \frac{c_0}{c_1 - c_2}.$$

Derive the above equations, compute optimum n^* that minimize $C(n)$ for λ/ω and $c_0/(c_1 - c_2)$, and compare them with Table 2.3.

2.11 Suppose that failed units from two factories arrive at a repair shop according to Poisson processes with rates λ_1 and λ_2 in Example 2.8. Show how many number of spare units should be provided in order to assure with probability α that the system will remain operating in time t in Example 2.8.

2.12 Suppose that failures are roughly classified into two kinds of failures; minor failure and major failure in Example 2.9. Minor and major failures occur at the rates of 0.90 and 0.10, respectively. When $\lambda t = 1, 2, 3$, compute the probabilities that the number of minor failures is k $(k = 0, 1, 2, 3)$.

2.13 When the objective function in Example 2.11 is [6, p. 41],

$$C(N) = AN + \frac{B}{N} \quad \text{or} \quad C(N) = A^N B^{1/N} \quad (N = 1, 2, \ldots)$$

for parameters $A, B > 0$, show that an optimum N^* is a unique minimum that satisfies

$$\sum_{k=1}^{N} k \geq \frac{B}{2A}.$$

2.14 Consider the multiple backward times T_k $(k = 1, 2, \ldots, N)$, where $S_k \equiv \sum_{j=1}^{k} T_j, S_0 \equiv 0$, and $S_N \equiv t$ in Example 2.12 [21].

2.15 Prove (2.39). When 10 failures have occurred in 100 h, derive the probabilities that $k (k = 0, 1, 2)$ failures have occurred in 10 h in Example 2.10, and the third failure has occurred until 50 h in (2.38).

2.16 Prove in p. 29 that when $H(t) \equiv \int_0^t h(u) \mathrm{d}u$, i.e., $h(t) \equiv H'(t)$,

$$\frac{[H(t)]^n}{n!} = \int_0^t \frac{[H(u)]^{n-1}}{(n-1)!} h(u) \mathrm{d}u \quad (n = 1, 2, \ldots).$$

2.17 Prove in (2.51) that when $h(t)$ increases strictly,

$$\frac{\sum_{k=0}^{N-1}\int_0^\infty P_k(t)\mathrm{d}t}{\int_0^\infty P_N(t)\mathrm{d}t} - (N-1) > \frac{\mu}{\int_0^\infty P_N(t)\mathrm{d}t} \quad (N=2,3,\ldots),$$

where $P_k(t) = \{[H(t)]^k/k!\}\mathrm{e}^{-H(t)}$ and $\mu \equiv \int_0^\infty \mathrm{e}^{-H(t)}\mathrm{d}t$.

2.18 Derive in Example 2.15 that

$$\int_0^\infty P_N(t)\mathrm{d}t = \frac{1}{\alpha}\frac{\Gamma(N+1/\alpha)}{\lambda^{1/\alpha}\Gamma(N+1)}, \quad \sum_{k=0}^{N-1}\int_0^\infty P_k(t)\mathrm{d}t = \frac{\Gamma(N+1/\alpha)}{\lambda^{1/\alpha}\Gamma(N)}.$$

2.19 Prove in Sect. 2.4.2 that the solution of the differential equation

$$\frac{\mathrm{d}H(t)}{\mathrm{d}t} = b(t)[a - H(t)],$$

is given by (2.54).

2.20 Derive $E\{Z(t)\}, V\{Z(t)\}$ from (2.57), and $E\{Y\}, V\{Y\}$ from (2.60) in Example 2.16.

2.21 Derive all equations in Example 2.17.

References

1. Kingman JFC (1993) Poisson processes. Oxford University Press, Oxford
2. Barlow RE, Proschan F (1965) Mathematical theory of reliability. Wiley, New York
3. Nakagawa T (2005) Maintenance theory of reliability. Springer, London
4. Osaki S (1992) Applied stochastic system modeling. Springer, Berlin
5. Çinlar E (1975) Introduction to stochastic processes. Prentice-Hall, Englewood Cliffs
6. Nakagawa T (2008) Advanced reliability models and maintenance policies. Springer, London
7. Havil J (2003) GAMMA: exploring Euler's constant. Princeton University Press, Princeton
8. Ross SM (1983) Stochastic processes. Wiley, New York
9. Beichelt FE, Fatti LP (2002) Stochastic processes and their applications. CRC Press, Boca Raton
10. Tijms HC (2003) A first course in stochastic models. Wiley, Chichester
11. Yamada S (1994) Software reliability models: fundamentals and applications. JUSE Press, Tokyo
12. Finkelstein M (2008) Failure rate modelling for reliability and risk. Springer, London
13. Murthy DNP, Xie M, Jiang R (2004) Weibull models. Wiley, Hoboken
14. Goel AL, Okumoto K (1979) Time-dependent error-detection rate model for software reliability and other performance measures. IEEE Trans Reliab R-28:206–211
15. Pham H (2006) System software reliability. Springer, London
16. Pham, H (ed) (2008) Recent advances in reliability and quality in design. Springer, London
17. Nakagawa T (2007) Shock and damage models in reliability theory. Springer, London
18. Cox DR (1962) Renewal theory. Methuen, London

19. Esary JD, Marshall AW, Proschan F (1973) Shock models and wear processes. Ann Probab 1:627–649
20. Yun WY, Nakagawa T (2010) Note on MTTF of a parallel system. In: 16th ISSAT International Conference on Reliability and Quality in Design, pp 235–237
21. Chen M, Mizutani S, Nakagawa T (2010) Optimal backward and backup policies in reliability theory. J Oper Res Soc Jpn 53:101–118

Chapter 3
Renewal Processes

In Chap. 2, we have defined a Poisson process and shown that the interarrival times are distributed exponentially. However, in general, such times have arbitrary distributed random variables, that is called a *renewal process*. In most reliability systems, they may deteriorate with age and break at last, and become like new by suitable replacement or maintenance. A renewal process arises from the study of self-renewing aggregates, and a renewal theory plays an important role in the analysis of stochastic models with the total sum of independent nonnegative random variables. The results can be applied to a wide variety of both theoretical and practical probability problems in engineering, natural, economical, and social sciences.

A renewal process in this chapter is the most fundamental process, and a renewal theory is an important theory in stochastic processes. Many basic stochastic systems form renewal processes essentially. Renewal processes are the largest and most crucial chapter in stochastic processes for studying reliability theory because learning a reliability theory is learning a renewal theory:

In Sect. 3.1, we define renewal processes by generalizing Poisson processes. Renewal processes are characterized by renewal functions. In Sect. 3.2, it is shown that a renewal function is obtained by solving a renewal equation, and its properties are given in the name of a renewal theory. Some theorems in a renewal theory are cited as theorems without mathematical proof because it needs more mathematical preparation. But, it would be sufficient on theorem forms to apply them to reliability theory.

The most fundamental policy in maintenance models is a replacement policy. In Sect. 3.3, we show how to apply a renewal theory to standard replacement policies. Using the techniques and results of a renewal theory, optimum policies for some replacement models are discussed analytically. In Sect. 3.4, we introduce an alternating renewal process that repeats on and off, up and down, or between operating and failure states. A one-unit system with repair maintenance which is the most basic reliability system is given as a typical example, and its properties are examined and the interval reliability is given by using the results of an alternating renewal process.

Finally, in Sect. 3.4, we take up a geometric and a discrete renewal processes as examples of modified renewal processes. A geometric process is the process whose

T. Nakagawa, *Stochastic Processes*, Springer Series in Reliability Engineering,
DOI: 10.1007/978-0-85729-274-2_3,

interarrival times decrease or increase geometrically, and a discrete process is defined to be a renewal process with discrete interarrival times.

3.1 Definition of Renewal Process

We define renewal processes by generalizing interarrival times X_k in a Poisson process: Consider a sequence of independent and nonnegative random variables $\{X_1, X_2, \ldots\}$ in Fig. 2.1, where it is assumed that $\Pr\{X_k = 0\} < 1$ for all k because of avoiding trivial cases. Suppose that $X_2, X_3, \ldots$ have an arbitrary identical distribution $F(t)$ with finite mean μ, however, X_1 possibly has a different distribution $F_1(t)$ with finite μ_1 from $F(t)$, in which both $F_1(t)$ and $F(t)$ are not degenerate at time $t = 0$ and $\overline{F}_1(t) \equiv 1 - F_1(t)$.

We introduce the total sum of random variables $S_n \equiv \sum_{k=1}^{n} X_k$, where $S_0 \equiv 0$ for convenience, and define $N(t) \equiv \max_n\{S_n \leq t\}$ that represents the number of renewals in the time interval $[0,t]$, where the process $\{N(t)\}$ is generally called a *counting process*. Clearly, from (2.16),

$$\{S_n \leq t\} \Leftrightarrow \{N(t) \geq n\}.$$

It is said that the nth renewal occurs at time S_n. We call epochs at which S_n ends *renewal points, renewal times* or *regeneration points*, and the time from the beginning of S_n to its end a *renewal cycle* with time interval X_n. We use properly two words of renewal point and regeneration point in the right place. Note that we have the memoryless property only at renewal points except $F_1(t)$ and $F(t)$ are exponential distributions. In other words, the process starts newly at each renewal point. We will use the terms of events and renewals interchangeably.

We define renewal processes according to three types of $F_1(t)$:

Definition 3.1

(i) If $F(t) = F_1(t)$, i.e., all random variables are identically distributed, then the counting process $\{N(t)\}$ is called an *ordinary renewal process* or renewal process for short.
(ii) If $F_1(t)$ and $F(t)$ are not the same, then the counting process $\{N(t)\}$ is called a *modified* or *delayed renewal process*.
(iii) If $F_1(t) \equiv \int_0^t \overline{F}_1(u)\mathrm{d}u/\mu$ that will be given in (3.24), then the counting process $\{N(t)\}$ is called an *equilibrium* or *stationary renewal process*.

Example 3.1 (Reliability systems) Consider a unit that is replaced with a new one upon failures: A unit begins to operate immediately after the replacement whose time is negligible. Suppose that the failure distribution of each new unit is $F(t)$. If a new unit is installed at time $t = 0$, then all failure times have the same distribution, and hence, we have an ordinary renewal process. On the other hand, if an old unit is in use at time $t = 0$, then X_1 represents its residual lifetime and could be different from the failure time of a new unit, and hence, we have a modified renewal process. In particular, if the time origin point at which we start observing the process is sufficiently large after the installation of an initial operating unit, it may be considered that the process has already entered the stationary state and the failure distribution of X_1 is given by $\int_0^t \overline{F}_1(u)\mathrm{d}u/\mu$, then we have an equilibrium renewal process.

We may consider an operating unit that is repaired upon failures and whose repair time may be assumed to be negligible because they are small usually compared with failure times. If we observe only the operating states that repeat successively, then it may be said that most reliability systems can form renewal processes. Therefore, renewal processes are the most fundamental stochastic process in reliability theory.

In general, if two positive random variables X and Y are independent and have the respective distributions $F(t) \equiv \Pr\{X \le t\}$ and $G(t) \equiv \Pr\{Y \le t\}$, then the sum of X and Y has the distribution

$$\Pr\{X+Y \le t\} = \int_0^t F(t-u)\mathrm{d}G(u) = \int_0^t G(t-u)\mathrm{d}F(u) \equiv F(t) * G(t),$$

and its LS transform is, from Appendix A,

$$\int_0^\infty \mathrm{e}^{-st}\mathrm{d}\Pr\{X+Y \le t\} = E\{\mathrm{e}^{-s(X+Y)}\} = E\{\mathrm{e}^{-sX}\}E\{\mathrm{e}^{-sY}\}$$

$$= \int_0^\infty \mathrm{e}^{-st}\mathrm{d}F(t) \int_0^\infty \mathrm{e}^{-st}\mathrm{d}G(t) = F^*(s)G^*(s),$$

i.e.,

$$\int_0^\infty \mathrm{e}^{-st}\mathrm{d}[F(t) * G(t)] = F^*(s)G^*(s),$$

where the asterisk is called the pairwise Stieltjes convolution. Denoting $F^{(0)}(t) \equiv 1$ for $t \ge 0$ and

$$\begin{aligned} F^{(n)}(t) &\equiv \Pr\{X_2 + X_3 + \cdots + X_{n+1} \le t\} \\ &= \int_0^t \Pr\{X_3 + X_4 + \cdots + X_{n+1} \le t-u\}\mathrm{d}\Pr\{X_2 \le u\} \\ &= \int_0^t F^{(n-1)}(t-u)\mathrm{d}F(u) \\ &= F(t) * \cdots * F(t) \quad (n = 1, 2, \ldots), \end{aligned} \tag{3.1}$$

$F^{(n)}(t)$ is called the n-fold Stieltjes convolution of $F(t)$ with itself. That is, $F^{(n)}(t)$ represents the distribution of the total sum of independent and identical random variables $\sum_{k=2}^{n+1} X_k$. In addition, the LS transform of $F^{(n)}(t)$ is, from Appendix A,

$$\int_0^\infty e^{-st} dF^{(n)}(t) = [F^*(s)]^n \quad (n = 0, 1, 2, \ldots), \tag{3.2}$$

where note that $\int_0^\infty e^{-st} dF^{(0)}(t) = 1$ for any $Re(s) \geq 0$.

Furthermore, from the relation of S_n and $N(t)$,

$$\begin{aligned} \Pr\{N(t) = 0\} &= \Pr\{X_1 > t\} = 1 - F_1(t), \\ \Pr\{N(t) = n\} &= \Pr\{S_n \leq t \text{ and } S_{n+1} > t\} = \Pr\{S_n \leq t\} - \Pr\{S_{n+1} \leq t\} \\ &= F_1(t) * F^{(n-1)}(t) - F_1(t) * F^{(n)}(t) \quad (n = 1, 2, \ldots). \end{aligned} \tag{3.3}$$

3.2 Renewal Function

We want to know properties of the number of renewals in $[0, t]$ and define its expected number as $M(t) \equiv E\{N(t)\}$, that is called the *renewal function*, and the derivative of $M(t)$ is $m(t) \equiv dM(t)/dt$, that is called the *renewal density*. From (3.3) (Problem 3.1), the renewal function is

$$\begin{aligned} M(t) &\equiv \sum_{n=1}^{\infty} n \Pr\{N(t) = n\} = \sum_{n=1}^{\infty} \Pr\{N(t) \geq n\} \\ &= \sum_{n=1}^{\infty} \Pr\{S_n \leq t\} = \sum_{n=1}^{\infty} F_1(t) * F^{(n-1)}(t). \end{aligned} \tag{3.4}$$

It is fairly easy to show that $M(t)$ is finite for all $t \geq 0$ because $\Pr\{X_k = 0\} < 1$ (Problem 3.1). Furthermore, from the notations of the convolutions in (3.1) and (3.4),

$$\begin{aligned} M(t) &= F_1(t) + \sum_{n=1}^{\infty} \int_0^t F^{(n)}(t-u) dF_1(u) = \int_0^t [1 + M_0(t-u)] dF_1(u), \\ m(t) &= f_1(t) + \int_0^t m_0(t-u) f_1(u) du, \end{aligned} \tag{3.5}$$

where $M_0(t)$ is the renewal function of an ordinary renewal process with distribution $F(t)$, i.e., $M_0(t) \equiv \sum_{n=1}^{\infty} F^{(n)}(t), m_0(t) \equiv dM_0(t)/dt = \sum_{n=1}^{\infty} f^{(n)}(t)$, and $f(t)$ and $f_1(t)$ are the respective density functions of $F(t)$ and $F_1(t)$. Such types of equations as (3.5) are called renewal-type equations and appear frequently in the analysis of reliability models, because most systems are renewed after maintenance as shown in Sect. 3.3. It is of great importance in reliability analysis to search for renewal points and set up renewal equations. Generally, solutions of renewal equations can be computed by only numerical methods, however, they are obtained explicitly in the form of LS transforms that will be shown in Sect. 3.2.3.

The LS transform of $M(t)$ is, from (3.4) or (3.5),

$$M^*(s) \equiv \int_0^\infty e^{-st} dM(t) = \frac{F_1^*(s)}{1 - F^*(s)}. \tag{3.6}$$

Thus, $M(t)$ is determined by $F_1(t)$ and $F(t)$. When $F_1(t)$ and $F(t)$ are identical, i.e., the process forms an ordinary renewal process, $M(t) = M_0(t)$, and (3.6) implies $F^*(s) = M^*(s)/[1 + M^*(s)]$. Therefore, $F(t)$ and $M(t)$ determines uniquely with each other because the LS transform determines its original function uniquely.

In particular, when the process forms a stationary renewal process, from (iii) of Definition 3.1,

$$F_1^*(s) = \int_0^\infty e^{-st} dF_1(t) = \frac{1 - F^*(s)}{\mu s},$$

and hence, from (3.6) and Table A.2 in Appendix A,

$$M^*(s) = \frac{1}{\mu s}, \quad \text{i.e.,} \quad M(t) = \frac{t}{\mu}.$$

After this, it is assumed that $F_1(t) = F(t)$, i.e., we deal with only an ordinary renewal process.

Example 3.2 (Gamma and normal distributions) Suppose that $F(t)$ is a gamma distribution, i.e., $F(t) = 1 - \sum_{j=0}^{k-1} [(\lambda t)^j / j!] e^{-\lambda t}$ $(k = 1, 2, \ldots)$. In particular, when $k = 1$, from Sect. 2.1.2,

$$M^*(s) = \frac{\lambda}{s}, \quad \text{i.e.,} \quad M(t) = \lambda t,$$

that is derived directly, as shown in (2.18).

When $k = 2, F(t) = 1 - (1 + \lambda t)e^{-\lambda t}$, and

$$F^*(s) = \left(\frac{\lambda}{s + \lambda}\right)^2,$$

and hence, from (3.6),

$$M^*(s) = \frac{\lambda^2}{s(s + 2\lambda)} = \frac{\lambda}{2s} - \frac{\lambda}{2(s + 2\lambda)}.$$

Using the LS transform method in Appendix A such that

$$\int_0^\infty e^{-st}d(at) = \frac{a}{s}, \quad \int_0^\infty e^{-st}d(1 - e^{-at}) = \frac{a}{s+a},$$

the inverse LS transform of $M^*(s)$ is

$$M(t) = \frac{\lambda t}{2} - \frac{1}{4}(1 - e^{-2\lambda t}).$$

In general, because the probability of n renewals in $[0, t]$ for a renewal process is equal to the probability of either $nk, nk+1, \ldots,$ or $nk+k-1$ events in $[0, t]$ for a Poisson process with rate λ (Fig. 3.1) [1, p. 57],

$$\Pr\{N(t) = n\} = \sum_{j=nk}^{nk+k-1} \frac{(\lambda t)^j}{j!} e^{-\lambda t},$$

and hence, because $m(t)dt$ is the probability of a renewal in $(t, t+dt]$,

$$m(t)dt = \sum_{n=1}^{\infty} \left[\frac{(\lambda t)^{nk-1}}{(nk-1)!} e^{-\lambda t} \right] \lambda dt.$$

Thus, a renewal function is

$$M(t) = \int_0^t m(u)du = \sum_{n=1}^{\infty} \int_0^t \frac{\lambda(\lambda u)^{nk-1}}{(nk-1)!} e^{-\lambda u} du$$
$$= \sum_{n=1}^{\infty} \sum_{j=nk}^{\infty} \frac{(\lambda t)^j}{j!} e^{-\lambda t}. \tag{3.7}$$

On the other hand, from Sect. 2.1.2 and (3.6), the LS transform of $M(t)$ for a gamma distribution with order k is

$$M^*(s) = \frac{\lambda^k}{(s+\lambda)^k - \lambda^k},$$

and inverting its LS transform [1, p. 57, 2, p. 121, 3]

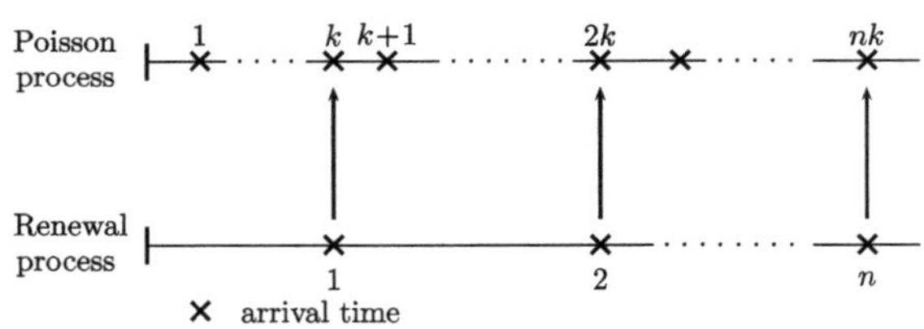

Fig. 3.1 Relationship between Poisson and renewal processes for a gamma distribution with order k

$$M(t) = \frac{\lambda t}{k} + \frac{1}{k}\sum_{j=1}^{k-1}\frac{\theta_j}{1-\theta_j}\left[1 - \mathrm{e}^{-\lambda t(1-\theta_j)}\right], \tag{3.8}$$

where $\theta_j \equiv \mathrm{e}^{2\pi ji/k}$ $(j = 1, 2, \ldots, k-1)$ is the distinct root of $s^k = 1$ and i is an imaginary order. In particular, when $k = 3$ [4, p. 162] (Problem 3.2),

$$M(t) = \frac{1}{3}\left[\lambda t - 1 + \frac{2}{\sqrt{3}}\mathrm{e}^{-1.5\lambda t}\sin\left(\frac{\sqrt{3}}{2}\lambda t + \frac{\pi}{3}\right)\right],$$

and when $k = 4$,

$$M(t) = \frac{1}{4}\left[\lambda t - \frac{3}{2} + \frac{1}{2}\mathrm{e}^{-2\lambda t} + \sqrt{2}\mathrm{e}^{-\lambda t}\sin\left(\lambda t + \frac{\pi}{4}\right)\right].$$

Figure 3.2 draws the renewal functions $M(t)$ for $k = 1, 2, 3$ when $\lambda = 1$.

Suppose generously that $F(t) = \int_0^t [\lambda(\lambda x)^{\alpha-1}\mathrm{e}^{-\lambda x}/\Gamma(\alpha)]\mathrm{d}x$ for parameters $\lambda > 0$ and $\alpha > 0$. Because the sum of two random variables with gamma distributions with the respective parameters (α, λ) and (β, λ) has a gamma distribution with parameter $(\alpha + \beta, \lambda)$ (Problem 3.3),

$$F^{(n)}(t) = \int_0^t \frac{\lambda(\lambda x)^{n\alpha-1}}{\Gamma(n\alpha)}\mathrm{e}^{-\lambda x}\mathrm{d}x, \tag{3.9}$$

and hence,

$$M(t) = \sum_{n=1}^{\infty}\int_0^t \frac{\lambda(\lambda x)^{n\alpha-1}}{\Gamma(n\alpha)}\mathrm{e}^{-\lambda x}\mathrm{d}x, \tag{3.10}$$

that agrees with (3.7) when $\alpha = k\,(k = 1, 2, \ldots)$.

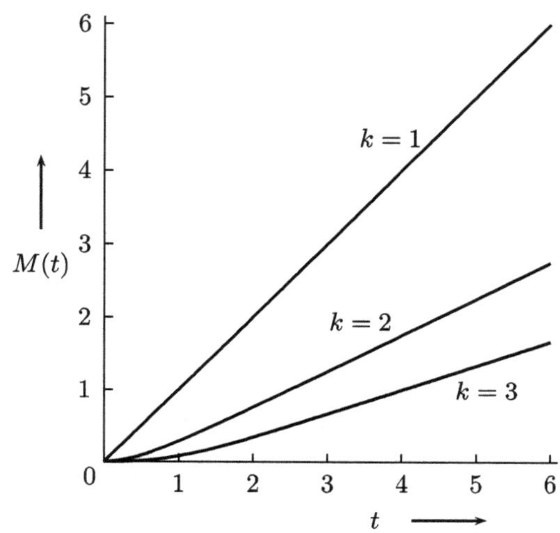

Fig. 3.2 Renewal functions $M(t)$ for $k = 1, 2, 3$

When X_k $(k = 1, 2, \ldots)$ is normally distributed with mean μ and variance σ^2 for $\mu \gg \sigma$, because S_n is normally distributed with mean $n\mu$ and variance $n\sigma^2$, from (3.4),

$$M(t) = \sum_{n=1}^{\infty} \Pr\{S_n \le t\} = \sum_{n=1}^{\infty} \Phi\left(\frac{t - n\mu}{\sqrt{n}\sigma}\right), \tag{3.11}$$

where $\Phi(x)$ is a standard normal distribution with mean 0 and variance 1 (Problems 3.4 and 3.5).

Next, we derive the variance $V\{N(t)\}$ [5, p. 89]: Because $n^2 = 2\sum_{j=1}^{n} j - n$,

$$\begin{aligned} E\{N(t)^2\} &= \sum_{n=1}^{\infty} n^2 \Pr\{N(t) = n\} = 2\sum_{n=1}^{\infty} n \Pr\{N(t) \ge n\} - M(t) \\ &= 2\sum_{n=1}^{\infty} n \Pr\{S_n \le t\} - M(t) = 2\sum_{n=1}^{\infty} nF^{(n)}(t) - M(t). \end{aligned}$$

Taking the LS transform on both sides above,

$$\begin{aligned} \int_0^{\infty} \mathrm{e}^{-st} \mathrm{d}E\left\{N(t)^2\right\} &= 2\sum_{n=1}^{\infty} n[F^*(s)]^n - M^*(s) \\ &= 2\left[\frac{F^*(s)}{1 - F^*(s)}\right]^2 + \frac{F^*(s)}{1 - F^*(s)} \\ &= 2[M^*(s)]^2 + M^*(s). \end{aligned} \tag{3.12}$$

Inverting (3.12),

$$E\left\{N(t)^2\right\} = 2M(t) * M(t) + M(t),$$

and hence,

$$V\{N(t)\} = 2M(t) * M(t) + M(t) - [M(t)]^2. \tag{3.13}$$

Because $M(t) * M(t) \le [M(t)]^2$,

$$V\{N(t)\} \le M(t) + [M(t)]^2.$$

In particular, when $F(t) = 1 - \mathrm{e}^{-\lambda t}$,

$$V\{N(t)\} = \lambda t,$$

that agrees with (2.18). Furthermore, when $F(t) = 1 - (1 + \lambda t)\mathrm{e}^{-\lambda t}$, from Example 3.2 and (3.13) (Problem 3.6),

$$V\{N(t)\} = -\frac{\lambda t}{4} + \frac{\lambda t}{2}(1 - \mathrm{e}^{-2\lambda t}) + \frac{1}{16}(1 - \mathrm{e}^{-4\lambda t}).$$

We summarize some important limiting theorems and results of renewal theory [5–7]:

Theorem 3.1 If $\mu_2 \equiv \int_0^\infty t^2 \mathrm{d}F(t) < \infty$ and $\sigma^2 \equiv \mu_2 - \mu^2$,

$$M(t) = \frac{t}{\mu} + \left(\frac{\sigma^2}{2\mu^2} - \frac{1}{2}\right) + o(1) \quad \text{as}\, t \to \infty, \tag{3.14}$$

and if $\mu_3 \equiv \int_0^\infty t^3 \mathrm{d}F(t) < \infty$,

$$V\{N(t)\} = \frac{\sigma^2 t}{\mu^3} + \left(\frac{5\sigma^4}{4\mu^4} + \frac{2\sigma^2}{\mu^2} + \frac{3}{4} - \frac{2\mu_3}{3\mu^3}\right) + o(1) \quad \text{as}\, t \to \infty, \tag{3.15}$$

where $o(h)/h \to 0$ as $h \to 0$.

Proof Expanding $F^*(s)$ with respect to s,

$$F^*(s) = 1 - \mu s + \frac{1}{2}(\sigma^2 + \mu^2)s^2 - \frac{1}{3!}\mu_3 s^3 + o(s^3).$$

Substituting $F^*(s)$ for (3.6) and arranging them,

$$\begin{aligned} M^*(s) &= \frac{1}{s\mu} + \left(\frac{\sigma^2 - \mu^2}{2\mu^2}\right) + o(1), \\ [M^*(s)]^2 &= \frac{1}{s^2\mu^2} + \frac{1}{s}\left(\frac{\sigma^2 - \mu^2}{\mu^3}\right) + \left(\frac{3\sigma^4}{4\mu^4} + \frac{\sigma^2}{2\mu^2} + \frac{3}{4} - \frac{\mu_3}{3\mu^3}\right) + o(1). \end{aligned} \tag{3.16}$$

Inverting $M^*(s)$ follows (3.14). Furthermore, substituting (3.16) for (3.12) and inverting it, we have (3.15) from (3.13) and (3.14) (Problem 3.6). □

From this theorem, $M(t)$ and $m(t)$ are approximately given by

$$\begin{aligned} M(t) &\approx \frac{t}{\mu} + \frac{\sigma^2}{2\mu^2} - \frac{1}{2}, \quad m(t) \approx \frac{1}{\mu}, \\ V\{N(t)\} &\approx \frac{\sigma^2 t}{\mu^3} + \frac{5\sigma^4}{4\mu^4} + \frac{2\sigma^2}{\mu^2} + \frac{3}{4} - \frac{2\mu_3}{3\mu^3} \end{aligned} \tag{3.17}$$

for large t. Furthermore, if $\sigma \ll \mu$, then

$$M(t) \approx \frac{t}{\mu} - \frac{1}{2}.$$

Theorem 3.2 If μ and σ^2 are finite,

$$\frac{M(t)}{t} \to \frac{1}{\mu}, \quad m(t) \to \frac{1}{\mu}, \quad \frac{V\{N(t)\}}{t} \to \frac{\sigma^2}{\mu^3} \quad \text{as}\, t \to \infty. \tag{3.18}$$

This theorem that is easily derived from Theorem 3.1 is called an *elementary renewal theorem* and is the most important one in a renewal process. This is also applied not only in the reliability theory but in also many other practical fields. Another proof of Theorem 3.2 that is more direct than that of the above was shown [1, p. 55].

A random variable X with distribution $F(t)$ is said to be lattice if there exists $h > 0$ such that $\sum_{n=0}^{\infty} \Pr\{X = nh\} = 1$. We give Blackwell's theorem [8, p. 347]:

Theorem 3.3 (Blackwell's theorem) If X is not lattice, then for $h > 0$,

$$\lim_{t\to\infty}[M(t+h) - M(t)] = \frac{h}{\mu}.$$

This theorem shows that the expected number of renewals in $[t, t+h]$ tends to h/μ approximately as $t \to \infty$, i.e., it is proportional to the time duration where the proportional constant is $1/\mu$.

Furthermore, we introduce a *renewal reward process* [6, p. 77]. For instance, if we consider the total reward produced by the successive production of some system, then the process is said to form a renewal reward process, where the successive production can be described by a renewal process and the total rewards caused by production may be additive. Chapter 6 will deal in detail with a *cumulative process* that is another name of a renewal reward process.

Define that some reward W_k is earned at the kth renewal point $(k = 1, 2, \dots)$. When a sequence of pairs $\{X_k, W_k\}$ is independent and identically distributed, $Z(t) \equiv \sum_{k=1}^{N(t)} W_k$ is the total reward earned in $[0, t]$.

Theorem 3.4 If $E\{W\} \equiv E\{W_k\} < \infty$,

$$\frac{E\{Z(t)\}}{t} \to \frac{E\{W\}}{\mu} \quad \text{as } t \to \infty.$$

This is proved by intuition because if $N(t)$ and W are independent,

$$E\{Z(t)\} = E\{N(t)\}E\{W\} = M(t)E\{W\}.$$

Thus, using Theorem 3.2,

$$\frac{E\{Z(t)\}}{t} = \frac{M(t)}{t}E\{W\} \to \frac{E\{W\}}{\mu} \quad \text{as } t \to \infty.$$

The detailed proof of Theorem 3.4 was shown [6, p. 78]. This theorem shows that if one cycle is denoted by the time interval between renewals, the expected reward per unit of time for an infinite time span is equal to the expected reward per cycle, divided by its mean time, i.e.,

$$\lim_{t\to\infty} \frac{E\{Z(t)\}}{t} = \frac{\text{Expected reward of one renewal cycle}}{\text{Mean time of one renewal cycle}}. \tag{3.19}$$

This function is adopted by appropriate objective functions of maintenance models where the reward of maintenance is given by costs. There will exist many optimization problems with aim to minimize them in reliability theory [1], as shown in Sect. 3.3.2

Applying the asymptotic formulae of (3.17) to the usual central limit theorem, we have a so-called central limit theorem for a renewal process:

Theorem 3.5

$$\lim_{t\to\infty} \Pr\left\{\frac{N(t)-t/\mu}{\sqrt{\sigma^2 t/\mu^3}} \le x\right\} = \frac{1}{\sqrt{2\pi}} \int_{-\infty}^{x} e^{-u^2/2} du, \tag{3.20}$$

i.e., $N(t)$ is asymptotically normally distributed with mean t/μ and variance $\sigma^2 t/\mu^3$ for large t. The detailed proof was found in [5, p. 93, 6, p. 62].

Example 3.3 (Asymptotic normal distribution) We compute approximately the number n when the probability that the number of failures is less than n in $t = 10,000$ h of operation is given by 0.90. The failure distribution of each unit is unknown, but, it is estimated from the sample data that the mean and variance are $\mu = 1,000$ h and $\sigma^2 = 200,000\,(\text{h})^2$, respectively. Then, from Theorem 3.5, when $t = 10,000$,

$$\frac{N(t)-t/\mu}{\sqrt{\sigma^2 t/\mu^3}} = \frac{N(10,000)-10}{\sqrt{2}}$$

is approximately normally distributed with mean 0 and variance 1. Thus,

$$\Pr\{N(10,000) \le n\} = \Pr\left\{\frac{N(10,000)-10}{\sqrt{2}} \le \frac{n-10}{\sqrt{2}}\right\}$$
$$\approx \frac{1}{\sqrt{2\pi}} \int_{-\infty}^{(n-10)/\sqrt{2}} e^{-u^2/2} du = 0.90.$$

Because $u_0 = 1.28$ that satisfies $(1/\sqrt{2\pi}) \int_{-\infty}^{u_0} e^{-u^2/2} du = 0.90$,

$$n \approx 10 + 1.28\sqrt{2} = 11.81.$$

In this case, it can be said that with probability more than 0.90, the system with 12 units, i.e., the MTTF is 1.2×10^4 h, including the first unit, will continue to operate for 10^4 hours without any supply (Problem 3.7).

Theorem 3.6 (Key or Fundamental renewal theorem) If $g(t)$ is bounded in $[0, \infty)$ and $F(t)$ is not lattice, then [2, p. 123, 5, p. 95, 9, p. 295],

$$\lim_{t\to\infty}\int_0^t g(t-u)\mathrm{d}M(u) = \frac{1}{\mu}\int_0^\infty g(t)\mathrm{d}t. \tag{3.21}$$

This is proved intuitively that from Theorem 3.3 [6, p. 65],

$$\lim_{h\to 0}\lim_{t\to\infty}\frac{M(t+h)-M(t)}{h} = \lim_{t\to\infty}\frac{\mathrm{dM(t)}}{\mathrm{d}t} = \frac{1}{\mu},$$

$$\lim_{t\to\infty}\int_0^t g(t-u)\mathrm{d}M(u) = \lim_{x\to\infty}\frac{1}{\mu}\int_0^t g(t-u)\mathrm{d}u = \frac{1}{\mu}\int_0^\infty g(t)\mathrm{d}t.$$

Recall that $m(t) = \mathrm{d}M(t)/\mathrm{d}t$ is called the renewal density and represents the expected number of renewals per unit of time.

3.2.1 Age and Residual Life Distributions

Let $\delta(t) \equiv t - S_{N(t)}$ and $\gamma(t) \equiv S_{N(t)+1} - t$ that denote the *shortage* and *excess* random variables (Fig. 3.3), respectively. In reliability theory, $\delta(t)$ represents the current age and $\gamma(t)$ represents the residual or remaining life of an operating unit. Thus, $\beta(t) \equiv \delta(t) + \gamma(t) = S_{N(t)+1} - S_{N(t)} = X_{N(t)+1}$ represents the total life of the unit.

Theorem 3.7

(i)

$$\Pr\{\gamma(t) \le x\} = F(t+x) - \int_0^t \overline{F}(t+x-u)\mathrm{d}M(u). \tag{3.22}$$

(ii)

$$\Pr\{\delta(t) \le x\} = \begin{cases} F(t) - \int_0^{t-x}\overline{F}(t-u)\mathrm{d}M(u) & \text{for } x \le t, \\ 1 & \text{for } x > t. \end{cases} \tag{3.23}$$

(iii) The limiting distribution is

$$\lim_{t\to\infty}\Pr\{\delta(t) \le x\} = \lim_{t\to\infty}\Pr\{\gamma(t) \le x\} = \frac{1}{\mu}\int_0^x \overline{F}(u)\mathrm{d}u. \tag{3.24}$$

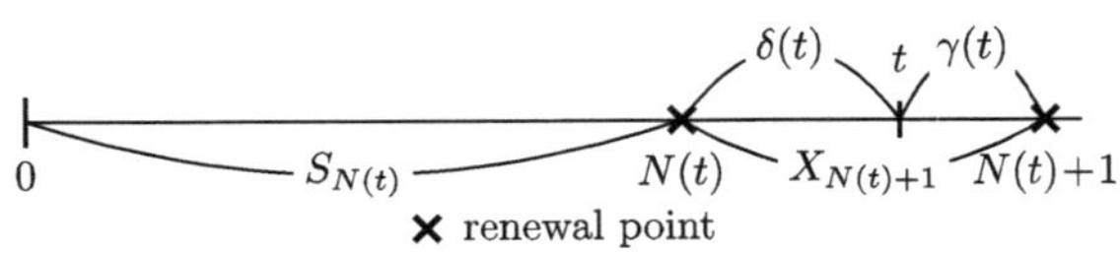

Fig. 3.3 Age $\delta(t)$ and residual life $\gamma(t)$ in a renewal process

Proof From Fig. 3.3 [1, p. 58, 5, p. 97],

$$
\begin{aligned}
\Pr\{\gamma(t) \le x\} &= \Pr\{S_{N(t)} \le t < S_{N(t)+1} \le t + x\} \\
&= \sum_{k=0}^{\infty} \Pr\{t < S_{k+1} \le t + x, N(t) = k\} \\
&= \sum_{k=0}^{\infty} \Pr\{S_k \le t < S_{k+1} \le t + x\} \\
&= \sum_{k=0}^{\infty} \int_0^t \Pr\{t - u < X_{k+1} \le t + x - u\} \mathrm{d}\Pr\{S_k \le u\} \\
&= \sum_{k=0}^{\infty} \int_0^t [F(t + x - u) - F(t - u)] \mathrm{d}F^{(k)}(u) \\
&= F(t + x) - F(t) + \int_0^t [F(t + x - u) - F(t - u)] \mathrm{d}M(u) \\
&= F(t + x) - \int_0^t \overline{F}(t + x - u) \mathrm{d}M(u).
\end{aligned}
$$

Furthermore, because both events $\{\delta(t) > x\}$ and $\{\gamma(t - x) > x\}$ mean that there is no renewal in $[t - x, t]$, we have the equality

$$\Pr\{\delta(t) > x\} = \Pr\{\gamma(t - x) > x\},$$

that implies (3.22) from (3.23). Applying Theorem 3.6 to (3.22) and (3.23),

$$\lim_{t \to \infty} \Pr\{\gamma(t) \le x\} = 1 - \frac{1}{\mu} \int_0^{\infty} \overline{F}(t + x) \mathrm{d}t = \frac{1}{\mu} \int_0^x \overline{F}(u) \mathrm{d}u,$$

and

$$\lim_{t \to \infty} \Pr\{\delta(t) \le x\} = \lim_{t \to \infty} \Pr\{\gamma(t - x) \le x\} = \lim_{t \to \infty} \Pr\{\gamma(t) \le x\},$$

that follows (3.24). This means that when the process started at $t = -\infty$, looking backwards and forwards from the present time t, the age and the residual life have the same distribution in a renewal process. □

From (i) of Theorem 3.7 (Problem 3.8),

$$
\begin{aligned}
E\{\gamma(t)\} &= \int_t^{\infty} \left[\overline{F}(x) + \int_0^t \overline{F}(x - u) \mathrm{d}M(u) \right] \mathrm{d}x \\
&= \mu[1 + M(t)] - t,
\end{aligned}
\tag{3.25}
$$

that is also easily derived because

$$E\{\gamma(t)\} = E\{S_{N(t)+1} - t\} = \mu[1 + M(t)] - t.$$

It is also of interest that the mean of the above limiting distribution in (3.24) is (Problem 3.8)

$$\frac{1}{\mu}\int_0^\infty t\overline{F}(t)\mathrm{d}t = \frac{\mu}{2} + \frac{\mu_2 - \mu^2}{2\mu} = \frac{\mu}{2} + \frac{\sigma^2}{2\mu}, \tag{3.26}$$

that is greater than half of the mean interval time μ.

In general [2, p. 128, 6 p. 69] (Problem 3.9),

$$\lim_{t\to\infty} \Pr\{\beta(t) \le x\} = \frac{1}{\mu}\int_0^x u\mathrm{d}F(u). \tag{3.27}$$

Example 3.4 (Exponential distribution) When $F(t) = 1 - \mathrm{e}^{-\lambda t}$,

$$\begin{aligned}\Pr\{\delta(t) \le x\} &= \Pr\{\gamma(t) \le x\} = \lim_{t\to\infty} \Pr\{\delta(t) \le x\} \\ &= \lim_{t\to\infty} \Pr\{\gamma(t) \le x\} = 1 - \mathrm{e}^{-\lambda x},\end{aligned}$$

and its mean time is $1/\lambda$. In addition,

$$\begin{aligned}\Pr\{\beta(t) \le x\} &= \Pr\{\delta(t) + \gamma(t) \le x\} = \lim_{t\to\infty} \Pr\{\beta(t) \le x\} \\ &= 1 - (1 + \lambda x)\mathrm{e}^{-\lambda x},\end{aligned}$$

i.e., the failure time of the unit that is operating at time t has a gamma distribution with order 2, and its mean life is twice as much as the usual unit.

Example 3.5 (Uniform distribution) Suppose that $F(t)$ is distributed uniformly on $[0, 10]$, i.e., $F(t) = t/10$ for $0 \le t \le 10$ and $\mu = 5$. In this case, from (3.24),

$$\lim_{t\to\infty} \Pr\{\gamma(t) \le x\} = \frac{1}{5}\int_0^x \left(1 - \frac{u}{10}\right)\mathrm{d}u = \frac{x}{5} - \frac{x^2}{100}$$

for $0 \le x \le 10$, and its mean is $10/3 > \mu/2 = 2.5$. From (3.27),

$$\lim_{t\to\infty} \Pr\{\beta(t) \le x\} = \frac{x^2}{100}$$

for $0 \le x \le 10$ and its mean is $20/3$.

3.2.2 Expected Number of Failures

The renewal function $M(t)$ in a renewal process and the cumulative hazard function $H(t)$ in a nonhomogeneous Poisson process play a great role in reliability theory. We summarize and compare the properties of $M(t)$ and $H(t)$: Suppose that the unit begins to operate at time 0 and fails according to a failure distribution $F(t)$ with finite mean μ and a density function $f(t) \equiv \mathrm{d}F(t)/\mathrm{d}t$.

3.2.2.1 Properties of $M(t)$

The unit is repaired immediately at failures and operates again like new.

(a) Renewal function $M(t)$ represents the expected number of failures in $[0, t]$ and is, from (3.5),

$$M(t) = F(t) + \int_0^t M(t-u)\mathrm{d}F(u)$$
$$= \sum_{n=1}^{\infty} F^{(n)}(t) = F(t) + F(t) * F(t) + \cdots.$$

The LS transform is, from (3.6),

$$M^*(s) = \frac{F^*(s)}{1 - F^*(s)}.$$

Thus, $F(t)$ and $M(t)$ determine uniquely with each other. However, it is not easy to compute $M(t)$ from $F(t)$ except $F(t)$ is a gamma distribution (Example 3.2).

(b) Renewal density is $m(t) = \mathrm{d}M(t)/\mathrm{d}t$, and $m(t)\mathrm{d}t$ represents the probability of a renewal occurring in $[t, t + \mathrm{d}t]$, that is given by a renewal equation

$$m(t) = f(t) + \int_0^t m(t-u)f(u)\mathrm{d}u,$$
$$= \sum_{n=1}^{\infty} f^{(n)}(t) = f(t) + f(t) * f(t) + \cdots.$$

The Laplace transform is

$$m^*(s) = \int_0^{\infty} \mathrm{e}^{-st} m(t)\mathrm{d}t = \frac{f^*(s)}{1 - f^*(s)},$$

where $f^*(s) \equiv \int_0^{\infty} \mathrm{e}^{-st} f(t)dt$.

3.2.2.2 Properties of $H(t)$

The unit is repaired at failures and undergoes minimal repair defined in Sect. 2.3.

(a) Cumulative hazard function $H(t)$ represents the expected number of failures in $[0,t]$ and is, from Sect. 2.3,

$$F(t)=1-\mathrm{e}^{-H(t)},\quad \text{i.e.,}\quad H(t)=-\log\overline{F}(t).$$

Hence,

$$F(t)=H(t)-\frac{[H(t)]^2}{2!}+\frac{[H(t)]^3}{3!}+\cdots,$$
$$H(t)=F(t)+\frac{[F(t)]^2}{2}+\frac{[F(t)]^3}{3}+\cdots.$$

Clearly,

$$h(t)\equiv H'(t)=f(t)[1+F(t)+F(t)^2+\cdots].$$

(b) Hazard rate or failure rate function is $h(t)\equiv \mathrm{d}H(t)/\mathrm{d}t$, and $h(t)\mathrm{d}t$ represents the probability of a failure occurring in $[t,t+\mathrm{d}t]$, given that the age is t, and the unit undergoes minimal repair at each failure that is given by

$$F(t)=1-\mathrm{e}^{-\int_0^t h(u)\mathrm{d}u},\quad \text{i.e.,}\quad h(t)=\frac{f(t)}{1-F(t)}.$$

When $F(t)=1-\mathrm{e}^{-\lambda t}$,

$$M(t)=H(t)=\lambda t,\quad m(t)=h(t)=\lambda.$$

3.2.2.3 Inequalities of $M(t)$ and $H(t)$

It is important to investigate $M(t)$ and $H(t)$ for maintenance models [11, p. 5] when $F(t)$ is IFR, i.e., $h(t)$ increases with t.

(a) When $F(t)$ with finite mean μ and variance σ^2 is IFR [1, p. 54] (Problem 3.10),

$$\mu\geq\sigma,$$
$$M(x)\leq M(t+x)-M(t),$$
$$M(x)\leq \lim_{t\to\infty}[M(t+x)-M(t)]=\frac{x}{\mu},$$
$$V[N(t)]\leq M(t)\leq\frac{t}{\mu},$$
$$\frac{t}{\mu}-1\leq\frac{t}{\int_0^t\overline{F}(u)\mathrm{d}u}-1\leq M(t)\leq\frac{tF(t)}{\int_0^t\overline{F}(u)\mathrm{d}u}\leq\frac{t}{\mu}.$$

Because $m(t)\mathrm{d}t$ is the probability of a renewal in $[t, t+\mathrm{d}t]$,

$$m(t) = \int_0^t h(x)\mathrm{d}\Pr\{\delta(t) \le x\}.$$

Thus, when $F(t)$ is IFR,

$$m(t) \le h(t), \quad \text{i.e.,} \quad M(t) \le H(t).$$

In particular, when $f(t) = \lambda^2 t\mathrm{e}^{-\lambda t}$, for $t > 0$,

$$m(t) = \frac{\lambda}{2}(1 - \mathrm{e}^{-2\lambda t}) < h(t) = \frac{\lambda^2 t}{1 + \lambda t},$$

$$M(t) = \frac{\lambda t}{2} - \frac{1}{4}(1 - \mathrm{e}^{-2\lambda t}) < H(t) = \lambda t - \log(1 + \lambda t).$$

(b) Because $\mathrm{e}^a \ge 1 + a$, and from (a) of Sect. 3.2.2.2,

$$\frac{H(t)}{1 + H(t)} \le F(t) \le H(t) \le \frac{F(t)}{\overline{F}(t)},$$

and hence, when $F(t)$ is IFR,

$$\frac{H(t)}{1 + H(t)} \le F(t) \le M(t) \le H(t) \le \frac{F(t)}{\overline{F}(t)}.$$

Furthermore, we have the inequality

$$\frac{H(t)}{t} \ge \frac{F(t)}{\int_0^t \overline{F}(u)\mathrm{d}u}, \tag{3.28}$$

that is proved as follows: Let

$$Q(t) \equiv H(t) \int_0^t \overline{F}(u)\mathrm{d}u - tF(t),$$

we have that $Q(0) = 0$ and

$$Q'(t) = h(t) \int_0^t \overline{F}(u)\mathrm{d}u - F(t) - \overline{F}(t)[th(t) - H(t)].$$

Furthermore, $Q'(0) = 0$ and

$$Q''(t) = h'(t) \int_0^t [F(t) - F(u)]\mathrm{d}u + f(t) \int_0^t [h(t) - h(u)]\mathrm{d}u \ge 0,$$

that implies (3.28). Therefore (Problem 3.11),

$$\frac{M(t)}{t} \leq \frac{F(t)}{\int_0^t \overline{F}(u)\mathrm{d}u} \leq \frac{H(t)}{t} \leq h(t) \leq \frac{\overline{F}(t)}{\int_t^\infty \overline{F}(u)\mathrm{d}u}. \tag{3.29}$$

3.2.2.4 Mean Time between Failures (MTBF)

Let X_k be the failure time between the $(k-1)$th and the kth failures $(k = 1, 2, \ldots)$. Then, for a renewal process,

$$E\{X_k\} = \int_0^\infty \overline{F}(t)\mathrm{d}t = \mu$$

for all $k \geq 1$, and for a nonhomogeneous Poisson process, from (2.44),

$$E\{X_k\} = \int_0^\infty \frac{[H(t)]^{k-1}}{(k-1)!} \mathrm{e}^{-H(t)} \mathrm{d}t \qquad (k = 1, 2, \ldots).$$

Thus, when $F(t)$ is IFR, from Sect. 2.3,

$$\mu = E\{X_1\} \geq E\{X_2\} \geq \cdots,$$

and

$$\mu \geq E\{X_k\} \to \frac{1}{h(\infty)} \quad \text{as } k \to \infty.$$

Furthermore, x_1 that satisfies $H(x_1) = 1$ is called *characteristic life* in the probability paper of a Weibull distribution, i.e., it represents the mean lifetime that about $1 - \mathrm{e}^{-1} \approx 63.2\%$ of units have failed until time x_1, as shown in (1) of Sect. 2.1.1. When $F(t)$ is IFR, t_1 that satisfies $M(t_1) = 1$ is greater than x_1 because $M(t) \leq H(t)$. This means that more than 63.2% of units have failed until time t_1 at which the expected number of failures is 1 for a renewal process.

3.2.3 Computation of Renewal Function

In general, it is not so easy to compute a renewal function except the interarrival distribution $F(t)$ is a gamma distribution. The following three methods of computing a renewal function $M(t)$ are presented [12, p. 310, 13]:

3.2.3.1 Gamma Approximation

Suppose that the distribution $F(t)$ of a random variable X is approximately replaced by a gamma distribution as follows: From (a) of Sect. 3.2.2.1,

$$M(t) = \sum_{n=1}^{\infty} F^{(n)}(t) \approx \sum_{n=1}^{k} F^{(n)}(t) + \sum_{n=k+1}^{\infty} F_G^{(n)}(t) \tag{3.30}$$

for $k = 2, 3, \ldots$, and $F_G(t)$ is a gamma distribution, i.e.,

$$F_G(t) = \frac{1}{\Gamma(\alpha)} \int_0^t \lambda^{\alpha} x^{\alpha-1} \mathrm{e}^{-\lambda x} \mathrm{d}x \quad \text{for } \lambda > 0 \text{ and } \alpha > 0,$$

and $F_G^{(n)}(t)$ is given in (3.9). Because the infinite series is truncated, this computation can be established by choosing that $E\{X\} = n\alpha/\lambda$ and $V\{X\} = n\alpha/\lambda^2$ and using the stopping criterion for the first k such that $F_G^{(k)}(t)F_G(t)/\overline{F}_G(t) < \varepsilon$ for some specified $\varepsilon > 0$.

3.2.3.2 Laplace Inversion

The LS transform of $M(t)$ is given in Example 3.2 and is obtained by inverting it. The inversion formula for computing the renewal density is

$$m(t) = \lim_{c \to \infty} \frac{1}{2\pi i} \int_{b-ic}^{b+ic} \mathrm{e}^{st} M^*(s) \mathrm{d}s, \tag{3.31}$$

where $i = \sqrt{-1}$, $b > \max\{\sigma, 0\}$, and σ is a radius of convergence. But, (3.31) would be hard to use actually.

Several methods of computing (3.31) are summarized [12, p. 462, 13]: The typical method is, applying the Fourier series to (3.31),

$$\begin{aligned} m(t) &= \frac{2\mathrm{e}^{\mathrm{bt}}}{\pi} \int_0^{\infty} \mathrm{Re}(M^*(b + iu)) \cos(xt) \mathrm{d}x \\ &= -\frac{2\mathrm{e}^{\mathrm{bt}}}{\pi} \int_0^{\infty} \mathrm{Im}(M^*(b + iu)) \sin(xt) \mathrm{d}x. \end{aligned} \tag{3.32}$$

3.2.3.3 Discretization Algorithm

From (3.5), the renewal function $M(t)$ is computed numerically by solving the renewal equation

$$M(t) = F(t) + \int_0^t M(t-u) \mathrm{d}F(u).$$

Xie [14] proposed the discretization algorithm for solving the above Volterra integral equation: Divide the time interval $[0, t]$ by n line segments with equal length d, and denote $M_i = M(id)$ and $K_i = F(id)$ $(i = 0, 1, 2, \ldots, n)$ at each point $i = 0, d, 2d, \ldots, nd = t$, where $F_i \equiv F((i - 0.5)d)$. Then, the following recursive formula gives the approximation of the renewal function M_n, starting from $M_0 \equiv 0$,

$$M_i = \frac{1}{1 - F_1}\left[K_i + \sum_{j=1}^{i-1}(M_j - M_{j-1})F_{i-j+1} - M_{i-1}F_1\right]. \tag{3.33}$$

This formula is easy to program and surprisingly shows accurate results. The renewal function of a Weibull distribution was presented in [12, p. 312] (Problem 3.4).

3.3 Age Replacement Policies

3.3.1 Renewal Equation

We can generalize the renewal equation in (3.5) to the following equation form [5, p. 95, 6, p. 65]

$$g(t) = h(t) + \int_0^t g(t - u)\mathrm{d}F(u), \tag{3.34}$$

where $h(t)$ and $F(t)$ are known, and $g(t)$ is unknown. This is called the *renewal-type equation*. Taking the LS transform of (3.34),

$$g^*(s) \equiv \int_0^\infty \mathrm{e}^{-st}\mathrm{d}g(t) = \frac{h^*(s)}{1 - F^*(s)} = h^*(s)[1 + M^*(s)].$$

Thus, inverting the LS transform $g^*(s)$,

$$g(t) = h(t) + \int_0^t h(t - u)\mathrm{d}M(u). \tag{3.35}$$

Such modified types of equations appear well in reliability theory, as shown in (3.22) and (3.23). We show how to set up renewal-type equations for a variety of reliability models by paying close attention to renewal points and making some suitable modifications: Consider an age replacement model where a unit is replaced at time $T (0 < T \le \infty)$ after its installation or at failure, whichever occurs first. The event $\{T = \infty\}$ represents that no planned replacement is done at all. It is assumed that failures are instantly detected and each failed unit is replaced with a new one, where its replacement time is negligible, and so, a newly installed unit begins to operate instantly. Furthermore, suppose that the failure time $X_k (k = 1, 2, \ldots)$ of each operating unit is independent and has an identical failure distribution $F(t) \equiv \Pr\{X_k \le t\}$ with finite mean $\mu \equiv \int_0^\infty \overline{F}(t)\mathrm{d}t < \infty$ throughout this section.

A new unit is installed at time $t = 0$. Then, an age replacement procedure generates a renewal process as follows: Let $\{X_k\}_{k=1}^\infty$ be the failure times of successive operating units. Define a new random variable $Z_k \equiv \min\{X_k, T\} (k = 1, 2, \ldots)$. Then, $\{Z_k\}_{k=1}^\infty$ represents the intervals between replacements caused by either failures or planned replacements (Fig. 3.4). In other words, because

$$\Pr\{Z_k = X_k\} = \Pr\{X_k \le T\} = F(T),$$
$$\Pr\{Z_k = T\} = \Pr\{X_k > T\} = \overline{F}(T),$$

such trials corresponds to a Bernoulli trial [5, p. 34] whose outcome can be classified as either *success* or *failure* with the respective probabilities $p \equiv \overline{F}(T)$ and $q \equiv F(T)$.

A sequence of random variables $\{Z_k\}_{k=1}^\infty$, where Z_k is independently and identically distributed, forms a renewal process with an interarrival distribution

$$\Pr\{Z_k \le t\} = \begin{cases} F(t) & \text{for } t \le T, \\ 1 & \text{for } t > T. \end{cases}$$

Thus, the mean time between replacements is

$$E\{Z_k\} = \int_0^T t\mathrm{d}F(t) + T\overline{F}(T) = \int_0^T \overline{F}(t)\mathrm{d}t.$$

First, let Y_1 be the first time at which the unit is replaced before time T by failure, and $G_1(t) \equiv \Pr\{Y_1 \le t\}$. Then, by paying attention to the renewal point of replacement times,

$$G_1(t) = \int_0^t \overline{D}(u)\mathrm{d}F(u) + \int_0^t G_1(t-u)\overline{F}(u)\mathrm{d}D(u),$$

where $D(t)$ is a degenerate distribution at time T, i.e., $D(t) \equiv 0$ for $t < T$ and 1 for $t \ge T$. Taking the LS transform and solving it for $G_1^*(s)$,

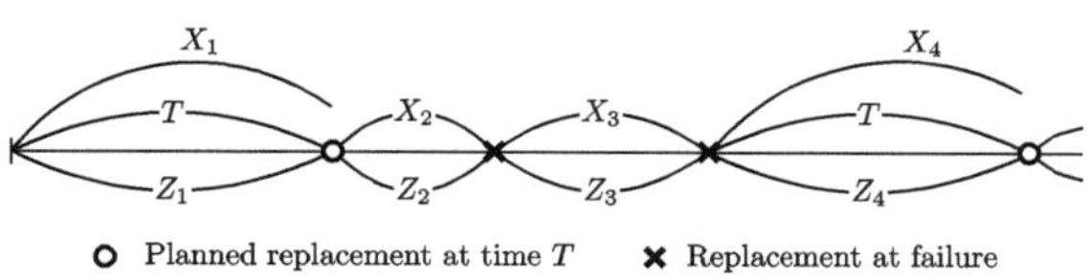

Fig. 3.4 Process of age replacement with planned time T

$$G_1^*(s) \equiv \int_0^\infty \mathrm{e}^{-st}\mathrm{d}G_1(t) = \frac{\int_0^\infty \mathrm{e}^{-st}\overline{D}(t)\mathrm{d}F(t)}{1-\int_0^\infty \mathrm{e}^{-st}\overline{F}(t)\mathrm{d}D(t)} = \frac{\int_0^T \mathrm{e}^{-st}\mathrm{d}F(t)}{1-\mathrm{e}^{-sT}\overline{F}(T)}. \tag{3.36}$$

Similarly, letting Y_2 be the first time at which the unit is replaced before failure by the planned replacement,

$$G_2(t) \equiv \Pr\{Y_2 \le t\} = \int_0^t \overline{F}(u)\mathrm{d}D(u) + \int_0^t G_2(t-u)\overline{D}(u)\mathrm{d}F(u),$$

and its LS transform is

$$G_2^*(s) \equiv \int_0^\infty \mathrm{e}^{-st}\mathrm{d}G_2(t) = \frac{\mathrm{e}^{-sT}\overline{F}(T)}{1-\int_0^T \mathrm{e}^{-st}\mathrm{d}F(t)}. \tag{3.37}$$

Therefore, the mean times of Y_1 and Y_2 are, respectively,

$$\begin{aligned} l_1 &\equiv E\{Y_1\} = \lim_{s\to 0}\frac{1-G_1^*(s)}{s} = \frac{1}{F(T)}\int_0^T \overline{F}(t)\mathrm{d}t, \\ l_2 &\equiv E\{Y_2\} = \lim_{s\to 0}\frac{1-G_2^*(s)}{s} = \frac{1}{\overline{F}(T)}\int_0^T \overline{F}(t)\mathrm{d}t. \end{aligned} \tag{3.38}$$

It can be easily seen that l_1 decreases from $1/h(0)$ to μ when $F(t)$ is IFR, l_2 increases from 0 to ∞, and when $F(T) = 1/2, l_1 = l_2$ (Problem 3.12).

The above results are easily derived by solving the respective renewal equations

$$l_1 = (T + l_1)\overline{F}(T) + \int_0^T t\mathrm{d}F(t),$$

$$l_2 = T\overline{F}(T) + \int_0^T (t + l_2)\mathrm{d}F(t),$$

and

$$E\{Z_k\} = 1/\left(\frac{1}{l_1} + \frac{1}{l_2}\right).$$

It is of interest that $2E\{Z_k\}$ is given by the harmonic mean of l_1 and l_2.

Next, let $M_1(t)$ be the expected number of replacements due to failures in $[0, t]$. We have a renewal equation for $M_1(t)$

$$M_1(t) = \int_0^t [1 + M_1(t-u)]\overline{D}(u)\mathrm{d}F(u) + \int_0^t M_1(t-u)\overline{F}(u)\mathrm{d}D(u).$$

Forming the LS transform and solving for $M_1(t)$,

$$M_1{}^*(s) = \int_0^\infty \mathrm{e}^{-st}\mathrm{d}M_1(t) = \frac{\int_0^T \mathrm{e}^{-st}\mathrm{d}F(t)}{\int_0^T s\mathrm{e}^{-st}\overline{F}(t)\mathrm{d}t}. \tag{3.39}$$

Similarly, let $M_2(t)$ be the expected number of replacements due to the planned replacement in $[0,t]$. We have a renewal equation for $M_2(t)$

$$M_2(t) = \int_0^t M_2(t-u)\overline{D}(u)\mathrm{d}F(u) + \int_0^t [1 + M_2(t-u)]\overline{F}(u)\mathrm{d}D(u).$$

Thus,

$$M_2{}^*(s) = \int_0^\infty \mathrm{e}^{-st}\mathrm{d}M_2(t) = \frac{\mathrm{e}^{-st}\overline{F}(T)}{\int_0^T s\mathrm{e}^{-st}\overline{F}(t)\mathrm{d}t}. \tag{3.40}$$

Then, by using a Tauberian Theorem in Appendix A.3,

$$\begin{aligned}
\lim_{t\to\infty}\frac{M_1(t)}{t} &= \lim_{s\to 0} sM_1{}^*(s) = \frac{F(T)}{\int_0^T \overline{F}(t)\mathrm{d}t} = \frac{1}{l_1},\\
\lim_{t\to\infty}\frac{M_2(t)}{t} &= \lim_{s\to 0} sM_2{}^*(s) = \frac{\overline{F}(T)}{\int_0^T \overline{F}(t)\mathrm{d}t} = \frac{1}{l_2}.
\end{aligned} \tag{3.41}$$

Therefore, letting $M(t)$ be $M(t) \equiv M_1(t) + M_2(t)$ that is the expected number of replacements in $[0,t]$,

$$\lim_{t\to\infty}\frac{M(t)}{t} = \frac{1}{\int_0^T \overline{F}(t)\mathrm{d}t} = \frac{1}{E\{Z_k\}} = \frac{1}{l_1} + \frac{1}{l_2}.$$

Three limiting results of $M_1(t), M_2(t)$ and $M(t)$ satisfy the elementary renewal theorem.

3.3.2 Optimum Replacement Policies

We give optimum policies for age and periodic replacements as applied examples of renewal processes [1, p. 84, 11, p. 70].

3.3.2.1 Age Replacement

We take up the age replacement model where the unit is replaced at time T or at failure, whichever occurs first, as shown in Sect. 3.3.1. The model forms a renewal process with an interarrival distribution $\Pr\{Z_k \leq t\}$ and its mean time $E\{Z_k\} = \int_0^T \overline{F}(t)\mathrm{d}t$.

We consider the problem of minimizing the expected cost per unit of time for an infinite time span. Introduce the following replacement costs: Cost c_1 is incurred for each failed unit that is replaced; this includes all costs resulting from the failure and its replacement. Cost c_2 $(<c_1)$ is incurred for each exchange of nonfailures at time T. Let $M_1(t)$ and $M_2(t)$ denote the expected replacement numbers of failures and nonfailures and the expected cost $\widehat{C}(t)$ denote $c_1M_1(t) + c_2M_2(t)$ in $[0, t]$. Then, from (3.41), the expected cost per unit of time for an infinite time span is

$$C(T) \equiv \lim_{t\to\infty} \frac{\widehat{C}(t)}{t} \equiv \lim_{t\to\infty} \left[c_1\frac{M_1(t)}{t} + c_2\frac{M_2(t)}{t}\right] \\ = \frac{c_1}{l_1} + \frac{c_2}{l_2} = \frac{c_1F(T) + c_2\overline{F}(T)}{\int_0^T \overline{F}(t)\mathrm{d}t}. \tag{3.42}$$

We consider the renewal cycle from one replacement to the next replacement. Then, the pairs of time and cost on each cycle are independently and identically distributed, and both have finite means. The expected cost of one cycle is

$$c_1 \Pr\{X_k \le T\} + c_2 \Pr\{X_k > T\} = c_1F(T) + c_2\overline{F}(T),$$

and the mean time of one cycle is $E\{Z_k\} = \int_0^T \overline{F}(t)\mathrm{d}t$. Thus, the expected cost per unit of time is easily given in (3.42) from (3.19). We call $C(T)$ the *expected cost rate* and generally adopt it as the objective function of the optimization problem for age replacement with an infinite time span [11, p. 71].

We derive an optimum planned replacement time T^* that minimizes the expected cost rate $C(T)$ in (3.42). It is assumed that the failure rate $h(t) \equiv f(t)/\overline{F}(t)$ is differentiable and increases strictly, i.e., $h'(t) \equiv \mathrm{d}h(t)/\mathrm{d}t > 0$, where $f(t)$ is a density function of $F(t)$. It is clearly seen that

$$C(0) \equiv \lim_{T\to 0} C(T) = \infty, \quad C(\infty) \equiv \lim_{T\to\infty} C(T) = \frac{c_1}{\mu}.$$

Thus, there exists a positive $T^*\,(0<T^* \le \infty)$. In addition, differentiating $C(T)$ with respect to T and setting it equal to zero,

$$h(T)\int_0^T \overline{F}(t)\mathrm{d}t - F(T) = \frac{c_2}{c_1 - c_2}. \tag{3.43}$$

Letting $Q_1(T)$ be the left-hand side of (3.43), $Q(0) = 0$ and

$$Q_1(\infty) = \mu h(\infty) - 1.$$

Differentiating $Q_1(T)$ with respect to T,

$$\frac{\mathrm{d}Q_1(\mathrm{T})}{\mathrm{d}T} = h'(T)\int_0^T \overline{F}(t)\mathrm{d}t > 0.$$

Therefore, we have the optimum policy when $K \equiv c_1/[\mu(c_1 - c_2)]$:

(i) If $h(\infty) > K$, then there exists a finite and unique T^* $(0<T^*<\infty)$ that satisfies (3.43), and the resulting cost rate is

$$C(T^*) = (c_1 - c_2)h(T^*).$$

(ii) If $h(\infty) \le K$, then $T^* = \infty$, i.e., the unit is replaced only at failure, and the expected cost rate is $C(\infty) = c_1/\mu$.

3.3.2.2 Three Periodic Replacement

We take up three periodic replacements where the unit is replaced at periodic times kT $(k = 1, 2, \ldots)$. These replacement models form the simplest renewal process with a constant time T $(0<T\le\infty)$.

First, we consider the block replacement where each failed unit between planned replacements is instantly detected and is replaced with a new one. That is, the model also forms a renewal process with a renewal function $M(t)$ between planned replacements. Then, because the expected number of failures in one planned replacement cycle is $M(T)$, the expected cost rate is, from (3.19),

$$C_1(T) = \frac{c_1 M(T) + c_2}{T}, \tag{3.44}$$

where $c_1 =$ cost of replacement at each failure and $c_2 =$ cost of planned replacement at time T. Clearly, from (3.18), $\lim_{T\to\infty} C_1(T) = c_1/\mu$.

Compare the expected costs with age and block replacements: Letting $A(T) \equiv c_1 F(T) + c_2 \overline{F}(T)$ and $B(T) \equiv c_1 M(T) + c_2$, we have the renewal equation (Problem 3.13)

$$B(T) = A(T) + \int_0^T B(T-t)\mathrm{d}F(t), \quad \text{or} \quad B(T) = A(T) + \int_0^T A(T-t)\mathrm{d}M(t),$$

i.e., $A(T)$ and $B(T)$ determine each other.

We seek an optimum planned replacement time T^* that minimizes $C_1(T)$ in (3.44). It is assumed that $M(t)$ is differentiable and define $m(t) \equiv \mathrm{d}M(t)/\mathrm{d}t$. Then, differentiating $C_1(T)$ with respect to T and setting it equal to zero,

$$Tm(T) - M(T) = \frac{c_2}{c_1}. \tag{3.45}$$

This is a necessary condition that there exists a finite T^*, and the resulting cost rate is

$$C_1(T^*) = c_1 m(T^*).$$

Next, we consider the periodic replacement where each failed unit between replacements undergo minimal repair, as shown in Sect. 2.4.1. It is assumed that the repair times are negligible and the failure rate remains undisturbed by any minimal repair. That is, the replacement model also forms a nonhomogeneous Poisson process with a cumulative hazard function $H(t)$ between planned replacements, where $F(t) = 1 - e^{-H(t)}$ and $H'(t) = h(t)$. Then, because the expected number of failures in one planned replacement cycle is $H(T)$ from (2.45), the expected cost rate is, from (3.19),

$$C_2(T) = \frac{c_1 H(T) + c_2}{T}, \tag{3.46}$$

where c_1 = cost of minimal at each failure, and c_2 is given in (3.44). When $F(t)$ is IFR, $C_2(T) \geq C_1(T)$ for the same replacement costs from $H(T) \geq M(T)$.

We seek an optimum planned time T^* that minimizes $C_2(T)$ in (3.46). Differentiating $C_2(T)$ with respect to T and setting it equal to zero,

$$Th(T) - H(T) = \frac{c_2}{c_1}, \tag{3.47}$$

and the resulting cost rate is

$$C_2(T^*) = c_1 h(T^*).$$

It is shown that if the failure rate $h(t)$ increases strictly, then the left-hand side of (3.47) also increases strictly [11, p. 102]. So that, if a solution T^* to (3.47) exists, then it is unique.

Finally, we consider the periodic replacement where each failed unit remains in a failed state for the time interval from failure to its planned replacement. Then, the mean time from a failure to the replacement is

$$\int_0^T (T - t)\mathrm{d}F(t) = \int_0^T F(t)\mathrm{d}t.$$

Thus, the expected cost rate is [11, p. 120]

$$C_3(T) = \frac{c_1 \int_0^T F(t)\mathrm{d}t + c_2}{T}, \tag{3.48}$$

where c_1 = downtime cost per unit of time elapsed between a failure and its replacement, and c_2 is given in (3.44).

Differentiating $C_3(T)$ with respect to T and setting it equal to zero,

$$TF(T) - \int_0^T F(t)\mathrm{d}t = \frac{c_2}{c_1}, \quad \text{or} \quad \int_0^T t\mathrm{d}F(t) = \frac{c_2}{c_1}. \tag{3.49}$$

Thus, if $\mu > c_2/c_1$, then there exists an optimum time $T_3{}^*$ that satisfies uniquely (3.49), and the resulting cost rate is

$$C_3(T_3{}^*) = c_1 F(T^*).$$

In general, the results of three periodic replacements are summarized as follows: The expected cost rate is [11, p. 125]

$$C(T) = \frac{c_1 \int_0^T \varphi(t)\mathrm{d}t + c_2}{T}, \tag{3.50}$$

where $\varphi(t)$ is $h(t), m(t)$, and $F(t)$, respectively. Differentiating $C(T)$ with respect to T and setting it equal to zero,

$$T\varphi(T) - \int_0^T \varphi(t)\mathrm{d}t = \frac{c_2}{c_1}, \quad \text{or} \quad \int_0^T t\mathrm{d}\varphi(t) = \frac{c_2}{c_1}. \tag{3.51}$$

If there exists T^* that satisfies (3.51), then the expected cost rate is

$$C(T^*) = c_1\varphi(T^*).$$

Conversely, if a finite T^* does not exists, i.e., $T^* = \infty$, and the unit does not undergo any planned replacement, then

$$C(\infty) = c_1\varphi(\infty).$$

3.4 Alternating Renewal Process

Alternating renewal processes are the process that combines two renewal processes $\{X_k\}$ and $\{Y_k\}$ generated alternately (Fig. 3.5) [7]. That is, each sequence of $\{X_k\}$ and $\{Y_k\}$ forms a renewal process, and the sequence of the sum of $\{X_k + Y_k\}$ also forms a renewal process. In this case, X_k and Y_k are allowed to be dependent. Most redundant systems with repair maintenance that repeat up and down states alternately, generate alternating renewal processes. Furthermore, the process is a special case of Markov renewal processes in Chap. 5.

3.4.1 Ordinary Alternating Renewal Process

We consider a simple system with one unit that repeats on and off states. Suppose that $\{X_k\}$ and $\{Y_k\}$ $(k = 1, 2, \ldots)$ are the respective on and off times of the kth renewal cycle and have the distribution $F(t) \equiv \Pr\{X_k \leq t\}$ with finite mean μ and

Fig. 3.5 Realization of alternating renewal process

the distribution $G(t) \equiv \Pr\{Y_k \le t\}$ with finite mean β. To analyze the system, we define the following states:

State 0: Unit is on.
State 1: Unit is off.

When the unit starts to State 0 at time 0, the unit repeats States 0 and 1 alternately (Fig. 3.5).

Let $M_{ij}(t)$ $(i,j=0,1)$ denote the expected number of State j in $[0,t]$ when the unit goes into State i at time 0, where the first visit to State j is not counted when $i=j$. Then, we have the following renewal equations:

$$M_{01}(t) = F(t) + \int_0^t M_{11}(t-u)\mathrm{d}F(u),$$

$$M_{10}(t) = G(t) + \int_0^t M_{00}(t-u)\mathrm{d}G(u), \tag{3.52}$$

$$M_{11}(t) = \int_0^t M_{01}(t-u)\mathrm{d}G(u), \quad M_{00}(t) = \int_0^t M_{10}(t-u)\mathrm{d}F(u).$$

Thus, taking the LS transforms of both sides of the above equations and solving them,

$$M_{01}^*(s) = \frac{F^*(s)}{1-F^*(s)G^*(s)}, \quad M_{10}^*(s) = \frac{G^*(s)}{1-F^*(s)G^*(s)}, \tag{3.53}$$
$$M_{11}^*(s) = G^*(s)M_{01}^*(s), \qquad M_{00}^*(s) = F^*(s)M_{10}^*(s),$$

where the asterisk of the function denotes the LS transform with itself, i.e., $\Phi^*(s) \equiv \int_0^\infty \mathrm{e}^{-st}\mathrm{d}\Phi(t)$ for any function $\Phi(t)$.

Next, let $Z(t)$ denote the state of the process at time t, and define $P_{ij}(t) \equiv \Pr\{Z(t)=j|Z(0)=i\}$ $(i,j=0,1)$. The probability $P_{ij}(t)$ is called the *transition probability* from State i to State j and represents the probability that the unit is in State j at time t if it starts from State i at time 0. By the similar method for obtaining (3.52),

$$P_{00}(t) = \overline{F}(t) + \int_0^t P_{10}(t-u)\mathrm{d}F(u),$$

$$P_{11}(t) = \overline{G}(t) + \int_0^t P_{01}(t-u)\mathrm{d}G(u), \tag{3.54}$$

$$P_{10}(t) = \int_0^t P_{00}(t-u)\mathrm{d}G(u), \quad P_{01}(t) = \int_0^t P_{11}(t-u)\mathrm{d}F(u).$$

Thus, again taking the LS transforms,

$$P_{00}^*(s) = \frac{1 - F^*(s)}{1 - F^*(s)G^*(s)}, \quad P_{11}^*(s) = \frac{1 - G^*(s)}{1 - F^*(s)G^*(s)},$$
$$P_{10}^*(s) = G^*(s)P_{00}^*(s), \qquad P_{01}^*(s) = F^*(s)P_{11}^*(s). \tag{3.55}$$

Thus, comparing (3.53) and (3.55), we have the following relations:

$$P_{01}(t) = \int_0^t \overline{G}(t-u)\mathrm{d}M_{01}(u) = M_{01}(t) - M_{00}(t),$$

$$P_{10}(t) = \int_0^t \overline{F}(t-u)\mathrm{d}M_{10}(u) = M_{10}(t) - M_{11}(t).$$

These relations between renewal functions and transition probabilities would be useful for the analysis of more complex systems.

Next, it is assumed that distributions $F(t)$ and $G(t)$ have finite variances ${\sigma_F}^2$ and ${\sigma_G}^2$, respectively. Then,

$$F^*(s) = 1 - \mu s + \frac{1}{2}(\mu^2 + \sigma_F^2)s^2 + o(s^2),$$
$$G^*(s) = 1 - \beta s + \frac{1}{2}(\beta^2 + \sigma_G^2)s^2 + o(s^2).$$

Substituting these equations for (3.53),

$$M_{01}^*(s) = \frac{1}{\mu+\beta}\frac{1}{s} - \frac{\mu}{\mu+\beta} + \frac{1}{2} + \frac{\sigma_F^2 + \sigma_G^2}{2(\mu+\beta)^2} + o(1).$$

The LS inversion of $M_{01}^*(s)$ gives, for large t,

$$M_{01}(t) = \frac{t}{\mu+\beta} - \frac{\mu}{\mu+\beta} + \frac{1}{2} + \frac{\sigma_F^2 + \sigma_G^2}{2(\mu+\beta)^2} + o(1). \tag{3.56}$$

Similarly,

$$M_{00}(t) = \frac{t}{\mu+\beta} - \frac{1}{2} + \frac{\sigma_F^2 + \sigma_G^2}{2(\mu+\beta)^2} + o(1),$$
$$P_{00}(t) = \frac{\mu}{\mu+\beta} + o(1), \quad P_{01}(t) = \frac{\beta}{\mu+\beta} + o(1). \tag{3.57}$$

Clearly,

$$\lim_{t\to\infty} \frac{M_{ij}(t)}{t} = \frac{1}{\mu+\beta} \quad (j = 0, 1),$$
$$\lim_{t\to\infty} P_{i0}(t) = 1 - \lim_{t\to\infty} P_{i1}(t) = \frac{\mu}{\mu+\beta} = \frac{E\{X_k\}}{E\{X_k\} + E\{Y_k\}}, \tag{3.58}$$

that are independent of an initial State i in the limit. Note that (3.58) represents the steady-state availability of a one-unit system [11, p. 9].

Example 3.6 (One-unit system with repair) Suppose that an operating unit is repaired when it fails, and it begins to operate again. In this case, States 0 and 1 represent that the unit is operating and under repair, respectively. It is assumed that the failure time has an exponential distribution $F(t) = 1 - e^{-\lambda t}$ and the repair time has an exponential distribution $G(t) = 1 - e^{-\theta t}$. Then, it is easy to invert the LS transforms as follows:

$$P_{01}(t) = \frac{\lambda}{\lambda+\theta}\left[1 - e^{-(\lambda+\theta)t}\right],$$
$$M_{01}(t) = \frac{\lambda\theta t}{\lambda+\theta} + \left(\frac{\lambda}{\lambda+\theta}\right)^2\left[1 - e^{-(\lambda+\theta)t}\right].$$

Thus, for large t,

$$P_{01}(t) \approx \frac{\lambda}{\lambda+\theta}, \quad M_{01}(t) \approx \frac{\lambda\theta}{\lambda+\theta}t + \left(\frac{\lambda}{\lambda+\theta}\right)^2.$$

Next, when $F(t) = 1 - (1+\lambda t)e^{-\lambda t}$ and the time for repair is constant β,

$$P^*_{01}(s) = \frac{\lambda^2(1 - e^{-\beta s})}{(s+\lambda)^2 - \lambda^2 e^{-\beta s}}, \quad M^*_{01}(s) = \frac{\lambda^2}{(s+\lambda)^2 - \lambda^2 e^{-\beta s}},$$

and for large t,

$$P_{01}(t) \approx \frac{\beta}{2/\lambda+\beta}, \quad M_{01}(t) \approx \frac{t}{2/\lambda+\beta}.$$

3.4.2 Interval Reliability

Interval reliability is defined as the probability that at a specified time, the unit is operating and will continue to operate for an interval duration [1, p. 7, 11, p. 11]. Then, the interval reliability $R(x;T_0)$ for an interval of duration x starting at time T_0 is

$$R(x;T_0) \equiv \Pr\{Z(u) = 0, T_0 \le u \le T_0 + x\}, \tag{3.59}$$

and its limit of $R(x;T_0)$ as $T_0 \to \infty$ is called the *limiting interval reliability*.

When the unit begins to operate at time 0 in Example 3.6, the interval reliability is

$$R(x;T_0) \equiv P_{00}(T_0)e^{-\lambda x} = \left[\frac{\theta}{\lambda+\theta} + \frac{\lambda}{\lambda+\theta}e^{-(\lambda+\theta)T_0}\right]e^{-\lambda x}, \tag{3.60}$$

and its limiting interval reliability is

$$R(x) \equiv \lim_{T_0 \to \infty} R(x; T_0) = \frac{\theta}{\lambda + \theta} e^{-\lambda x}. \tag{3.61}$$

Suppose that the unit has a failure distribution $F(t)$ with mean μ and a repair distribution $G(t)$ with mean β in Sect. 3.4.1. When the unit begins to operate at time 0, the interval reliability is [11, p. 48]

$$R(x; T_0) = \overline{F}(T_0 + x) + \int_0^{T_0} \overline{F}(T_0 + x - u) \mathrm{d}M_{00}(u), \tag{3.62}$$

where $M_{00}(t)$ is given in (3.53). Hence, its Laplace transform is

$$R^*(x; s) \equiv \int_0^\infty e^{-sT_0} R(x; T_0) \mathrm{d}T_0 = \frac{e^{sx} \int_x^\infty e^{-st} \overline{F}(t) \mathrm{d}t}{1 - F^*(s) G^*(s)}, \tag{3.63}$$

and its limiting interval reliability is

$$R(x) \equiv \lim_{T_0 \to \infty} R(x; T_0) = \lim_{s \to 0} s R^*(x; s) = \frac{\int_x^\infty \overline{F}(t) \mathrm{d}t}{\mu + \beta}. \tag{3.64}$$

Next, suppose that we set the preventive maintenance (PM) at time $T\,(0 < T \le \infty)$ for the operating unit [11, p. 141]. However, the PM is not done in the interval $[T_0, T_0 + x]$ even if it is time for PM. It is assumed that the distribution of time to the PM completion is the same as the repair distribution $G(t)$.

Let $D(t)$ be a degenerate distribution placing unit mass at T, i.e., $D(t) \equiv 0$ for $t < T$ and 1 for $t \ge T$. Then, in a similar way of obtaining (3.52),

$$M_{00}(t) = \int_0^t M_{10}(t - u)[\overline{D}(u) \mathrm{d}F(u) + \overline{F}(u) \mathrm{d}D(u)],$$

$$M_{10}(t) = \int_0^t [1 + M_{00}(t - u)] \mathrm{d}G(u).$$

Taking the LS transforms,

$$M_{00}^*(s) = \frac{G^*(s)\left[1 - \int_0^T s e^{-st} \overline{F}(t) \mathrm{d}t\right]}{1 - G^*(s)\left[1 - \int_0^T s e^{-st} \overline{F}(t) \mathrm{d}t\right]}. \tag{3.65}$$

By the similar method of obtaining (3.62), the interval reliability is

$$R(T; x, T_0) = \overline{F}(T_0 + x) \overline{D}(T_0) + \int_0^{T_0} \overline{F}(T_0 + x - u) \overline{D}(T_0 - u) \mathrm{d}M_{00}(u). \tag{3.66}$$

Taking the Laplace transform of (3.66) and substituting it for (3.65),

$$R^*(T;x,s) \equiv \int_0^\infty \mathrm{e}^{-sT_0} R(T;x,T_0)\mathrm{d}T_0$$
$$= \frac{\mathrm{e}^{sx}\int_x^{T+x}\mathrm{e}^{-st}\overline{F}(t)\mathrm{d}t}{1-G^*(s)+G^*(s)\int_0^T s\mathrm{e}^{-st}\overline{F}(t)\mathrm{d}t}, \tag{3.67}$$

that agrees with (3.63) when $T=\infty$ (Problem 3.14). Thus, the limiting interval reliability is

$$R(T;x) \equiv \lim_{T_0\to\infty} R(T;x,T_0) = \lim_{s\to 0} sR^*(T;x,s)$$
$$= \frac{\int_x^{T+x}\overline{F}(t)\mathrm{d}t}{\int_0^T \overline{F}(t)\mathrm{d}t+\beta}, \tag{3.68}$$

that also agrees with (3.64) when $T=\infty$ (Problem 3.15).

We seek an optimum PM time T^* that maximizes $R(T;x)$ for a fixed $x>0$. Let $\lambda(t;x)\equiv[F(t+x)-F(t)]/\overline{F}(t)$ for $t\geq 0$ and $\lambda(\infty;x)\equiv \lim_{t\to\infty}\lambda(t;x)\leq 1$. Then, note that both $\lambda(t;x)$ and the failure rate $h(t)\equiv f(t)/\overline{F}(t)$ have the same property [1, p. 23, 11, p. 6].

It is assumed that $\lambda(t;x)$ increases strictly with t. Then, differentiating $R(T;x)$ in (3.68) with respect to T and setting it equal to zero,

$$\lambda(T;x)\left[\int_0^T \overline{F}(t)\mathrm{d}t+\beta\right] - \int_0^T [F(t+x)-F(t)]\mathrm{d}t = \beta, \tag{3.69}$$

whose left-hand side $Q(T;x)$ increases strictly from $Q(0)=\beta F(x)$ to

$$Q(\infty;x) = \lambda(\infty;x)(\mu+\beta) - \int_0^x \overline{F}(t)\mathrm{d}t.$$

Therefore, if $Q(\infty;x)>\beta$, then there exists a finite and unique T^* $(0<T^*<\infty)$ that satisfies (3.69), and the resulting interval reliability is

$$R(T^*;x) = \frac{\overline{F}(T^*+x)}{\overline{F}(T^*)},$$

that represents the probability that the unit with age T^* does not fail in $[T^*, T^*+x]$.

Example 3.7 When $F(t)=1-(1+\lambda t)\mathrm{e}^{-\lambda t}$,

$$\lambda(t;x) = 1-\left(1+\frac{\lambda x}{1+\lambda t}\right)\mathrm{e}^{-\lambda x},$$

that increases strictly with t to $(1 - e^{-\lambda x})$. Thus, if $Q(\infty; x) > \beta$, i.e., $x > \beta$, an optimum T^* is a unique solution of the equation

$$\lambda T(x - \beta) - x(1 - e^{-\lambda T}) = (1 + \lambda x)\beta,$$

and the interval reliability is

$$R(T^*; x) = \left(1 + \frac{\lambda x}{1 + \lambda T^*}\right) e^{-\lambda x}.$$

In this case, an optimum T^* decreases with x to a unique solution of

$$T - \frac{1 - e^{-\lambda T}}{\lambda} = \beta.$$

Conversely, if $x \leq \beta$, then $T^* = \infty$ and the interval reliability in (3.68) is given in (3.64).

3.4.3 Off Distribution

Let $N(t)$ and $D(t)$ be the number of on states and the total amount of time spent in off states in $[0, t]$, respectively. It is said in reliability theory that $D(t)$ is called a *downtime* [1, p. 79, 11, p. 45]. Then, from the formula of the total sum of independent random variables,

$$\begin{aligned}
&\Pr\{Y_1 + Y_2 + \cdots + Y_n \leq x | N(t) = n\} \Pr\{N(t) = n\} \\
&\quad = \Pr\{Y_1 + Y_2 + \cdots + Y_n \leq x\} \Pr\{X_1 + \cdots + X_n \leq t - x < X_1 + \cdots + X_{n+1}\} \\
&\quad = G^{(n)}(x)\left[F^{(n)}(t - x) - F^{(n+1)}(t - x)\right].
\end{aligned}$$

Thus, the downtime distribution is

$$\Pr\{D(t) \leq x\} = \begin{cases} \sum_{n=0}^{\infty} G^{(n)}(x)[F^{(n)}(t - x) - F^{(n+1)}(t - x)] & \text{for } t > x, \\ 1 & \text{for } t \leq x. \end{cases} \tag{3.70}$$

The above equation can be written as

$$\Pr\{D(t + x) \leq x\} = \sum_{n=0}^{\infty} G^{(n)}(x)\left[F^{(n)}(t) - F^{(n+1)}(t)\right],$$

$$\Pr\{D(t) > x\} = \begin{cases} \sum_{n=0}^{\infty}\left[G^{(n)}(x) - G^{(n+1)}(x)\right] F^{(n+1)}(t - x) & \text{for } t > x, \\ 0 & \text{for } t \leq x. \end{cases}$$

Next, we investigate the properties of $D(t)$. Denote $L_\delta \equiv \min_t\{D(t) > \delta\}$ as the first time that the total time spent in off states, i.e., the total downtime, exceeds a threshold level δ. We have the following relation:

$$\Pr\{L_\delta \le t\} = \Pr\{D(t) > \delta\}$$
$$= \sum_{n=0}^{\infty}\left[G^{(n)}(\delta) - G^{(n+1)}(\delta)\right]F^{(n+1)}(t-\delta). \tag{3.71}$$

Thus, the mean time that the total downtime first exceeds δ is

$$E\{L_\delta\} = \delta + \mu\sum_{n=0}^{\infty} G^{(n)}(\delta). \tag{3.72}$$

The mean downtime in $[0, t]$ is

$$E\{D(t)\} = \int_0^t x\mathrm{d}\Pr\{D(t) \le x\}. \tag{3.73}$$

In general, it is very difficult to compute (3.73). It is noted easily from Sect. 3.4.1 that

$$E\{D(t)\} = \int_0^t P_{01}(u)\mathrm{d}u, \tag{3.74}$$

and its LS transform is

$$\int_0^\infty \mathrm{e}^{-st}P_{01}(t)\mathrm{d}t = \frac{F^*(s)[1-G^*(s)]}{s[1-F^*(s)G^*(s)]}. \tag{3.75}$$

Thus, from (3.57)

$$\lim_{t\to\infty}\frac{E\{D(t)\}}{t} = \lim_{t\to\infty}P_{01}(t) = \lim_{s\to 0} s\int_0^\infty \mathrm{e}^{-st}P_{01}(t)\mathrm{d}t = \frac{\beta}{\mu+\beta}. \tag{3.76}$$

Example 3.8 (One-unit system with repair) We consider the same one-unit system in Example 3.6. The downtime distribution is [1, p. 80]

$$\Pr\{D(t) > x\} = \sum_{n=0}^{\infty}\frac{(\lambda x)^n}{n!}\mathrm{e}^{-\lambda x}\sum_{k=n+1}^{\infty}\frac{[\theta(t-x)]^k}{k!}\mathrm{e}^{-\theta(t-x)},$$

or

$$\Pr\{D(t) \le x\} = \mathrm{e}^{-\theta(t-x)}\left[1+\sqrt{\lambda\theta(t-x)}\int_0^x \mathrm{e}^{-\lambda y}y^{-1/2}I_1\left(2\sqrt{\lambda\theta(t-x)y}\right)\mathrm{d}y\right],$$

where $I_1(x)$ is the Bessel function of order 1 for the imaginary argument defined by

$$I_1(x) \equiv \sum_{j=0}^{\infty} \frac{(x/2)^{2j+1}}{j!(j+1)!}.$$

In addition, from (3.72) and (3.74),

$$E\{L_\delta\} = \delta + \frac{1}{\lambda}(1 + \theta\delta),$$

$$E\{D(t)\} = \frac{\lambda}{\lambda + \theta}t - \frac{\lambda}{(\lambda + \theta)^2}\left[1 - e^{-(\lambda+\theta)t}\right].$$

Takács [10] also proved the following important theorem:

Theorem 3.8 If $\sigma_F^2 < \infty$ and $\sigma_G^2 < \infty$, then

$$\lim_{t\to\infty} \Pr\left\{\frac{D(t) - \beta t/(\mu + \beta)}{\sqrt{\left[(\beta\sigma_F)^2 + (\mu\sigma_G)^2\right]t/(\mu + \beta)^3}} \le x\right\} = \frac{1}{\sqrt{2\pi}} \int_{-\infty}^{x} e^{-u^2/2} du. \tag{3.77}$$

That is, if the means and variances of $F(t)$ and $G(x)$ are statistically estimated, then the probability of the total time spent in off states, i.e., downtime $D(t)$ is approximately computed for large t, by using a standard normal distribution.

Example 3.9 (Asymptotic downtime distribution) [11, p. 50] We wish to compute the time δ when the probability that the system is down more than δ in $t = 10,000$ hours of operation is given by 0.90. The failure and repair distributions are unknown, but from the sample data, the estimates of means and variances are

$$\mu = 1,000, \quad {\sigma_F}^2 = 100,000, \quad \beta = 100, \quad {\sigma_G}^2 = 400.$$

Then, from Theorem 3.8, when $t = 10,000$,

$$\frac{D(t) - \beta t/(\mu + \beta)}{\sqrt{\left[(\beta\sigma_F)^2 + (\mu\sigma_G)^2\right]t/(\mu + \beta)^3}} = \frac{D(10,000) - 909.09}{102.56}$$

is approximately normally distributed with mean 0 and variance 1. Hence,

$$\Pr\{D(t) > \delta\} = \Pr\left\{\frac{D(10,000) - 909.09}{102.56} > \frac{\delta - 909.09}{102.56}\right\}$$

$$\approx \frac{1}{\sqrt{2\pi}} \int_{(\delta-909.09)/102.56}^{\infty} e^{-u^2/2} du = 0.90.$$

Because $u_0 = -1.28$ such that $(1/\sqrt{2\pi}) \int_u^\infty e^{-u^2/2} du_0 = 0.90$,

$$\delta \approx 909.09 - 102.56 \times 1.28 = 777.81,$$

i.e., the system is down more than 778 h in $t = 10{,}000$ h with probability 0.90 (Problem 3.16).

3.4.4 Terminating Renewal Process

Consider the first time that an amount of the time spent in off states exceeds a threshold level U with distribution $W(t)$. We define that if $Y_k \le U$ then $\widetilde{Y}_k \equiv Y_k$, and conversely, if $U < Y_k$ then $\widetilde{U}_k \equiv U$ $(k = 1, 2, \ldots)$. The process ends with the first event of $\{U < Y_N\}$ that is called the terminating process $\{X_1, \widetilde{Y}_1, X_2, \widetilde{Y}_2, \ldots, X_{N-1}, \widetilde{Y}_{N-1}, X_N, \widetilde{U}_N\}$, and its interest quantities are the total sum of random variables $Z_N \equiv \sum_{k=1}^{N-1}(X_k + \widetilde{Y}_k) + X_N + \widetilde{U}_N$ $(N = 1, 2, \ldots)$ and its distribution $L(t) \equiv \sum_{N=1}^\infty \Pr\{Z_N \le t\}$.

The probability that Y_k is not greater than U and $Y_k \le t$ is

$$B(t) \equiv \Pr\{Y_k \le U, Y_k \le t\} = \int_0^t \overline{W}(u) dG(u),$$

and the probability that Y_k is greater than U and $U \le t$ is

$$\widetilde{B}(t) \equiv \Pr\{U < Y_k, U \le t\} = \int_0^t \overline{G}(u) dW(u).$$

Thus, from the formula of the total sum of independent random variables,

$$L(t) = \sum_{N=1}^\infty \Pr\{Z_N \le t\} = \sum_{k=0}^\infty F^{(k)}(t) * B^{(k)}(t) * F(t) * \widetilde{B}(t). \tag{3.78}$$

Therefore, the LS transform of $L(t)$ is

$$L^*(s) \equiv \int_0^\infty e^{-st} dL(t) = \frac{F^*(s)\widetilde{B}^*(s)}{1 - F^*(s)B^*(s)}, \tag{3.79}$$

and its mean time is

$$l \equiv \int_0^\infty t dL(t) = \lim_{s \to 0} \frac{1 - L^*(s)}{s} = \frac{\mu + \int_0^\infty \overline{G}(t)\overline{W}(t) dt}{\int_0^\infty W(t) dG(t)}. \tag{3.80}$$

In particular, when a random variable U is constant $c (c > 0)$, i.e., $W(t) \equiv 0$ for $t < c$, and 1 for $t \geq c$,

$$L^*(s) = \frac{F^*(s)\mathrm{e}^{-sc}\overline{G}(c)}{1 - F^*(s)\int_0^c \mathrm{e}^{-st}\mathrm{d}G(t)}, \quad l = \frac{\mu + \int_0^c \overline{G}(t)\mathrm{d}t}{\overline{G}(c)}, \tag{3.81}$$

where c is called a *critical allowed time* for maintenance [15].

Example 3.10 (Repair limit policy) [11, p. 51] Consider a one-unit system that is repaired or replaced if it fails. Let μ denote the mean failure time of the unit and $G(t)$ denote the repair distribution of the failed unit with finite mean β. When the unit fails, its repair is started immediately, and when the repair is not completed within time T $(0 \leq T < \infty)$, that is called the repair limit time, it is replaced with a new one. If $Z_k \equiv \min\{Y_k, T\}$, then the sequence of $\{X_k, Z_k\}$ $(k = 1, 2, \ldots)$ forms an alternating renewal process.

Let c_1 be the replacement cost of a failed unit, caused by failure and replacement, and $c_r(t)$ be the expected repair cost during $[0, t]$, that includes all costs incurred due to repair and downtime during $[0, t]$. Consider one cycle from the beginning of an operating unit to the repair or replacement completion. Then, the expected cost of one cycle is

$$[c_1 + c_r(T)]\overline{G}(T) + \int_0^T c_r(t)\mathrm{d}G(t) = c_1\overline{G}(T) + \int_0^T \overline{G}(t)\mathrm{d}c_r(t),$$

and the mean time of one cycle is

$$\mu + T\overline{G}(T) + \int_0^T t\mathrm{d}G(t) = \mu + \int_0^T \overline{G}(t)\mathrm{d}t.$$

Thus, from (3.19), the expected cost rate is

$$C(T) = \frac{c_1\overline{G}(T) + \int_0^T \overline{G}(t)\mathrm{d}c_r(t)}{\mu + \int_0^T \overline{G}(t)\mathrm{d}t}. \tag{3.82}$$

Clearly,

$$C(0) \equiv \lim_{T\to 0} C(T) = \frac{c_1}{\mu}, \quad C(\infty) \equiv \lim_{T\to\infty} C(T) = \frac{\int_0^\infty \overline{G}(t)\mathrm{d}c_r(t)}{\mu + \beta},$$

that represent the expected cost rates with only replacement and only repair maintenance, respectively.

Suppose that $c_r(t) = at$ for $a > 0$. Then, the expected cost rate is, from (3.82),

$$C(T) = \frac{c_1\overline{G}(T) + a\int_0^T \overline{G}(t)\mathrm{d}t}{\mu + \int_0^T \overline{G}(t)\mathrm{d}t}. \tag{3.83}$$

We find an optimum repair limit time T^* that minimizes $C(T)$. It is assumed that $r(t) \equiv g(t)/\overline{G}(t)$, where $g(t)$ is a density function of $G(t)$. Then, differentiating $C(T)$ with respect to T and setting it equal to zero,

$$r(T)\left[\mu + \int_0^T \overline{G}(t)\mathrm{d}t\right] + \overline{G}(T) = \frac{a\mu}{c_1}. \tag{3.84}$$

Letting $Q(T)$ be the left-hand side of (3.84),

$$Q(0) \equiv \mu r(0) + 1, \quad Q(\infty) = (\mu + \beta) r(\infty).$$

Thus, if $r(t)$ decreases strictly and $Q(0) > a\mu/c_1 > Q(\infty)$, there exists uniquely a finite and positive $T^* (0 < T^* < \infty)$ that satisfies (3.84), and the resulting cost rate is

$$C(T^*) = a - c_1 r(T^*).$$

If $Q(0) \le a\mu/c_1$ then $T^* = 0$, i.e., no repair should be done. If $Q(\infty) \ge a\mu/c_1$ then $T^* = \infty$, i.e., no replacement should be done (Problem 3.17).

3.5 Modified Renewal Processes

This section takes up two modified renewal processes.

3.5.1 Geometric Renewal Process

Suppose that a renewal process is defined by a sequence of random variables $\{X_1, X_2/a, \ldots, X_k/a^{k-1}, \ldots\}$ for $a > 0$ with distribution $F(a^{k-1}t) \equiv \Pr\{X_k/a^{k-1} \le t\}$, where $F(t)$ has finite mean μ and variance σ^2 [16]. It is shown in a one-unit system that each operating time decreases geometrically with the number of repairs for $a > 1$. Then, the mean time and variance are

$$E\left\{\frac{X_k}{a^{k-1}}\right\} = \frac{\mu}{a^{k-1}}, \quad V\left\{\frac{X_k}{a^{k-1}}\right\} = \frac{\sigma^2}{a^{2(k-1)}}.$$

When $S_n \equiv \sum_{k=1}^{n} X_k/a^{k-1}$,

$$E\{S_n\} = \frac{\mu(a - a^{-n+1})}{a-1}, \quad V\{S_n\} = \frac{\sigma^2\left(a^2 - a^{-2(n-1)}\right)}{a^2 - 1},$$

and for $a > 1$,

$$E\{S_n\} \to \frac{\mu a}{a-1}, \quad V\{S_n\} \to \frac{\sigma^2 a^2}{a^2 - 1} \quad \text{as } n \to \infty.$$

Furthermore, a renewal function $M(t,a) \equiv \sum_{n=1}^{\infty} \Pr\{S_n \le t\}$ is, from (3.5),

$$M(t,a) = F(t) + \int_0^t M(a(t-u),a)\mathrm{d}F(u), \tag{3.85}$$

because

$$\Pr\{S_n \le t\} = \int_0^t \Pr\{S_{n-1} \le a(t-u)\}\mathrm{d}F(u) \qquad (n=1,2,\ldots).$$

The distributions of the age $\delta(t)$ and the residual life $\gamma(t)$ at time t in (3.22) and (3.23) are rewritten as, respectively,

$$\Pr\{\gamma(t) \le x\} = F(t+x) - \sum_{n=1}^{\infty} \int_0^t \overline{F}(a^n(t+x-u))\mathrm{d}F^{(n)}(u),$$

$$\Pr\{\delta(t) \le x\} = \begin{cases} F(t) - \sum_{n=1}^{\infty} \int_0^{t-x} \overline{F}(a^n(t-u))\mathrm{d}F^{(n)}(u) & \text{for } x \le t, \\ 1 & \text{for } x > t. \end{cases} \tag{3.86}$$

In general, it is very difficult to derive $M(t,a)$ explicitly even if $F(t)$ is exponential, and so, there is nothing but to compute it numerically. For example, when $F(t) = 1 - \mathrm{e}^{-\lambda t}$ [17, p. 156] (Problem 3.18),

$$F^{(n)}(t) = 1 - \sum_{k=1}^{n} \left[\prod_{j=1, j\neq k}^{n} \frac{1}{1-a^{k-j}} \right] \exp(-a^{k-1}\lambda t) \quad (n=1,2,\ldots). \tag{3.87}$$

The renewal process with interarrival times $\{X_1, aX_1, a^2X_1, \ldots\}$ is also called the quasi-renewal process [18, p. 53].

Example 3.11 (Replacement policy) [15, p. 201] It is assumed in Sect. 2.4.1 that the time between the $(k-1)$th and kth failures is denoted by X_k/a^{k-1} and its mean time is μ/a^{k-1} $(k=1,2,\ldots)$ for $a>1$. Then, the expected cost rate is (Problem 3.19)

$$C_1(N) = \frac{c_1(N-1)+c_2}{\mu(a-a^{-N+1})/(a-1)} \quad (N=1,2,\ldots), \tag{3.88}$$

where costs c_1 and c_2 are given in (3.44). From the inequality $C_1(N+1) - C_1(N) \ge 0$,

$$\frac{a^{N+1}-a}{a-1} - (N-1) \ge \frac{c_2}{c_1} \quad (N=1,2,\ldots). \tag{3.89}$$

It is easily seen that the left-hand side of (3.89) increases strictly from a to ∞. Thus, there exists a finite and unique N^* $(1 \le N^* < \infty)$ that satisfies (3.89). If $a \ge c_2/c_1$, then $N^* = 1$. In addition, because the left-hand side of (3.89) also increases with a, N^* decreases with a.

Next, consider an alternating renewal process defined by two sequences of $\{X_k/a^{k-1}\}$ and $\{Y_k/b^{k-1}\}$ for $a, b > 0$.

Example 3.12 (One-unit system) Consider a one-unit system in Example 3.6 where the failure time for the kth operation is X_k/a^{k-1} with mean time μ/a^{k-1} for $a \ge 1$ and the repair time for the kth repair is Y_k/b^{k-1} with mean time β/b^{k-1} for $0 < b \le 1$. That is, the mean failure times decrease with their number, and conversely, the mean repair times increase with their number.

Suppose that the unit is replaced with a new one at the Nth failure. Then, the expected cost rate is (Problem 3.20)

$$C_2(N) = \frac{c_1(N-1) + c_2}{\sum_{k=1}^{N}(\mu/a^{k-1}) + \sum_{k=1}^{N-1}(\beta/b^{k-1})} \quad (N = 1, 2, \ldots), \tag{3.90}$$

where c_1 = cost of repair and c_2 = replacement cost at the Nth failure. Then, from $C_2(N+1) - C_2(N) \ge 0$,

$$\frac{\sum_{k=1}^{N}(\mu/a^{k-1}) + \sum_{k=1}^{N-1}(\beta/b^{k-1})}{\mu/a^N + \beta/b^{N-1}} - (N-1) \ge \frac{c_2}{c_1} \quad (N = 1, 2, \ldots), \tag{3.91}$$

where $\sum_1^0 \equiv 0$. Thus, if

$$\frac{\mu}{a^k} + \frac{\beta}{b^{k-1}} > \frac{\mu}{a^{k+1}} + \frac{\beta}{b^k},$$

i.e., the mean times between failures decrease, then there exists a finite and unique N^* $(1 \le N^* < \infty)$ that satisfies (3.91). If $\mu/(\mu/a + \beta) \ge c_2/c_1$, then $N^* = 1$.

3.5.2 Discrete Renewal Process

Until now, it has been assumed in a renewal process that the sequence of random variables $\{X_k\}$ has an identical continuous distribution $F(t)$. However, the times to some event in actual fields are often measured by the number of cycles, might not be recorded at its exact instant time and be observed statistically per day, per month, per year, and so on. In failure studies, the failure times of units such as tires of jet fighters, switching devices, railroad tracks, and ball bearings might be measured discretely.

In general, the conversion from continuous models to discrete ones is possible. However, the results are often too complicated to be useful, unless skillful treatment is used. This section simply shows how to make the conversion.

Consider the time over an infinitely long cycle $j\,(j=1,2,\ldots)$ and renewal cycle $k\,(k=1,2,\ldots)$. Suppose that X_k has an identical probability function $p_j\,(j=1,2,\ldots)$ with finite mean μ and a probability distribution P_j, i.e., $p_j \equiv \Pr\{X_k=j\}, P_j \equiv \sum_{i=1}^{j} p_i\,(j=1,2,\ldots)$, and

$$\mu \equiv \sum_{j=1}^{\infty} jp_j = \sum_{j=1}^{\infty}(1-P_j)<\infty,$$

where $p_0 \equiv P_0 \equiv 0$. The generating function of a probability function is defined as $p^*(z) \equiv \sum_{j=1}^{\infty} z^j p_j$ for $|z|\le 1$ from Appendix A.4, and hence,

$$P^*(z) \equiv \sum_{j=1}^{\infty} z^j P_j = \frac{p^*(z)}{1-z}. \tag{3.92}$$

Let the total sum of random variables $S_n \equiv \sum_{k=1}^{n} X_k$ and its probability function $p_j^{(n)} \equiv \Pr\{S_n=j\}$, where $p_j^{(1)}=p_j$ and $p_j^{(0)} \equiv 1$. Then,

$$\begin{aligned} p_j^{(2)} &= \Pr\{S_2=j\} = \Pr\{X_1+X_2=j\} \\ &= \sum_{i=1}^{j} \Pr\{X_2=j-i|X_1=i\}\Pr\{X_1=i\} = \sum_{i=1}^{j} p_{j-i}p_i, \end{aligned}$$

and its generating function is

$$\sum_{j=1}^{\infty} z^j p_j^{(2)} = \sum_{j=1}^{\infty} z^j \sum_{i=1}^{j} p_{j-i}p_i = [p^*(z)]^2.$$

Similarly, by using the mathematical induction,

$$\sum_{j=1}^{\infty} z^j p_j^{(n)} = [p^*(z)]^n \quad (n=0,1,2,\ldots). \tag{3.93}$$

The generating function of $P_j^{(n)} \equiv \Pr\{S_n \le j\}$, where $P_j^{(0)} \equiv 1$, is

$$\sum_{j=1}^{\infty} z^j P_j^{(n)} = \sum_{j=1}^{\infty} z^j \sum_{i=1}^{j} p_i^{(n)} = \frac{[p^*(z)]^n}{1-z} \quad (n=0,1,2,\ldots). \tag{3.94}$$

Thus, the probability function of $N_j \equiv \max_n\{S_n \le j\}$ is

$$\begin{aligned} \Pr\{N_j=0\} &= \Pr\{X_1>j\} = 1-P_j, \\ \Pr\{N_j=n\} &= \Pr\{S_n \le j \text{ and } S_{n+1}>j\} = \Pr\{S_n \le j\} - \Pr\{S_{n+1}\le j\} \\ &= P_j^{(n)} - P_j^{(n+1)} \quad (n=0,1,2,\ldots). \end{aligned} \tag{3.95}$$

The renewal function $M_j \equiv E\{N_j\}\,(j=1,2,\ldots)$ that represents the expected number of renewals in cycle j is

$$\begin{aligned} M_j &= \sum_{n=1}^{\infty} n \Pr\{N_j = n\} = \sum_{n=1}^{\infty} P_j^{(n)} \\ &= P_j + \sum_{i=1}^{j} p_i M_{j-i} \quad (j = 1, 2, \ldots), \end{aligned} \tag{3.96}$$

and its generating function is, from (3.92),

$$M^*(z) \equiv \sum_{j=1}^{\infty} z^j M_j = \frac{P^*(z)}{1 - p^*(z)} = \frac{1}{1-z} \frac{p^*(z)}{1 - p^*(z)}. \tag{3.97}$$

In particular, when $p_j = pq^{j-1}$ $(j = 1, 2, \ldots)$, where $q \equiv 1 - p$,

$$p^*(z) = \frac{pz}{1 - qz}, \quad P^*(z) = \frac{1}{1-z} \frac{pz}{1 - qz},$$

and hence, from (3.97),

$$M^*(z) = \frac{pz}{(1-z)^2} = p \sum_{j=1}^{\infty} j z^j, \quad \text{i.e.,} \quad M_j = jp = \frac{j}{E\{X_k\}},$$

that is the same result as the exponential case of the continuous renewal process shown in Example 3.2.

Example 3.13 (Age replacement policy) [11, p. 80] Consider the unit with a failure probability function p_j that should be operating for the time over an infinitely long cycle j $(j = 1, 2, \ldots)$. The unit is replaced at a planned cycle N $(N = 1, 2, \ldots)$ after its installation or at failure, whichever occurs first. Letting $Z_k \equiv \min\{X_k, N\}$ $(k = 1, 2, \ldots)$, a sequence of $\{Z_k\}_{k=1}^{\infty}$ forms a discrete renewal process with an independent and identical distribution

$$\Pr\{Z_k \le j\} = \begin{cases} \sum_{i=1}^{j} p_i & \text{for } j \le N, \\ 1 & \text{for } j > N, \end{cases}$$

and its mean time is

$$E\{Z_k\} = \sum_{j=1}^{N} j p_j + N \sum_{j=N+1}^{\infty} p_j = \sum_{i=1}^{N} \sum_{j=i}^{\infty} p_j.$$

Thus, by the similar method of obtaining (3.42), the expected cost rate is

$$\begin{aligned} C(N) &= \frac{c_1 \Pr\{X_k \le N\} + c_2 \Pr\{X_k > N\}}{E\{Z_k\}} \\ &= \frac{c_1 \sum_{j=1}^{N} p_j + c_2 \sum_{j=N+1}^{\infty} p_j}{\sum_{i=1}^{N} \sum_{j=i}^{\infty} p_j} \quad (N = 1, 2, \ldots), \end{aligned} \tag{3.98}$$

where replacement costs are given in (3.42).

We seek an optimum replacement cycle N^* that minimizes $C(N)$. Let $h_j \equiv p_j / \sum_{i=j}^{\infty} p_i \; (j = 1, 2, \ldots)$ be the discrete failure rate of the probability function p_j with finite mean μ. Suppose that h_j increases strictly with j. From the inequality $C(N+1) - C(N) \geq 0$,

$$h_{N+1} \sum_{i=1}^{N} \sum_{j=i}^{\infty} p_j - \sum_{j=1}^{N} p_j \geq \frac{c_2}{c_1 - c_2} \quad (N = 1, 2, \ldots). \tag{3.99}$$

Letting $L(N)$ be the right-hand side of (3.99),

$$L(N+1) - L(N) = (h_{N+2} - h_{N+1}) \sum_{i=1}^{N+1} \sum_{j=i}^{\infty} p_j > 0,$$

$$L(\infty) \equiv \lim_{N \to \infty} L(N) = \mu h_\infty - 1.$$

Therefore, if $L(\infty) > c_2/(c_1 - c_2)$, i.e., $h_\infty > K$, where $K \equiv c_1/[\mu(c_1 - c_2)]$, then there exists a finite unique minimum N^* $(1 \leq N^* < \infty)$ that satisfies (3.99), and the resulting cost rate is

$$(c_1 - c_2) h_{N^*} \leq C(N^*) < (c_1 - c_2) h_{N^*+1}.$$

Conversely, if $h_\infty \leq K$, then $N^* = \infty$, i.e., no planned replacement should be done. Note that $0 < h_j \leq 1$ from the definition of the failure rate h_j. Thus, if $K \geq 1$, then $N^* = \infty$.

In particular, assume that the failure distribution is a negative binomial distribution $p_j = jp^2 q^{j-1} \; (j = 1, 2, \ldots)$, where $q \equiv 1 - p \; (0 < p < 1)$. Then, $\mu = (1+q)/p$ and $h_j = jp^2/(jp + q)$ that increases strictly from p^2 to p. Therefore, from the above results, if $c_1 q > c_2(1+q)$, an optimum N^* $(1 \leq N^* < \infty)$ is given by a finite and unique minimum that satisfies

$$\frac{(N+1)p(1+q) + q^{N+2}}{Np + 1} \geq \frac{c_1}{c_1 - c_2}.$$

For example, when $c_1 = 10, c_2 = 1$, and $\mu = 9$, i.e., $p = 1/5, N^* = 4$ and $C(4) = 0.92$. In this case, $C(\infty) = c_1/\mu = 1.11$.

Consider the discrete alternating renewal process with $\{X_k\}$ and $\{Y_k\}$ that have the respective probability functions $p_j \equiv \Pr\{X_k = j\}$ and $q_j \equiv \Pr\{Y_k = j\}, P_j \equiv \sum_{i=1}^{j} p_i$, and $Q_j \equiv \sum_{i=1}^{j} q_i$. Other assumptions are the same as those of Sect. 3.4.

Let $M_{ij}(n) \; (i, j = 0, 1)$ denote the expected number of visits to State j in n cycles, starting from State i at cycle 1, and its generating function is $M_{ij}^*(z) = \sum_{n=1}^{\infty} z^n M_{ij}(n)$ for $|z| < 1$. Then, the results in Sect. 3.4 are rewritten as

$$M_{01}(n) = P_n + \sum_{k=1}^{n} p_k M_{11}(n-k), \quad M_{10}(n) = Q_n + \sum_{k=1}^{n} q_k M_{00}(n-k),$$

$$M_{11}(n) = \sum_{k=1}^{n} q_k M_{01}(n-k), \qquad M_{00}(n) = \sum_{k=1}^{n} p_k M_{10}(n-k).$$

Forming the generating functions as the above equations and solving them for $M_{ij}^*(z)$,

$$\begin{aligned} M_{00}^*(z) &= p^*(z) M_{10}^*(z), \quad M_{01}^*(z) = \frac{1}{1-z} \frac{p^*(z)}{1-p^*(z)q^*(z)}, \\ M_{11}^*(z) &= q^*(z) M_{01}^*(z), \quad M_{10}^*(z) = \frac{1}{1-z} \frac{q^*(z)}{1-p^*(z)q^*(z)}. \end{aligned} \tag{3.100}$$

Next, let $Z(n)$ denote the state of the process at cycle n, and define $P_{ij}(n) = \Pr\{Z(n) = j | Z(1) = i\}$. Then, from (3.54), the results in continuous time are rewritten as

$$P_{00}(n) = 1 - P_n + \sum_{k=1}^{n} p_k P_{10}(n-k),$$

$$P_{11}(n) = 1 - Q_n + \sum_{k=1}^{n} q_k P_{01}(n-k),$$

$$P_{10}(n) = \sum_{k=1}^{n} q_k P_{00}(n-k), \quad P_{01}(n) = \sum_{k=1}^{n} p_k P_{11}(n-k),$$

and their generating functions $P_{ij}^*(z) \equiv \sum_{n=1}^{\infty} z^n P_{ij}(n)$ are

$$\begin{aligned} P_{01}^*(z) &= p^*(z) P_{11}^*(z), \quad P_{00}^*(z) = \frac{1}{1-z} \frac{1-p^*(z)}{1-p^*(z)q^*(z)}, \\ P_{10}^*(z) &= q^*(z) P_{00}^*(z), \quad P_{11}^*(z) = \frac{1}{1-z} \frac{1-q^*(z)}{1-p^*(z)q^*(z)}. \end{aligned} \tag{3.101}$$

Clearly,

$$P_{01}(n) = M_{01}(n) - M_{00}(n), \quad P_{10}(n) = M_{10}(n) - M_{11}(n).$$

Example 3.14 (One-unit system with repair) Consider a one-unit system with repair maintenance whose failure and repair distributions are geometric, i.e., $p_k = pq^{k-1}$ and $q_k = \alpha\beta^{k-1}$ $(k = 1, 2, \ldots)$, where $p + q = 1$ and $\alpha + \beta = 1$. Then, the results in Example 3.6 are rewritten as

$$P_{01}^*(z) = \frac{pz}{(1-z)[1-(\beta-p)z]},$$

$$M_{01}^*(z) = \frac{pz}{1-z} \frac{1-\beta z}{(1-z)[1-(\beta-p)z]}.$$

Inverting the generating functions for $1 > \beta - p > 0$ (Problem 3.21),

$$P_{01}(n) = \frac{p}{p+\alpha}[1-(\beta-p)^n],$$
$$M_{01}(n) = \frac{p}{p+\alpha}\left[n\alpha + \frac{p[1-(\beta-p)^n]}{p+\alpha}\right],$$

and as $n \to \infty$,

$$P_{01}(n) \to \frac{p}{p+\alpha}, \quad \frac{M_{01}(n)}{n} \to \frac{p\alpha}{p+\alpha}.$$

3.6 Problems 3

3.1 Prove that $M(t) < \infty$ for all $t < \infty$ [6, p. 57, 9, p. 287].

3.2 In Example 3.2, show that when $k = 3$ and $k = 4$ [4, p. 162], respectively,

$$M(t) = \frac{1}{3}\left[\lambda t - 1 + \frac{2}{\sqrt{3}}e^{-1.5\lambda t}\sin\left(\frac{\sqrt{3}}{2}\lambda t + \frac{\pi}{3}\right)\right],$$
$$M(t) = \frac{1}{4}\left[\lambda t - \frac{3}{2} + \frac{1}{2}e^{-2\lambda t} + \sqrt{2}e^{-\lambda t}\sin\left(\lambda t + \frac{\pi}{4}\right)\right].$$

3.3 Prove that when X_1 and X_2 have a gamma distribution with the respective parameters (α, λ) and (β, λ), $X \equiv X_1 + X_2$ has a gamma distribution $(\alpha + \beta, \lambda)$.

3.4 When $F(t) = 1 - e^{-(\lambda t)^\alpha}$, show that

$$E\{X\} = \frac{1}{\lambda}\Gamma\left(1 + \frac{1}{\alpha}\right), \quad V\{X\} = \frac{1}{\lambda^2}\left\{\Gamma\left(1 + \frac{2}{\alpha}\right) - \left[\Gamma\left(1 + \frac{1}{\alpha}\right)\right]^2\right\},$$

and using the discretization algorithm, compute $M(t)$ for α = 1, 1.5, 2.0, 2.5, 3.0 and t = 1, 2, ..., 10 when λ = 1, and compare them with $t/E\{X\}$ given by the approximation of an elementary renewal theorem in Theorm 2.3

3.5 For a two-unit parallel system with an exponential failure distribution $(1 - e^{-\lambda t})$, show that the failure distribution of the system is $F(t) = (1 - e^{-\lambda t})^2$ and a renewal function is [24, p. 161]

$$M(t) = \frac{2}{3}\left[\lambda t - \frac{1}{3}(1 - e^{-3\lambda t})\right].$$

3.6 When $F(t) = 1 - (1 + \lambda t)e^{-\lambda t}$, derive $V\{N(t)\}$ and (3.15) in Theorem 3.1.

3.7 In Example 3.3, compute the number n of units when the probability that the number of failures is less than n in $t = 10^4$ hours is given by $\alpha = 0.95, 0.99$.

3.8 Show (3.25) and (3.26) [19, p. 259].

3.9 Prove (3.27).

3.10 In inequalities of $M(t)$ and $H(t)$, prove that

$$\frac{t}{\int_0^t \overline{F}(u)\mathrm{d}u} - 1 \le M(t) \le \frac{tF(t)}{\int_0^t \overline{F}(u)\mathrm{d}u}.$$

3.11 In (3.29), prove when $F(t)$ is IFR,

$$\frac{F(t)}{\int_0^t \overline{F}(u)\mathrm{d}u} \le h(t) \le \frac{\overline{F}(t)}{\int_t^\infty \overline{F}(u)\mathrm{d}u}.$$

3.12 In (3.38), prove that when $F(t)$ is IFR, l_1 decreases from $1/h(0)$ to μ, and l_2 increases from 0 to ∞.

3.13 In Sect. 3.3.2.2, prove that when $A(T) = c_1F(T) + c_2\overline{F}(T)$ and $B(T) \equiv c_1M(T) + c_2$,

$$B(T) = A(T) + \int_0^T B(T-t)\mathrm{d}F(t), \quad B(T) = A(T) + \int_0^T A(T-t)\mathrm{d}M(t).$$

3.14 Show that the interval reliability in (3.67), when T_0 is a random variable with an exponential distribution $(1-\mathrm{e}^{-\alpha t})(0<\alpha<\infty)$, is $R(T;x,\alpha) = \alpha R^*(T;x,\alpha)$, and derive an optimum policy that maximizes $R(T;x,\alpha)$ [11, p. 143] when the failure rate increases strictly.

3.15 If the time for PM has a distribution $G_2(t)$ with mean β_2 and the time for repair has $G_1(t)$ with mean β_1, then show that the limiting interval reliability in (3.68) is [11, p. 143]

$$R(T;x) = \frac{\int_x^{T+x} \overline{F}(t)\mathrm{d}t}{\int_0^T \overline{F}(t)\mathrm{d}t + \beta_1 F(T) + \beta_2\overline{F}(T)}.$$

3.16 In Example 3.9, compute the time δ when the probability that the system is down more than δ in $t = 10{,}000$ h of operation is given by $\alpha = 0.95, 0.99$.

3.17 In Example 3.10, derive optimum policies when (i) $c_r(t) = at, G(t) = 1-\mathrm{e}^{-\theta\sqrt{t}}$, and (ii) $c_r(t) = at^2, G(t) = 1-\mathrm{e}^{-\theta t}$.

3.18 Derive (3.87).

3.19 Derive (3.88) and (3.89).

3.20 Derive (3.90) and (3.91).

3.21 In Example 3.14, derive $P_{01}(n)$ and $M_{01}(n)$.

3.22 A generalized renewal process in which

$$S_{n+1} = S_n + A_{n+1}X_{n+1} \qquad (n = 0, 1, 2, \ldots)$$

is applied to a imperfect repair model [20]. When $A_n = a$, derive

$$\Pr\{X_{n+1} \le x | S_n = t\}.$$

References

1. Barlow RE, Proschan F (1965) Mathematical theory of reliability. Wiley, New York
2. Prabhu NU (2007) Stochastic processes. World Scientific, Singapore
3. Parzen E (1962) Stochastic process. Holden-Day, San Francisco
4. Beichelt FE (2006) Stochastic processes in science, engineering and finance. Chapman & Hall, Boca Raton
5. Osaki S (1992) Applied stochastic system modeling. Springer, Berlin
6. Ross SM (1983) Stochastic processes. Wiley, New York
7. Cox DR (1962) Renewal theory. Methuen, London
8. Feller W (1968) An introduction to probability theory and its application, vol I, 3rd edn. Wiley, New York
9. Çinlar E (1975) Introduction to stochastic processes. Prentice-Hall, Englewood Cliffs
10. Takács L (1960) Stochastic processes. Wiley, New York
11. Nakagawa T (2005) Maintenance theory of reliability. Springer, London
12. Tijms HC (2003) A first course in stochastic models. Wiley, Chichester
13. Dohi T, Kaio N, Osaki S (2002) Renewal processes and their computational aspects. In: Osaki S (ed) Stochastic models in reliability and maintenance. Springer, Berlin
14. Xie M (1989) On the solution of renewal-type integral equations. Commun Statist B 18:281–293
15. Calabro SR (1962) Reliability principles and practices. McGraw-Hill, New York
16. Lam Yeh (2007) Geometric process and its applications. World Scientific, Singapore
17. Nakagawa T (2007) Shock and damage models in reliability theory. Springer, London
18. Wang H, Pham H (2006) Reliability and optimal maintenance. Springer, London
19. Rausand M, Høyland A (2004) System reliability theory. Wiley, Hoboken
20. Kijima M (2002) Generalized renewal processes and general repair models. In: Osaki S (ed) Stochastic models in reliability and maintenance. Springer, Berlin

Chapter 4
Markov Chains

In Sect. 3.4, we consider the system that repeats up and down such as operating and failed states alternately. Next, as one example of extended models, we take up the system with repair maintenance: The system begins to operate at time 0 and undergoes repair according to two types of failures such as minor and major failures. After the repair completion, the system becomes like new and begins to operate again.

To form the stochastic modeling of the above system, we define the following three states:

State 0: System is operating.
State 1: System is under repair according to minor failure.
State 2: System is under repair according to major failure.

In general, the repair time and cost required for major failure might be greater than those for minor failure. The system repeats transitions between the states of $0 \to 1 \to 0$ or $0 \to 2 \to 0$ (Fig. 4.1).

We explain simply and clearly Markov processes on the subject of the system: Suppose that the failure occurs according to a general distribution $F(t)$, and when it has occurred, the probability of minor failure is $\alpha\,(0 \le \alpha \le 1)$ and the probability of major failure is $1-\alpha$, where both probabilities are independent of failure times. The distribution of time to the repair completion of minor and major failures are $G_1(t)$ and $G_2(t)$, respectively. If we pay no regard to transition times and are concerned only with transition probabilities among three states, we express the process in the following matrix form:

$$\boldsymbol{P} = \begin{matrix} & \begin{matrix} 0 & 1 & 2 \end{matrix} \\ \begin{matrix} 0 \\ 1 \\ 2 \end{matrix} & \begin{pmatrix} 0 & \alpha & 1-\alpha \\ 1 & 0 & 0 \\ 1 & 0 & 0 \end{pmatrix} \end{matrix}. \tag{4.1}$$

T. Nakagawa, *Stochastic Processes*, Springer Series in Reliability Engineering,
DOI: 10.1007/978-0-85729-274-2_4,

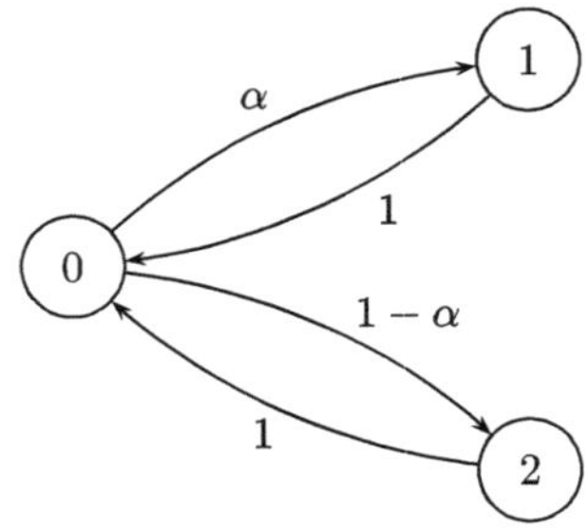

Fig. 4.1 Process of the system with two types of failures

The stochastic process with discrete-state space such as (4.1) is called a *discrete-time Markov chain*. Using the transition matrix in (4.1), we can obtain stochastic behaviors of the system in Sect. 4.1.

Next, assume that $F(t) = 1 - \mathrm{e}^{-\lambda t}$ and $G_j(t) = 1 - \mathrm{e}^{-\theta_j t}$ $(j = 1, 2)$. We investigate the states of the system at time t. Let $Z(t)$ denote the state of the system at time t when it starts from State 0, and define that $P_j(t) \equiv \Pr\{Z(t) = j\}$ $(j = 0, 1, 2)$, that is called a *transition probability*. Note that the process has the memoryless property irrespective of elapsed operating and repair times because of exponential distribution in Sect. 2.1.1. Thus, we have the following equations: For small $h > 0$,

$$
\begin{aligned}
P_0(t+h) &= P_0(t)\mathrm{e}^{-\lambda h} + P_1(t)(1 - \mathrm{e}^{-\theta_1 h}) + P_2(t)(1 - \mathrm{e}^{-\theta_2 h}) + o(h), \\
P_1(t+h) &= P_1(t)\mathrm{e}^{-\theta_1 h} + P_0(t)\alpha(1 - \mathrm{e}^{-\lambda h}) + o(h), \\
P_2(t+h) &= P_2(t)\mathrm{e}^{-\theta_2 h} + P_0(t)(1-\alpha)(1 - \mathrm{e}^{-\lambda h}) + o(h),
\end{aligned}
$$

where $o(h)/h \to 0$ as $h \to 0$. From these equations,

$$
\begin{aligned}
\frac{P_0(t+h) - P_0(t)}{h} &= -P_0(t)\frac{1 - \mathrm{e}^{-\lambda h}}{h} + P_1(t)\frac{1 - \mathrm{e}^{-\theta_1 h}}{h} + P_2(t)\frac{1 - \mathrm{e}^{-\theta_2 h}}{h} + \frac{o(h)}{h}, \\
\frac{P_1(t+h) - P_1(t)}{h} &= -P_1(t)\frac{1 - \mathrm{e}^{-\theta_1 h}}{h} + P_0(t)\alpha\frac{1 - \mathrm{e}^{-\lambda h}}{h} + \frac{o(h)}{h}, \\
\frac{P_2(t+h) - P_2(t)}{h} &= -P_2(t)\frac{1 - \mathrm{e}^{-\theta_2 h}}{h} + P_0(t)(1-\alpha)\frac{1 - \mathrm{e}^{-\lambda h}}{h} + \frac{o(h)}{h}.
\end{aligned}
$$

As $h \to 0$, we have the differential equations

$$
\begin{aligned}
P_0'(t) &= -\lambda P_0(t) + \theta_1 P_1(t) + \theta_2 P_2(t), \\
P_1'(t) &= -\theta_1 P_1(t) + \alpha\lambda P_0(t), \\
P_2'(t) &= -\theta_2 P_2(t) + (1-\alpha)\lambda P_0(t).
\end{aligned}
$$

In a matrix form,

$$
\boldsymbol{P}'(t) \equiv \begin{pmatrix} P_0'(t) \\ P_1'(t) \\ P_2'(t) \end{pmatrix} = \begin{pmatrix} -\lambda & \theta_1 & \theta_2 \\ \alpha\lambda & -\theta_1 & 0 \\ (1-\alpha)\lambda & 0 & -\theta_2 \end{pmatrix} \boldsymbol{P}(t). \tag{4.2}
$$

Therefore, by solving the above differential equations mathematically and numerically, we can obtain the probabilities $P_j(t)$ $(j = 0, 1, 2)$.

The stochastic process, whose conditional probability of transition to the *future* states depends only on the *present* state and is independent of the *past* history, is called a *Markov property*. The stochastic process with Markov property at any time is a *Markov process*. Many real-world phenomena can be modeling stochastically by Markov processes because we may observe the phenomena only at Markov property points without keeping track of the time elapsed since the last point. In other words, if we know the state for any specific value of time, we can predict the stochastic behavior beyond the point. Stochastic processes with independent increments defined in Sect. 2.2 always have the Markov property.

Markov processes can be analyzed by forming differential equations such as (4.2) and are also called a *continuous-time Markov chain*. More detailed discussions for Markov processes are done in Sect. 4.2.

4.1 Discrete-Time Markov Chain

4.1.1 Transition Probabilities

Consider a discrete-time stochastic process $\{X_n, n = 0, 1, 2, \ldots\}$ with a finite discrete-state set $\{0, 1, 2, \ldots, m\}$. Note that the event $\{X_n = j\}$ represents that the process is in State j $(j = 0, 1, \ldots, m)$ at time n $(n = 0, 1, 2, \ldots)$. That is, the process moves among $(m + 1)$ states at some unit of time, time unit, or jump according to a given probability law. Then, we want to know mainly in which state the process is at time n and converges as n becomes larger.

If we suppose that

$$\Pr\{X_{n+1} = j | X_0 = i_0,\ X_1 = i_1, \ldots, X_n = i_n\} = \Pr\{X_{n+1} = j | X_n = i_n\} \tag{4.3}$$

for all $i_0, i_1, \ldots, i_n$ and all $n \geq 0$, then the process $\{X_n, n = 0, 1, 2, \ldots\}$ is said to be a *Markov chain* (Markov 1806). This property indicates that, given the value of X_n, the future value of X_{n+1} does not depend on the value of X_k for $0 \leq k \leq n-1$. If the probability of X_{n+1} being in State j, given that X_n is in State i, is independent of n, i.e.,

$$\Pr\{X_{n+1} = j | X_n = i\} \equiv P_{ij},$$

then the process has a *stationary* or *homogeneous* (*one-step*) *transition probability*, i.e., the process transits from State i to State j with probability P_{ij} and remains in State i with probability P_{ii} regardless of n. In other words, the transition probability from State i to State j does not change even if time varies. We restrict ourselves only to discrete-time Markov chains with stationary transition probabilities and a finite state space, because most reliability systems form such processes. A matrix

$\boldsymbol{P} \equiv \{P_{ij}\}$ is called a *transition probability matrix*, and manifestly, P_{ij} satisfies $P_{ij} \geq 0$ and $\sum_{j=0}^{m} P_{ij} = 1$ for any i, j. Let $\boldsymbol{P}$ denote the *transition matrix* given by

$$\boldsymbol{P} \equiv (P_{ij}) = \begin{pmatrix} P_{00} & P_{01} & \dots & P_{0m} \\ P_{10} & P_{10} & \dots & P_{1m} \\ \vdots & \vdots & & \vdots \\ P_{i0} & P_{i1} & \dots & P_{im} \\ \vdots & \vdots & & \vdots \\ P_{m0} & P_{m1} & \dots & P_{mm} \end{pmatrix}.$$

Example 4.1 (Alternating renewal process) Consider an alternating renewal process in Sect. 3.4, where a unit begins to operate at time 0 and undergoes repair at failures. After the repair completion, a unit becomes like new and begins to operate again. Then, we define

State 0: unit is operating.
State 1: unit is under repair.

The process has the simplest Markov chain with two states $\{0, 1\}$, and a transition probability matrix is

$$\boldsymbol{P} \equiv \begin{pmatrix} P_{00} & P_{01} \\ P_{10} & P_{10} \end{pmatrix} = \begin{pmatrix} 0 & 1 \\ 1 & 0 \end{pmatrix}.$$

The process repeats alternately between operating and repair. When the repair is made, the unit operates like new with probability 0.8, however, it undergoes repair again because of its imperfectness with probability 0.2. Then, a transition probability matrix is

$$\boldsymbol{P} = \begin{pmatrix} 0 & 1 \\ 0.8 & 0.2 \end{pmatrix}.$$

It is helpful to understand the process whose behavior is illustrated by using *state transition diagram* as shown in Figs. 4.1 and 4.2, where the number *circled* denotes a state and the number appearing on an *arc* is a transition probability or a transition rate (Fig. 4.3). It is indispensable to draw a state transition diagram and to make clear of the behavior for modeling reliability systems stochastically.

Example 4.2 (One-unit system) Consider a one-unit system with repair in Example 3.6, where the failure and repair times have exponential distributions

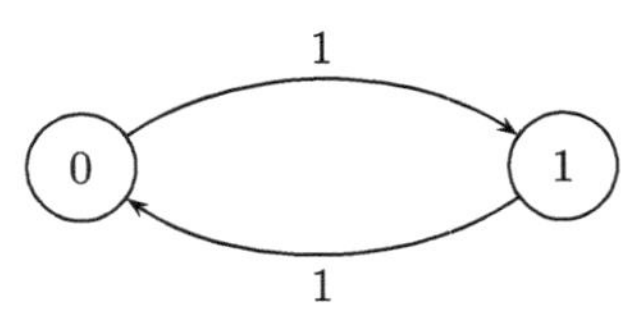

Fig. 4.2 Diagram of one-unit system

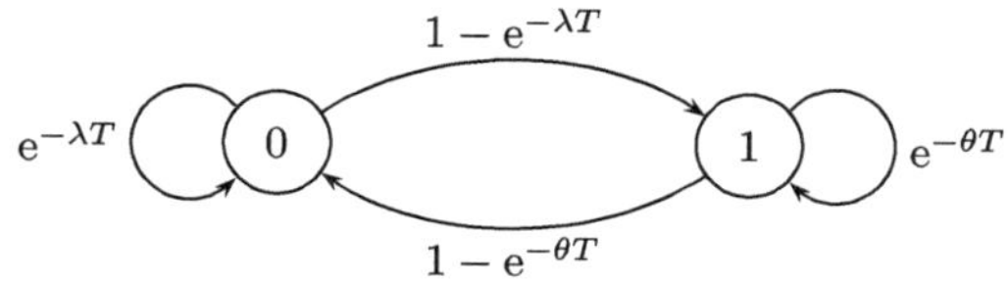

Fig. 4.3 Diagram of one-unit system with exponential failure and repair times

$(1-e^{-\lambda t})$ and $(1-e^{-\theta t})$, respectively. To know the behavior of the system, we check the unit at periodic times kT $(k=1,2,\ldots)$. If the unit has failed at some check, it undergoes repair immediately, and otherwise, leaves alone. Similarly, if the repair is completed at some check, it begins to operate, and otherwise, continues to make the repair. Then, because of the memoryless property of exponential distributions, the transition matrix is

$$\boldsymbol{P}=\begin{pmatrix} P_{00} & P_{01} \\ P_{10} & P_{10} \end{pmatrix}=\begin{pmatrix} e^{-\lambda T} & 1-e^{-\lambda T} \\ 1-e^{-\theta T} & e^{-\theta T} \end{pmatrix},$$

and the state transition diagram is shown in Fig. 4.3. When $T\to\infty$, and $e^{-\lambda T}=0$ and $e^{-\theta T}=0.2$, the transition matrices correspond to those in Example 4.1, respectively.

A Markov chain is completely specified by the transition probabilities P_{ij} and an initial probability $\Pr\{X_0=i\}$. Let P_{ij}^n denote the probability that the process goes from State i to State j in n transitions that is called the *n-step transition probability*, i.e.,

$$\begin{aligned} &P_{ij}^n\equiv\Pr\{X_{n+k}=j|X_k=i\} \quad (n=1,2,\ldots), \\ &P_{ij}^0\equiv 0 \quad \text{for } i\neq j, \\ &P_{ii}^0\equiv 1, \end{aligned}$$

$P_{ij}^n\geq 0$, $\sum_{j=0}^m P_{ij}^n=1$ for all $n\geq 0$, and $P_{ij}^1\equiv P_{ij}$ for any i, j. Note that P_{ij}^n is independent of the present time k and depends only on the time duration n. Then,

$$\begin{aligned} P_{ij}^n&=\Pr\{X_{n+k}=j|X_k=i\}=\sum_{l=0}^m\Pr\{X_{n+k}=j,X_{r+k}=l|X_k=i\} \\ &=\sum_{l=0}^m\Pr\{X_{n+k}=j|X_{r+k}=l\}\Pr\{X_{r+k}=l|X_k=i\} \\ &=\sum_{l=0}^m P_{il}^r P_{lj}^{n-r} \quad (r=0,1,2,\ldots,n), \end{aligned} \tag{4.4}$$

that is called the *Chapman–Kolmogorov equation*.

In a matrix form, from (4.4),

$$\boldsymbol{P}^{(n)}\equiv\left(P_{ij}^n\right)=\boldsymbol{P}^{(1)}\cdot\boldsymbol{P}^{(n-1)}=\boldsymbol{P}\cdot\boldsymbol{P}^{(n-1)}=\cdots=\boldsymbol{P}^n, \tag{4.5}$$

i.e., $\boldsymbol{P}^{(n)}$ can be computed by the nth power of matrix $\boldsymbol{P}$. Thus, (4.4) is written in the following matrix form

$$\boldsymbol{P}^n = \boldsymbol{P}^r \cdot \boldsymbol{P}^{n-r} \quad (r = 0, 1, 2, \ldots, n), \tag{4.6}$$

where note that $\boldsymbol{P}^{(0)} = \boldsymbol{P}^0 = \boldsymbol{I}$ is the identity matrix where all diagonal elements are 1. The Chapman–Kolmogorov equation in (4.4) plays a great role in analyzing Markov chains that will be shown in (4.25).

Therefore, the probability that the process is in State j at time n, irrespective of an initial state, is

$$\begin{aligned}
\pi_j(n) &\equiv \Pr\{X(n) = j\} = \sum_{i=0}^{m} \Pr\{X(n) = j, \quad X(0) = i\} \\
&= \sum_{i=0}^{m} \Pr\{X(n) = j | X(0) = i\} \Pr\{X(0) = i\} \\
&= \sum_{i=0}^{m} \pi_i(0) P_{ij}^n \quad (j = 0, 1, 2, \ldots, m),
\end{aligned} \tag{4.7}$$

where note that $\sum_{j=0}^{m} \pi_j(n) = 1$ for all $n \geq 0$.

Letting $\boldsymbol{\pi}(n) \equiv \{\pi_0(n), \pi_1(n), \ldots, \pi_m(n)\}$, (4.7) is written in a matrix form

$$\boldsymbol{\pi}(n) = \boldsymbol{\pi}(0)\boldsymbol{P}^{(n)} \quad (n = 0, 1, 2, \ldots). \tag{4.8}$$

Example 4.3 In Example 4.2, setting $a \equiv 1 - \mathrm{e}^{-\lambda T}$ and $b \equiv 1 - \mathrm{e}^{-\theta T}$ [1, p. 101],

$$\boldsymbol{P} = \begin{pmatrix} 1-a & a \\ b & 1-b \end{pmatrix},$$

$$\boldsymbol{P}^2 = \begin{pmatrix} (1-a)^2 + ab & (1-a)a + (1-b)a \\ (1-a)b + (1-b)b & ab + (1-b)^2 \end{pmatrix}.$$

In general (Problem 4.1),

$$\boldsymbol{P}^n = \frac{1}{a+b}\begin{pmatrix} b & a \\ b & a \end{pmatrix} + \frac{(1-a-b)^n}{a+b}\begin{pmatrix} a & -a \\ -b & b \end{pmatrix}.$$

If the initial state is $\boldsymbol{\pi}(0) = (1, 0)$, *i.e.*, the unit starts to operate at time 0,

$$\begin{aligned}
\boldsymbol{\pi}(n) &= \frac{1}{a+b}(b, a) + \frac{(1-a-b)^n}{a+b}(a, -a) \\
&= \left(\frac{b + a(1-a-b)^n}{a+b}, \quad \frac{a[1-(1-a-b)^n]}{a+b}\right).
\end{aligned}$$

For example, when $\lambda T = 0.1$ and $\theta T = 0.2$, i.e., $a = 0.095$ and $b = 0.181$,

$$\boldsymbol{\pi}(n) = (0.656 + 0.344(0.724)^n, \ 0.344[1 - (0.724)^n])$$

that goes to $\pi(\infty) \equiv \lim_{n\to\infty} \pi(n) = (0.656, 0.344)$, and the values of 0.656 and 0.344 are approximately equal to the availability $\theta/(\lambda+\theta)$ and the unavailability $\lambda/(\lambda+\theta)$ in Example 4.2.

4.1.2 Classification of States

Two States i and j *communicate* if and only if there exist integers $n \geq 0$ and $k \geq 0$ such that $P_{ij}^k > 0$ and $P_{ji}^n > 0$. By using a communication relation, all states are classified into several classes. Furthermore, we define the first-passage distribution as

$$F_{ij}^n \equiv \Pr\{X_n = j,\ X_k \neq j,\ k = 1, 2, \ldots, n-1 | X_0 = i\} \quad (n = 1, 2, \ldots),$$

that is the probability that starting in State i, the first transition into State j occurs at the nth transition, where $F_{ij}^0 \equiv 0$ for all i, j. In addition, we define $F_{ij} \equiv \sum_{n=1}^{\infty} F_{ij}^n$ that represents the eventual transition probability from State i to State j.

All states in a finite Markov chain are classified into two kinds of states:

(i) If $F_{ii} = 1$, then State i is said to be *recurrent*,
(ii) If $F_{ii} < 1$, then State i is said to be *transient*.

A recurrent set of states is one in which all states communicate and which cannot be left, once it is entered. A transient set of states is one in which all states communicate and which might be left even if it is entered.

We have the following relation between a transition probability P_{ij}^n and a first-passage distribution F_{ij}^n

$$P_{ij}^n = \sum_{k=0}^{n} F_{ij}^k P_{jj}^{n-k} \quad (n = 1, 2, \ldots), \tag{4.9}$$

because P_{ij}^n is denoted by the probability that the process transits from State i to State j for the first time at the kth step ($k = 1, 2, \ldots, n$) and returns to State j at the $(n-k)$th step, where note that $P_{ij}^0 = 0$ for $i \neq j$, and $P_{jj}^0 = 1$.

Let $P_{ij}^*(z)$ and $F_{ij}^*(z)$ be the generating functions of P_{ij}^n and F_{ij}^n, i.e., $P_{ij}^*(z) \equiv \sum_{n=0}^{\infty} z^n P_{ij}^n$ and $F_{ij}^*(z) \equiv \sum_{n=0}^{\infty} z^n F_{ij}^n$ for $|z| < 1$ from Appendix A.4. Then, (4.9) is rewritten by the generating function, for $i \neq j$,

$$\begin{aligned} P_{ij}^*(z) &= \sum_{n=0}^{\infty} z^n P_{ij}^n = \sum_{n=0}^{\infty} z^n \sum_{k=0}^{n} F_{ij}^k P_{ij}^{n-k} \\ &= \sum_{k=0}^{\infty} z^k F_{ij}^k \sum_{n=0}^{\infty} z^n P_{jj}^n = F_{ij}^*(z) P_{jj}^*(z), \end{aligned} \tag{4.10}$$

and for $i = j$,

$$P_{ii}^*(z) = 1 + F_{ii}^*(z)P_{ii}^*(z), \quad \text{i.e.,} \quad P_{ii}^*(z) = \frac{1}{1 - F_{ii}^*(z)}. \tag{4.11}$$

Next, let M_{ij}^n denote the expected number of visits to State j in the nth transition if the process starts from State i, not including the first at time 0. Then, clearly

$$M_{ij}^n = \begin{cases} \sum_{k=1}^n P_{ij}^k & (n = 1, 2, \ldots), \\ 0 & (n = 0). \end{cases} \tag{4.12}$$

Forming the generating function of (4.12), for $i \neq j$,

$$M_{ij}^*(z) \equiv \sum_{n=0}^{\infty} z^n M_{ij}^n = \frac{P_{ij}^*(z)}{1 - z},$$
$$M_{ii}^*(z) \equiv \sum_{n=0}^{\infty} z^n M_{ii}^n = \frac{P_{ii}^*(z) - 1}{1 - z}. \tag{4.13}$$

Using Theorem A.5 of Appendix A, when State i is recurrent,

$$\lim_{z \to 1} \sum_{n=0}^{\infty} z^n F_{ii}^n = \lim_{z \to 1} F_{ii}(z) = \sum_{n=0}^{\infty} F_{ii}^n = 1,$$

and hence, from (4.11),

$$\lim_{z \to 1} P_{ii}^*(z) = \lim_{z \to 1} \sum_{n=0}^{\infty} z^n P_{ii}^n = \infty.$$

Thus, using Theorem A.5,

$$\sum_{n=0}^{\infty} P_{ii}^n = \infty.$$

Next, when $\sum_{n=0}^{\infty} P_{ii}^n = \infty$, we assume that $F_{ii} < 1$. Then, from (4.11), $\lim_{z \to 1} P_{ii}^*(z) < \infty$ using Theorem A.5. In addition, using Theorem A.5 again, $\sum_{n=0}^{\infty} P_{ii}^n < \infty$ that contradicts the assumption. Thus, if $\sum_{n=0}^{\infty} P_{ii}^n = \infty$, then $F_{ii} = 1$. When State i is transient, we can make similar discussions.

Therefore, we have the following results:

Theorem 4.1

(i) *State i is recurrent if and only if* $\sum_{n=0}^{\infty} P_{ii}^n = \infty$.
(ii) *State i is transient if and only if* $\sum_{n=0}^{\infty} P_{ii}^n < \infty$.

The above results show that for any recurrent State i, the process can revisit State i infinitely. For any transient State i, the process transits to states of another class after it revisits to State i finitely, and cannot return to State i.

Example 4.4 In Example 4.3, it is assumed that $0<a<1$ and $b=0$, i.e., any failed units remain a failed state without making repair. Then, a transition probability matrix is

$$\boldsymbol{P} = \begin{pmatrix} 1-a & a \\ 0 & 1 \end{pmatrix},$$

and

$$\sum_{n=0}^{\infty} P_{00}^{n} = 1 + (1-a) + (1-a)^2 + \cdots = \frac{1}{a},$$

$$\sum_{n=0}^{\infty} P_{11}^{n} = 1 + 1 + \cdots = \infty,$$

that implies that State 0 is transient and State 1 is recurrent.

In general, the transition probability of a finite Markov chain can be expressed in a matrix form [1, p. 124, 2, p. 36]

$$\boldsymbol{P} = \begin{array}{c} \boldsymbol{C}_1 \\ \boldsymbol{C}_2 \\ \vdots \\ \boldsymbol{C}_k \\ \boldsymbol{T} \end{array} \left(\begin{array}{cccc|c} \boldsymbol{P}_1 & & \boldsymbol{0} & & \\ & \boldsymbol{P}_2 & & & \\ & \boldsymbol{0} & \ddots & & \boldsymbol{0} \\ & & & \boldsymbol{P}_k & \\ \hline \boldsymbol{R}_1 & \boldsymbol{R}_2 & \ldots & \boldsymbol{R}_k & \boldsymbol{Q} \end{array} \right), \tag{4.14}$$

where $\boldsymbol{C}_1, \boldsymbol{C}_2, \ldots, \boldsymbol{C}_k$ are all the sets of recurrent classes and $\boldsymbol{T}$ is a set of transient states.

4.1.3 Limiting Probabilities

Recall that if State i is recurrent, then $F_{ii}=1$ and $\sum_{n=0}^{\infty} P_{ii}^{n} = \infty$, and conversely, if State i is transient, then $F_{ii}<1$ and $\sum_{n=0}^{\infty} P_{ii}^{n} < \infty$. Define that $\mu_j \equiv \sum_{n=1}^{\infty} nF_{jj}^{n}$ is the mean recurrence time for State j ($j=0,1,\ldots,m$).

Theorem 4.2
(i) *If State jis recurrent, then*

$$P_{jj}^{n} \to \frac{1}{\mu_j} > 0 \quad \text{as } n \to \infty. \tag{4.15}$$

(ii) *If State jis transient, then*

$$P_{jj}^{n} \to 0 \quad \text{as } n \to \infty. \tag{4.16}$$

(iii) *If State jis recurrent, then*

$$P_{ij}^n \to \frac{F_{ij}}{\mu_j}. \tag{4.17}$$

Proof From (4.12),

$$M_{jj}^n - M_{jj}^{n-1} = P_{jj}^n \to \frac{1}{\mu_j} \quad \text{as } n \to \infty.$$

Furthermore, because $\sum_{j=0}^{m} P_{jj}^n = 1$ for any $n \geq 0$, $1/\mu_j > 0$. If State j is transient, $\sum_{n=1}^{\infty} P_{jj}^n < \infty$, and hence, $P_{jj}^n \to 0$ as $n \to \infty$. If a Markov chain starts from State i, visits State j in a recurrent class with probability F_{ij}^n and results infinitely State j. Thus, noting that F_{ij} represents the eventual transition probability from State i to State j, we get easily (iii). □

Theorem 4.3 *If States i and j* $(i, j = 0, 1, \ldots, m)$ *are in the same recurrent class,* $\pi_j \equiv \lim_{n\to\infty} P_{ij}^n > 0$, *that is a unique and positive solution to satisfy*

$$\pi_j = \sum_{i=0}^{m} \pi_i P_{ij} \quad (j = 0, 1, \ldots, m), \quad \sum_{j=0}^{m} \pi_j = 1. \tag{4.18}$$

Proof From the assumption that both States i and j are in the same recurrent class, $F_{ij} = 1$, and hence, $\lim_{n\to\infty} P_{ij}^n = 1/\mu_j > 0$ from (i) and (iii) of Theorem 4.2. Furthermore, from (4.7),

$$\pi_j(n) = \sum_{i=0}^{m} \pi_i(n-1) P_{ij}, \quad \sum_{j=0}^{m} \pi_j(n) = 1 \quad \text{for all } n \geq 1.$$

Letting $n \to \infty$, we easily have (4.18). □

Two methods of solving (4.18) were proposed [3, p. 107]; direct method and iterative method. However, because the number of states is not so large in most reliability models, it would be sufficient to solve (4.18) by the direct method. Note that (4.18) is expressed in a matrix form

$$\boldsymbol{\pi} \equiv (\pi_0, \pi_1, \ldots, \pi_m) = \boldsymbol{\pi} \boldsymbol{P}. \tag{4.19}$$

Therefore, we summarize the liming probabilities $P_{ij}^{\infty} \equiv \lim_{n\to\infty} P_{ij}^n$ of a finite Markov chain in (4.14)

$$\begin{aligned}
&P_{ij}^{\infty} = \frac{1}{\mu_j} \quad (i,j \in \boldsymbol{C}_l \quad l = 1,2,\ldots,k),\\
&P_{ij}^{\infty} = \frac{F_{ij}}{\mu_j} \quad (i \in \boldsymbol{T}, j \in \boldsymbol{C}_l \quad l = 1,2,\ldots,k),\\
&P_{ij}^{\infty} = 0 \quad (i \in \boldsymbol{C}_l, j \in \mathbf{C}_r \quad l \neq r),\\
&P_{ij}^{\infty} = 0 \quad (i,j \in \boldsymbol{T}),\\
&P_{ij}^{\infty} = 0 \quad (i \in \boldsymbol{C}_l \quad l = 1,2,\ldots,k, j \in \boldsymbol{T}).
\end{aligned} \tag{4.20}$$

Example 4.5 When $\lambda T = 0.1$ and $\theta T = 0.2$, i.e., $a = 0.095$ and $b = 0.181$ in Example 4.3, a transition probability matrix is

$$\boldsymbol{P} = \begin{matrix} & \begin{matrix} 0 & \quad 1 \end{matrix} \\ \begin{matrix} 0 \\ 1 \end{matrix} & \begin{pmatrix} 0.905 & 0.095 \\ 0.181 & 0.819 \end{pmatrix} \end{matrix},$$

that consists of one recurrent class. Then, the limiting probabilities π_0 and π_1 are, from Theorem 4.3,

$$(\pi_0, \pi_1) = (\pi_0, \pi_1)\begin{pmatrix} 0.905 & 0.095 \\ 0.181 & 0.819 \end{pmatrix}.$$

Thus, solving two equations

$$\pi_0 = 0.905\pi_0 + 0.181\pi_1, \quad \pi_0 + \pi_1 = 1,$$

$\pi_0 \approx 0.656$ and $\pi_1 \approx 0.344$ that are equal to $\pi(\infty)$ in Example 4.3.

In a transition probability matrix in (4.1), the limiting transition probabilities are given by solving simultaneous equations

$$\pi_1 = \pi_0\alpha, \quad \pi_2 = \pi_0(1-\alpha), \quad \pi_0 + \pi_1 + \pi_2 = 1.$$

Thus, $\pi_0 = 1/2$, $\pi_1 = \alpha/2$, and $\pi_2 = (1-\alpha)/2$ (Problem 4.2).

Example 4.6 Suppose that a transition probability matrix is given by

$$\boldsymbol{P} = \begin{matrix} & \begin{matrix} 0 & 1 & 2 & 3 & 4 & 5 & 6 \end{matrix} \\ \begin{matrix} 0 \\ 1 \\ 2 \\ 3 \\ 4 \\ 5 \\ 6 \end{matrix} & \left(\begin{array}{c:cc:cc:cc} 1 & 0 & 0 & 0 & 0 & 0 & 0 \\ \hdashline 0 & \frac{1}{2} & \frac{1}{2} & 0 & 0 & 0 & 0 \\ 0 & \frac{1}{4} & \frac{3}{4} & 0 & 0 & 0 & 0 \\ \hdashline 0 & 0 & 0 & \frac{2}{3} & \frac{1}{3} & 0 & 0 \\ 0 & 0 & 0 & 1 & 0 & 0 & 0 \\ \hdashline \frac{1}{5} & \frac{1}{10} & 0 & \frac{1}{10} & 0 & 0 & \frac{3}{5} \\ \frac{1}{2} & 0 & 0 & 0 & \frac{1}{6} & \frac{1}{3} & 0 \end{array}\right) \end{matrix}.$$

Then, the recurrent classes are $\boldsymbol{C}_1 = \{0\}, \boldsymbol{C}_2 = \{1, 2\}, \boldsymbol{C}_3 = \{3, 4\}$, and a transient class is $T = \{5, 6\}$. The transition diagram among seven states is shown in Fig. 4.4

For a recurrent class $\boldsymbol{C}_2 = \{1, 2\}$, the transition probability matrix is

$$\boldsymbol{P}_2 = \begin{pmatrix} \frac{1}{2} & \frac{1}{2} \\ \frac{1}{4} & \frac{3}{4} \end{pmatrix},$$

and the limiting probabilities are, from (4.18),

$$\pi_1 = \frac{1}{3}, \qquad \pi_2 = \frac{2}{3}.$$

Similarly, $\boldsymbol{C}_3 = \{3, 4\}$,

$$\boldsymbol{P} = \begin{pmatrix} \frac{2}{3} & \frac{1}{3} \\ 1 & 0 \end{pmatrix}, \quad \pi_3 = \frac{3}{4}, \quad \pi_4 = \frac{1}{4}.$$

We derive the first-passage distributions F_{ij} $(i \in \boldsymbol{T}, j \in \boldsymbol{C}_k\ (k = 2, 3))$ defined in Sect. 4.1.2 [1, p. 127].

Theorem 4.4 *The first-passage distribution F_{ij} from State i $(i \in \boldsymbol{T})$ to State j $(j \in \boldsymbol{C}_k)$ $(k = 1, 2, \ldots)$ satisfies*

$$F_{ij} = \sum_{l \in \boldsymbol{C}_k} P_{il} + \sum_{l \in \boldsymbol{T}} P_{il} F_{lj} \qquad (i \in \boldsymbol{T}, j \in \boldsymbol{C}_k), \tag{4.21}$$

and its matrix form is

$$(F_{ij}) = (\boldsymbol{I} - \boldsymbol{Q})^{-1} \boldsymbol{R}_k \cdot \mathbf{1}, \tag{4.22}$$

where $\mathbf{1}$ is a column vector where all elements are 1.

Proof The first-passage distribution F_{ij} has two possibilities: one is to transit from State i to State l $(l \in \boldsymbol{C}_k)$ directly, and the other is to transit from State i to State l $(l \in \boldsymbol{T})$ once including itself and then to transit to State j $(j \in \boldsymbol{C}_k)$. Thus, we have (4.21). Rewriting (4.21) in a matrix form,

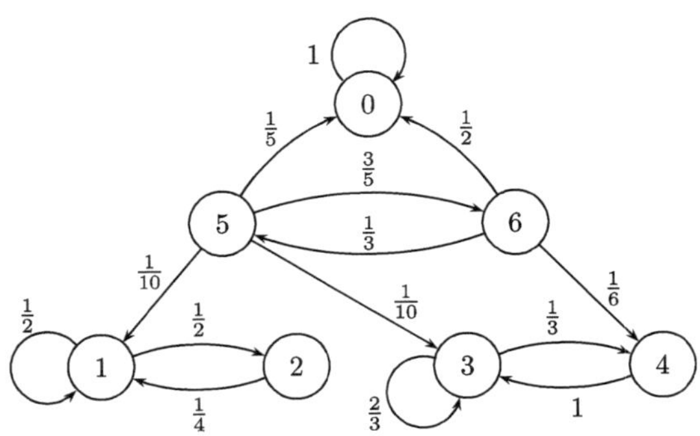

Fig. 4.4 Transition diagram of Example 4.6

$$F_{ij} = \boldsymbol{R}_k \cdot \mathbf{1} + \mathbf{Q}(F_{ij}),$$

i.e.,

$$(F_{ij}) = (\boldsymbol{I} - \boldsymbol{Q})^{-1} \boldsymbol{R}_k \cdot \mathbf{1}.$$

□

Example 4.7 In Example 4.6,

$$\boldsymbol{Q} = \begin{pmatrix} 0 & \frac{3}{5} \\ \frac{1}{3} & 0 \end{pmatrix}, \quad \text{i.e.,} \quad (\boldsymbol{I} - \boldsymbol{Q}) = \begin{pmatrix} 1 & -\frac{3}{5} \\ -\frac{1}{3} & 1 \end{pmatrix}.$$

Hence, the inverse matrix is

$$(\boldsymbol{I} - \boldsymbol{Q})^{-1} = \begin{pmatrix} \frac{5}{4} & \frac{3}{4} \\ \frac{5}{12} & \frac{5}{4} \end{pmatrix}.$$

Thus, using Theorem 4.4,

$$(\boldsymbol{I} - \boldsymbol{Q})^{-1} \boldsymbol{R}_1 \cdot \mathbf{1} = \begin{pmatrix} \frac{5}{4} & \frac{3}{4} \\ \frac{5}{12} & \frac{5}{4} \end{pmatrix} \begin{pmatrix} \frac{1}{5} \\ \frac{1}{2} \end{pmatrix} (1) = \begin{pmatrix} \frac{5}{8} \\ \frac{17}{24} \end{pmatrix},$$

$$(\boldsymbol{I} - \boldsymbol{Q})^{-1} \boldsymbol{R}_2 \cdot \mathbf{1} = \begin{pmatrix} \frac{5}{4} & \frac{3}{4} \\ \frac{5}{12} & \frac{5}{4} \end{pmatrix} \begin{pmatrix} \frac{1}{10} & 0 \\ 0 & 0 \end{pmatrix} \begin{pmatrix} 1 \\ 1 \end{pmatrix} = \begin{pmatrix} \frac{1}{8} \\ \frac{1}{24} \end{pmatrix},$$

$$(\boldsymbol{I} - \boldsymbol{Q})^{-1} \boldsymbol{R}_3 \cdot \mathbf{1} = \begin{pmatrix} \frac{5}{4} & \frac{3}{4} \\ \frac{5}{12} & \frac{5}{4} \end{pmatrix} \begin{pmatrix} \frac{1}{10} & 0 \\ 0 & \frac{1}{6} \end{pmatrix} \begin{pmatrix} 1 \\ 1 \end{pmatrix} = \begin{pmatrix} \frac{1}{4} \\ \frac{1}{4} \end{pmatrix}.$$

Consequently,

$$F_{50} = \frac{5}{8}, \qquad F_{5j} = \frac{1}{8} \quad (j = 1, 2), \qquad F_{5j} = \frac{1}{4} \quad (j = 3, 4),$$
$$F_{60} = \frac{17}{24}, \qquad F_{6j} = \frac{1}{24} \quad (j = 1, 2), \qquad F_{6j} = \frac{1}{4} \quad (j = 3, 4),$$

where note that $\frac{5}{8} + \frac{1}{8} + \frac{1}{4} = 1$ and $\frac{17}{24} + \frac{1}{24} + \frac{1}{4} = 1$. Therefore, from (4.20),

$$P_{50}^{\infty} = \frac{5}{8}, \qquad P_{51}^{\infty} = \frac{1}{8} \times \frac{1}{3} = \frac{1}{23}, \qquad P_{52}^{\infty} = \frac{1}{8} \times \frac{2}{3} = \frac{1}{11},$$
$$P_{53}^{\infty} = \frac{1}{4} \times \frac{3}{4} = \frac{3}{16}, \qquad P_{54}^{\infty} = \frac{1}{4} \times \frac{1}{4} = \frac{1}{16},$$
$$P_{60}^{\infty} = \frac{17}{24}, \qquad P_{61}^{\infty} = \frac{1}{24} \times \frac{1}{3} = \frac{1}{72}, \qquad P_{62}^{\infty} = \frac{1}{24} \times \frac{2}{3} = \frac{1}{36},$$
$$P_{63}^{\infty} = \frac{1}{4} \times \frac{3}{4} = \frac{3}{16}, \qquad P_{64}^{\infty} = \frac{1}{4} \times \frac{1}{4} = \frac{1}{16},$$

where note that $\sum_{j=0}^{4} P_{5j}^{\infty} = \sum_{j=0}^{4} P_{6j}^{\infty} = 1$.

Collecting these results, a matrix of limiting probabilities is

$$
(P_{ij}^{\infty}) = \begin{array}{c} \\ 0 \\ 1 \\ 2 \\ 3 \\ 4 \\ 5 \\ 6 \end{array}
\begin{array}{c} \begin{array}{ccccccc} 0 & 1 & 2 & 3 & 4 & 5 & 6 \end{array} \\
\left(\begin{array}{c|cc|cc|cc}
1 & 0 & 0 & 0 & 0 & 0 & 0 \\ \hline
0 & \frac{1}{2} & \frac{1}{2} & 0 & 0 & 0 & 0 \\
0 & \frac{1}{4} & \frac{3}{4} & 0 & 0 & 0 & 0 \\ \hline
0 & 0 & 0 & \frac{2}{3} & \frac{1}{3} & 0 & 0 \\
0 & 0 & 0 & 1 & 0 & 0 & 0 \\ \hline
\frac{5}{8} & \frac{1}{24} & \frac{1}{12} & \frac{3}{16} & \frac{1}{16} & 0 & 0 \\
\frac{17}{24} & \frac{1}{72} & \frac{1}{36} & \frac{3}{16} & \frac{1}{16} & 0 & 0
\end{array}\right)\end{array}.
$$

4.1.4 Absorbing Markov Chain

We are sometimes interested in stochastic behaviors before system failure. An absorbing Markov chain consists of one in which all recurrent states are absorbing. The transition probability matrix of an absorbing Markov chain with k absorbing states and transient states can be expressed, from (4.14),

$$
\boldsymbol{P} = \begin{array}{c} 1 \\ 2 \\ \vdots \\ k \\ \boldsymbol{T} \end{array}
\left(\begin{array}{cccc|c}
1 & & \boldsymbol{0} & & \\
 & 1 & & & \\
 & & \ddots & & \boldsymbol{0} \\
\boldsymbol{0} & & & 1 & \\ \hline
\boldsymbol{R}_1 & \boldsymbol{R}_2 & \ldots & \boldsymbol{R}_k & \boldsymbol{Q}
\end{array}\right), \tag{4.23}
$$

where $\boldsymbol{0}$ is zero matrix in which all elements are zero.

We obtain the expected number of visits to State j, starting from State i $(i, j \in \boldsymbol{T})$, i.e.,

$$M_{ij} = E\{N_j(\infty)|X(0) = i\} \qquad (i, j \in \boldsymbol{T}).$$

Theorem 4.5 *The expected number of visits from State i to State j* is [2, p. 46]

$$(M_{ij}) = (\boldsymbol{I} - \boldsymbol{Q})^{-1}. \tag{4.24}$$

Proof Letting U_j^n be the function that takes 1 when the process is State j at time n, and otherwise, takes 0, it can be easily seen that $N_j(\infty) = \sum_{n=0}^{\infty} U_j^n$. Thus,

$$
\begin{aligned}
(M_{ij}) &= \left(E\left\{\sum_{n=0}^{\infty} U_j^n \middle| X(0)=i\right\}\right) = \left(\sum_{n=0}^{\infty} E\left\{U_j^n | X(0)=i\right\}\right) \\
&= \left(\sum_{n=0}^{\infty}\left[P_{ij}^{(n)} \times 1 + \left(1-P_{ij}^{(n)}\right) \times 0\right]\right) = \left(\sum_{n=0}^{\infty} P_{ij}^{(n)}\right) \\
&= \sum_{n=0}^{\infty} \boldsymbol{Q}^n = (\boldsymbol{I}-\boldsymbol{Q})^{-1}.
\end{aligned}
$$

Note that at least one of sums of each row in $\boldsymbol{Q}$ is <1 because $\boldsymbol{Q}$ is a matrix of transient states, and hence, the inverse of a matrix $(\boldsymbol{I}-\boldsymbol{Q})$ exists. □

Example 4.8 In Example 4.7,

$$
(\boldsymbol{I}-\boldsymbol{Q})^{-1} = \begin{matrix} 5 \\ 6 \end{matrix}\begin{pmatrix} \frac{5}{4} & \frac{3}{4} \\ \frac{5}{12} & \frac{5}{4} \end{pmatrix}.
$$

Thus, the expected numbers of visits to transient States 5 and 6 before absorbing are $5/4 + 3/4 = 2$ from State 5 and $5/12 + 5/4 = 5/3$ from State 6, where note that 2 and $5/3$ are the mean times from States 5 and 6 to an absorbing state, respectively, because each transition time is a unit of time in a Markov chain.

Example 4.9 (Two-unit parallel system with spare units) Consider a two-unit parallel system with no repair, however, spare units in inventory: Two units begin to operate at time 0. If one of operating unit fails before time T, the spare unit is ordered immediately and is placed into service as soon as it is delivered after a lead time L. If two operating units do not fail up to time T, the spare unit is ordered at time T and is delivered after time L. In this case, when two units have failed, the spare unit is put into service and the spare unit is ordered immediately. When only one unit has failed, the spare unit is put into service, and when two units are operating, the spare unit is put into inventory.

It is assumed that each unit fails according to an identical exponential distribution $(1 - e^{-\lambda t})$. We define the states:

State 0: Two units are operating and there is no spare.
State 1: Two units are operating and spare is ordered.
State 2: One unit is operating and spare is ordered.
State 3: Two units are operating and spare is delivered.
State 4: One unit is operating and spare is delivered.
State 5: Two units have failed and spare is delivered.

We draw a state transition diagram of the above system in Fig. 4.5. Referring to the diagram, transition probabilities are

$$
\begin{aligned}
&P_{01} = e^{-2\lambda T}, && P_{02} = 1 - e^{-2\lambda T}, && P_{13} = e^{-2\lambda L}, \\
&P_{14} = 2e^{-\lambda L}(1 - e^{-\lambda L}), && P_{15} = (1 - e^{-\lambda L})^2, && \\
&P_{24} = e^{-\lambda L}, && P_{25} = 1 - e^{-\lambda L}, && P_{30} = P_{40} = P_{52} = 1,
\end{aligned}
$$

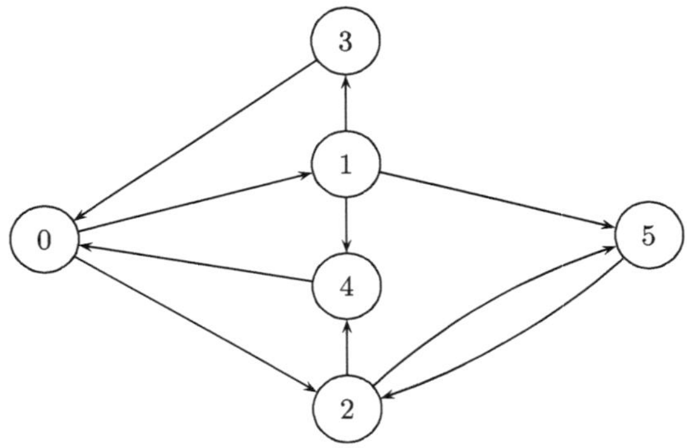

Fig. 4.5 Diagram of two-unit parallel system with spare units in inventory

and its matrix form is

$$P = \begin{matrix} & \begin{matrix} 0 & 1 & 2 & 3 & 4 & 5 \end{matrix} \\ \begin{matrix} 0 \\ 1 \\ 2 \\ 3 \\ 4 \\ 5 \end{matrix} & \begin{pmatrix} 0 & e^{-2\lambda T} & 1-e^{-2\lambda T} & 0 & 0 & 0 \\ 0 & 0 & 0 & e^{-2\lambda L} & 2e^{-\lambda L}(1-e^{-\lambda L}) & (1-e^{-\lambda L})^2 \\ 0 & 0 & 0 & 0 & e^{-\lambda L} & 1-e^{-\lambda L} \\ 1 & 0 & 0 & 0 & 0 & 0 \\ 1 & 0 & 0 & 0 & 0 & 0 \\ 0 & 0 & 1 & 0 & 0 & 0 \end{pmatrix} \end{matrix}.$$

When $\lambda T = 0.1$ and $\lambda L = 0.05$, the transition probability matrix is

$$P = \begin{matrix} & \begin{matrix} 0 & 1 & 2 & 3 & 4 & 5 \end{matrix} \\ \begin{matrix} 0 \\ 1 \\ 2 \\ 3 \\ 4 \\ 5 \end{matrix} & \begin{pmatrix} 0 & 0.8187 & 0.1813 & 0 & 0 & 0 \\ 0 & 0 & 0 & 0.9048 & 0.0928 & 0.0024 \\ 0 & 0 & 0 & 0 & 0.9512 & 0.0488 \\ 1 & 0 & 0 & 0 & 0 & 0 \\ 1 & 0 & 0 & 0 & 0 & 0 \\ 0 & 0 & 1 & 0 & 0 & 0 \end{pmatrix} \end{matrix}.$$

Suppose that State 5 is in a failed state of the system and is absorbing. Striking out row 5 and column 5 of $\boldsymbol{I} - \boldsymbol{P}$ and inverting it, we can compute the expected number of visits to State j before system failure

$$\begin{matrix} & \begin{matrix} 0 & 1 & 2 & 3 & 4 \end{matrix} \\ \begin{matrix} 0 \\ 1 \\ 2 \\ 3 \\ 4 \end{matrix} & \begin{pmatrix} 488.944 & 400.303 & 88.360 & 392.387 & 95.557 \\ 488.800 & 401.185 & 88.604 & 393.241 & 95.549 \\ 484.079 & 396.319 & 88.749 & 388.483 & 95.597 \\ 488.944 & 400.303 & 88.630 & 393.387 & 95.557 \\ 488.944 & 400.303 & 88.630 & 392.387 & 96.557 \end{pmatrix} \end{matrix}.$$

Thus, the mean time from Sate 0 to State 5 is $1,465.551 \times (1/\lambda)$ by summing the number of column 0. For example, when $1/\lambda = 100$, the mean time is 146, 555.1 (Problem 4.3).

4.2 Continuous-Time Markov Chain

4.2.1 Transition Probabilities

Consider a continuous-time stochastic process $\{X(t), t \geq 0\}$ with a finite state set $\{0, 1, 2, \ldots, m\}$. Note that the event $\{X(t) = j\}$ represents that the process is in State j ($j = 0, 1, 2, \ldots, m$) at time t. Then, we want to know mainly in which state the process is at time t and the process converges as t becomes larger.

If we suppose that

$$\Pr\{X(t) = j | X(t_0) = i_0,\ X(t_1) = i_1, \ldots, X(t_n) = i_n\} = \Pr\{X(t) = j | X(t_n) = i_n\}$$

for all i_0, i_1, ..., i_n and $0 \leq t_0 < t_1 < \cdots < t_n < t$, then the process $\{X(t)\}$ is said to be a *continuous-time Markov chain*. This property indicates that, given the value of $X(t_n)$, the future value of $X(t)$ for $t > t_n$ does not depend on the value of $X(t_{n-1})$ for any $t_{n-1} < t_n$ and depends only on the present state. If the probability of $X(t+u)$ being in State j, given that $X(u)$ is in State i, is independent of $u \geq 0$, i.e.,

$$\Pr\{X(t+u) = j | X(u) = i\} \equiv P_{ij}(t),$$

then the process has a *stationary* or *homogeneous transition probability* that depends only on the time difference t.

We restrict ourselves to continuous-time Markov chains with stationary transition probabilities and a finite state space $\{0, 1, 2, \ldots, m\}$ like discrete-time Markov chains. A continuous-time Markov chain is a stochastic process in which the process moves among states in accordance with a discrete-time Markov chain, however, its sojourn time spent in each state to visit the next state is independent and distributed exponentially.

A matrix $\boldsymbol{P}(t) = \{P_{ij}(t)\}$ is called a *transition probability matrix* at time t, and satisfies $P_{ij}(t) \geq 0$ and $\sum_{j=0}^{m} P_{ij}(t) = 1$ for any $t \geq 0$ and i, j. In a similar way of obtaining (4.4),

$$\begin{aligned} P_{ij}(t+u) &= \Pr\{X(t+u) = j | X(0) = i\} \\ &= \sum_{k=0}^{m} \Pr\{X(t+u) = j, X(u) = k | X(0) = i\} \\ &= \sum_{k=0}^{m} \Pr\{X(t+u) = j | X(u) = k\} \Pr\{X(u) = k | X(0) = i\} \\ &= \sum_{k=0}^{m} P_{ik}(u) P_{kj}(t) \end{aligned} \tag{4.25}$$

for t, $u \geq 0$, that is called the *Chapman–Kolmogorov equation*. In a matrix form, (4.25) is rewritten as

$$\boldsymbol{P}(t+u) = \boldsymbol{P}(t)\boldsymbol{P}(u). \tag{4.26}$$

However, we cannot calculate the transition probability by the matrix operation as shown in Sect. 4.1.1.

4.2.2 *Pure Birth Process and Birth and Death Process*

Recall that a Poisson process is a counting process in which the process has stationary and independent increments, and the interarrival times have an identical exponential distribution with mean $1/\lambda$. Expanding the Poisson process, the law of probability of arrival events depends on the total number of events that has already arrived, i.e., the interarrival times are independent and distributed exponentially with mean $1/\lambda_k$ that depends on the present State k. In this case, states are denoted by the total number of events.

Definition 4.1 If $\{X(t)\}$ is a continuous-time Markov chain with

(i) $X(0) = 0$,
(ii) $\Pr\{X(t+h) - X(t) = 1|X(t) = k\} = \lambda_k h + o(h)$,
(iii) $\Pr\{X(t+h) - X(t) \geq 2|X(t) = k\} = o(h)$,

then the process $\{X(t)\}$ is said to be a *pure birth process* with parameters λ_k and transits only from k to $k+1$.

Using this definition, we derive the stationary transition probability

$$P_k(t) \equiv \Pr\{X(t) = k|X(0) = 0\} = \Pr\{X(t) = k\} \quad (k = 0, 1, \ldots, m),$$

that represents that k events occur in the time interval $[0, t]$. By the similar method of obtaining (2.25), from Definition 4.1,

$$\begin{aligned} P_0(t+h) &= \Pr\{X(t+h) = 0\} \\ &= \Pr\{X(t) = 0\}\Pr\{X(t+h) - X(t) = 0|X(t) = 0\} \\ &= P_0(t)[1 - \lambda_0 h + o(h)]. \end{aligned}$$

Hence, as $h \to 0$,

$$\frac{\mathrm{d}P_0(t)}{\mathrm{d}t} = -\lambda_0 P_0(t).$$

By integration and setting $P_0(0) = 1$ from (i),

$$P_0(t) = \mathrm{e}^{-\lambda_0 t}. \tag{4.27}$$

Similarly,

$$\begin{aligned}P_k(t+h) &= \Pr\{X(t+h)=k\}\\ &= \Pr\{X(t)=k-1, X(t+h)-X(t)=1\}\\ &\quad+\Pr\{X(t)=k,\ X(t+h)-X(t)=0\}\\ &\quad+\sum_{j=2}^{k}\Pr\{X(t)=k-j,\ X(t+h)-X(t)=j\}\\ &= P_{k-1}(t)[\lambda_{k-1}h+o(h)]+P_k(t)[1-\lambda_k h+o(h)]+o(h).\end{aligned}$$

Hence, as $h \to 0$,

$$\frac{\mathrm{d}P_k(t)}{\mathrm{d}t} = -\lambda_k P_k(t) + \lambda_{k-1}P_{k-1}(t),$$

i.e.,

$$\mathrm{e}^{\lambda_k t}[P_k'(t)+\lambda_k P_k(t)] = \frac{\mathrm{d}[\mathrm{e}^{\lambda_k t}P_k(t)]}{\mathrm{d}t} = \lambda_{k-1}\mathrm{e}^{\lambda_k t}P_{k-1}(t).$$

By integration under $P_k(0)=0$ for $k \geq 1$,

$$P_k(t) = \lambda_{k-1}\,\mathrm{e}^{-\lambda_k t}\int_0^t \mathrm{e}^{\lambda_k u}P_{k-1}(u)\mathrm{d}u \quad (k=1,2,\ldots). \tag{4.28}$$

Therefore, from (4.27) and (4.28), we can compute $P_k(t)$ successively. Furthermore, interarrival times are independent and have an exponential distribution with mean $1/\lambda_k$ $(k=0,1,2,\ldots,m)$, as shown in Sect. 2.2.

Example 4.10 (Parallel system) Consider a parallel redundant system with n identical units each of which has an independent exponential distribution $(1-\mathrm{e}^{-\lambda t})$. Then, setting that $\lambda_k \equiv (n-k)\lambda$ $(k=0,1,\ldots,n-1)$, the system forms a pure birth process. Solving (4.28) under $P_0(t)=\mathrm{e}^{-n\lambda t}$,

$$P_k(t) = \binom{n}{k}(1-\mathrm{e}^{-\lambda t})^k\mathrm{e}^{-(n-k)\lambda t} \quad (k=0,1,\ldots,n),$$

that is easily derived by using a binomial distribution (Problem 4.4). In addition, it is shown that the mean time to system failure is given in (2.8).

Next, consider the stochastic process that states increase or decrease according to arrivals or departures of events.

Definition 4.2 If the process $\{X(t)\}$ is a continuous-time Markov chain with

(i) $X(0)=i$,
(ii) $\Pr\{X(t+h)-X(t)=1|X(t)=k\} = \lambda_k h + o(h)$,

(iii) $\Pr\{X(t+h) - X(t) = -1|X(t) = k\} = \mu_k h + o(h)$,
(iv) $\Pr\{|X(t+h) - X(t)| > 1|X(t) = k\} = o(h)$,

then the process $\{X(t)\}$ is said to be a *birth and death process* with parameters $\{\lambda_k, \mu_{k+1}\}$ and transits only from k to $k-1$ or $k+1$, where λ_k and μ_{k+1} $(k = 0, 1, 2, \ldots)$ are called the birth and death rates, respectively. It follows easily from (ii–iv) that

$$\Pr\{X(t+h) - X(t) = 0|X(t) = 0\} = 1 - \lambda_0 h + o(h),$$
$$\Pr\{X(t+h) - X(t) = 0|X(t) = k\} = 1 - (\lambda_k + \mu_k)h + o(h) \quad (k = 1, 2, \ldots).$$

Using this definition, we derive the stationary transition probability

$$P_{ij}(t) \equiv \Pr\{X(t) = j|X(0) = i\} \qquad (i, j = 0, 1, \ldots, m).$$

From the Chapman–Kolmogorov equation in (4.25),

$$P_{ij}(t+h) = \sum_{k=0}^{m} P_{ik}(t)P_{kj}(h).$$

Thus, from (ii–iv),

$$\begin{aligned}
P_{ij}(t+h) &= P_{ij-1}(t)\Pr\{X(t+h) - X(t) = 1|X(t) = j-1\} \\
&\quad + P_{ij}(t)\Pr\{X(t+h) - X(t) = 0|X(t) = j\} \\
&\quad + P_{ij+1}(t)\Pr\{X(t+h) - X(t) = -1|X(t) = j+1\} \\
&\quad + \sum_{k=0, k \neq j-1, j, j+1}^{m} P_{ik}(t)\Pr\{X(t+h) - X(t) = j-k|X(t) = k\} \\
&= P_{ij-1}(t)[\lambda_{j-1}h + o(h)] + P_{ij}(t)[1 - (\lambda_j + \mu_j)h + o(h)] \\
&\quad + P_{ij+1}(t)[\mu_{j+1}h + o(h)] + o(h).
\end{aligned}$$

Hence, as $h \to 0$,

$$\frac{\mathrm{d}P_{ij}}{\mathrm{d}t} = \lambda_{j-1}P_{ij-1}(t) - (\lambda_j + \mu_j)P_{ij}(t) + \mu_{j+1}P_{ij+1}(t) \quad (j = 1, 2, \ldots, m-1). \tag{4.29}$$

In particular, for the respective cases of $j = 0$ and $j = m$,

$$\begin{aligned}
\frac{\mathrm{d}P_{i0}(t)}{\mathrm{d}t} &= -\lambda_0 P_{i0}(t) + \mu_1 P_{i1}(t), \\
\frac{\mathrm{d}P_{im}(t)}{\mathrm{d}t} &= \lambda_{m-1}P_{im-1}(t) - \mu_m P_{im}(t).
\end{aligned} \tag{4.30}$$

Therefore, from (4.29) and (4.30), we can compute $P_{ij}(t)$ under the condition of

$$P_{ii}(0) = \Pr\{X(0) = i\} = 1, \quad P_{ij}(0) = 0 \quad \text{for } i \neq j.$$

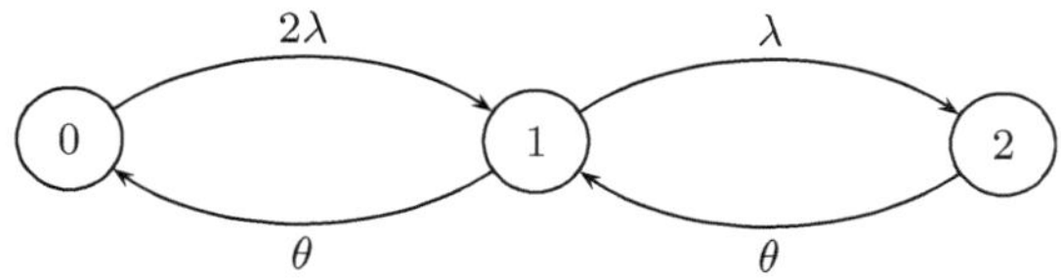

Fig. 4.6 Transition diagram among three states

Example 4.11 (two-unit parallel system) Consider a two-unit parallel system with one repair person, i.e., only one failed unit is repaired by the repair person. It is assumed that two units begin to operate at time 0 and fail according to an identical exponential distribution $(1 - e^{-\lambda t})$. Failed units are repaired by one repair person according to an exponential distribution $(1 - e^{-\theta t})$, become like new ones, and start to operate again immediately upon the repair completion. Then, we define the states:

State 0: Two units are operating.
State 1: One unit is operating and the other is under repair.
State 2: One unit is under repair and the other waits for repair.

Note that states represent the number of failed units. Setting that $\lambda_0 = 2\lambda$, $\lambda_1 = \lambda$, and $\mu_1 = \mu_2 = \theta$, the transition diagram among three states is shown in Fig. 4.6.

Setting that $P_{00}(0) \equiv 1$, (4.29) and (4.30) for the system become

$$\begin{aligned} P'_{00}(t) &= -2\lambda P_{00}(t) + \theta P_{01}(t), \\ P'_{01}(t) &= 2\lambda P_{00}(t) - (\lambda + \theta) P_{01}(t) + \theta P_{02}(t), \\ P'_{02}(t) &= \lambda P_{01}(t) - \theta P_{02}(t). \end{aligned}$$

From Appendix A, using the Laplace–Stieltjes (LS) transformation of

$$\int_0^\infty e^{-st} dP'_{ij}(t) = s[P^*_{ij}(s) - P_{ij}(0)],$$

the above differential equations are

$$\begin{aligned} sP^*_{00}(s) - s &= -2\lambda P^*_{00}(s) + \theta P^*_{01}(s), \\ sP^*_{01}(s) &= 2\lambda P^*_{00}(s) - (\lambda + \theta) P^*_{01}(s) + \theta P^*_{02}(s), \\ sP^*_{02}(s) &= \lambda P^*_{01}(s) - \theta P^*_{02}(s). \end{aligned}$$

Solving these equations,

$$P^*_{00}(s) = \frac{s^2 + (\lambda + 2\theta)s + \theta^2}{s^2 + (3\lambda + 2\theta)s + 2\lambda^2 + 2\lambda\theta + \theta^2},$$

$$P^*_{01}(s) = \frac{2\lambda s + 2\lambda\theta}{s^2 + (3\lambda + 2\theta)s + 2\lambda^2 + 2\lambda\theta + \theta^2},$$

$$P_{02}^*(s) = \frac{2\lambda^2}{s^2 + (3\lambda + 2\theta)s + 2\lambda^2 + 2\lambda\theta + \theta^2},$$

where note that $P_{00}^*(s) + P_{01}^*(s) + P_{02}^*(s) = 1$. Expanding the right-hand side of $P_{00}^*(s)$ in partial fractions,

$$P_{00}^*(s) = C + \frac{As}{s + w_1} + \frac{Bs}{s + w_2},$$

where

$$C \equiv \frac{\theta^2}{2\lambda^2 + 2\lambda\theta + \theta^2},$$
$$(s + w_1)(s + w_2) = s^2 + (3\lambda + 2\theta)s + 2\lambda^2 + 2\lambda\theta + \theta^2,$$

i.e.,

$$w_1 \equiv \frac{1}{2}\left(3\lambda + 2\theta + \sqrt{\lambda^2 + 4\lambda\theta}\right), \quad w_2 \equiv \frac{1}{2}\left(3\lambda + 2\theta - \sqrt{\lambda^2 + 4\lambda\theta}\right).$$

From the following identity equation

$$C(s + w_1)(s + w_2) + As(s + w_2) + Bs(s + w_1) = s^2 + (\lambda + 2\theta)s + \theta^2,$$
$$A \equiv \frac{2\lambda(w_1 - \lambda - \theta)}{w_1(w_1 - w_2)}, \quad B \equiv \frac{2\lambda(w_2 - \lambda - \theta)}{w_2(w_2 - w_1)}.$$

Thus, using the inverse LS transform of

$$\int_0^\infty e^{-st}\, de^{-wt} = \frac{s}{s + w},$$

the transition probability is

$$P_{00}(t) = \frac{\theta^2}{2\lambda^2 + 2\lambda\theta + \theta^2} + \frac{2\lambda}{w_1 - w_2}\left(\frac{w_1 - \lambda - \theta}{w_1} e^{-w_1 t} - \frac{w_2 - \lambda - \theta}{w_2} e^{-w_2 t}\right).$$

Similarly,

$$P_{01}(t) = \frac{2\lambda\theta}{2\lambda^2 + 2\lambda\theta + \theta^2} + \frac{2\lambda}{w_1 - w_2}\left(\frac{-w_1 + \theta}{w_1} e^{-w_1 t} - \frac{-w_2 + \theta}{w_2} e^{-w_2 t}\right),$$
$$P_{02}(t) = \frac{2\lambda^2}{2\lambda^2 + 2\lambda\theta + \theta^2} + \frac{2\lambda^2}{w_1 - w_2}\left(\frac{1}{w_1} e^{-w_1 t} - \frac{1}{w_2} e^{-w_2 t}\right),$$

where note that $P_{00}(t) + P_{01}(t) + P_{02}(t) = 1$. Thus, the limiting probabilities $P_j \equiv \lim_{t\to\infty} P_{ij}(t)$ are

$$P_0 \equiv \frac{\theta^2}{2\lambda^2 + 2\lambda\theta + \theta^2}, \quad P_1 \equiv \frac{2\lambda\theta}{2\lambda^2 + 2\lambda\theta + \theta^2},$$

and $P_2 = 1 - P_0 - P_1$ that represents the unavailability for a two-unit parallel system with one repair person.

Next, suppose that there exists two repair persons. In this case, any units are repaired immediately upon failures. Then, setting $\lambda_0 = 2\lambda$, $\lambda_1 = \lambda$, $\mu_1 = \theta$, and $\mu_2 = 2\theta$,

$$\begin{aligned}
P'_{00}(t) &= -2\lambda P_{00}(t) + \theta P_{01}(t), \\
P'_{01}(t) &= 2\lambda P_{00}(t) - (\lambda + \theta)P_{01}(t) + 2\theta P_{02}(t), \\
P'_{02}(t) &= \lambda P_{01}(t) - 2\theta P_{02}(t).
\end{aligned}$$

By the same method of the previous model,

$$\begin{aligned}
P^*_{00}(s) &= \frac{s^2 + (\lambda + 3\theta)s + 2\theta^2}{(s + 2\lambda + 2\theta)(s + \lambda + \theta)}, \\
P^*_{01}(s) &= \frac{2\lambda(s + 2\theta)}{(s + 2\lambda + 2\theta)(s + \lambda + \theta)}, \\
P^*_{02}(s) &= \frac{2\lambda^2}{(s + 2\lambda + 2\theta)(s + \lambda + \theta)}.
\end{aligned}$$

Forming the inverse LS transforms of $P^*_{0j}(s)$,

$$\begin{aligned}
P_{00}(t) &= \left[\frac{\theta + \lambda e^{-(\lambda+\theta)t}}{\lambda + \theta}\right]^2, \\
P_{01}(t) &= \frac{2\lambda}{\lambda + \theta}\left[1 - e^{-(\lambda+\theta)t}\right]\frac{1}{\lambda + \theta}\left[\theta + \lambda e^{-(\lambda+\theta)t}\right], \\
P_{02}(t) &= \left\{\frac{\lambda}{\lambda + \theta}\left[1 - e^{-(\lambda+\theta)t}\right]\right\}^2
\end{aligned}$$

(Problem 4.5).

Example 4.12 We compute explicitly $P_j(t)$ $(j = 0, 1, 2)$ given in (4.2). Forming the LS transforms of (4.2),

$$\begin{aligned}
sP^*_0(s) - s &= -\lambda P^*_0(s) + \theta_1 P^*_1(s) + \theta_2 P^*_2(s), \\
sP^*_1(s) &= -\theta_1 P^*_1(s) + \alpha\lambda P^*_0(s), \\
sP^*_2(s) &= -\theta_2 P^*_2(s) + (1 - \alpha)\lambda P^*_0(s).
\end{aligned}$$

Solving these equations,

$$P_0^*(s) = \frac{(s+\theta_1)(s+\theta_2)}{s^2 + s(\lambda+\theta_1+\theta_2) + (1-\alpha)\lambda\theta_1 + \alpha\lambda\theta_2 + \theta_1\theta_2},$$
$$P_1^*(s) = \frac{\alpha\lambda}{s+\theta_1} P_0^*(s),$$
$$P_2^*(s) = \frac{(1-\alpha)\lambda}{s+\theta_2} P_0^*(s).$$

Forming the inverse LS transforms of $P_j^*(s)$,

$$\begin{aligned} P_0(t) = {} & \frac{\theta_1\theta_2}{(1-\alpha)\lambda\theta_1 + \alpha\lambda\theta_2 + \theta_1\theta_2} + \frac{(w_1-\theta_1)(w_1-\theta_2)}{w_1(w_1-w_2)} \mathrm{e}^{-w_1 t} \\ & + \frac{(w_2-\theta_1)(w_2-\theta_2)}{w_2(w_2-w_1)} \mathrm{e}^{-w_2 t}, \\ P_1(t) = {} & \frac{\alpha\lambda\theta_2}{(1-\alpha)\lambda\theta_1 + \alpha\lambda\theta_2 + \theta_1\theta_2} - \frac{\alpha\lambda(w_1-\theta_2)}{w_1(w_1-w_2)} \mathrm{e}^{-w_1 t} \\ & - \frac{\alpha\lambda(w_2-\theta_2)}{w_2(w_2-w_1)} \mathrm{e}^{-w_2 t}, \\ P_2(t) = {} & \frac{(1-\alpha)\lambda\theta_1}{(1-\alpha)\lambda\theta_1 + \alpha\lambda\theta_2 + \theta_1\theta_2} - \frac{(1-\alpha)\lambda(w_1-\theta_1)}{w_1(w_1-w_2)} \mathrm{e}^{-w_1 t} \\ & - \frac{(1-\alpha)\lambda(w_2-\theta_1)}{w_2(w_2-w_1)} \mathrm{e}^{-w_2 t}, \end{aligned}$$

where

$$w_1 \equiv \frac{1}{2}\left\{\lambda+\theta_1+\theta_2 + \sqrt{(\lambda+\theta_1+\theta_2)^2 - 4[(1-\alpha)\lambda\theta_1 + \alpha\lambda\theta_2 + \theta_1\theta_2]}\right\},$$
$$w_2 \equiv \frac{1}{2}\left\{\lambda+\theta_1+\theta_2 - \sqrt{(\lambda+\theta_1+\theta_2)^2 - 4[(1-\alpha)\lambda\theta_1 + \alpha\lambda\theta_2 + \theta_1\theta_2]}\right\}.$$

From (4.29) and (4.30), the transition probability matrix satisfies the differential equation

$$\boldsymbol{P}'(t) = \boldsymbol{P}(t)\boldsymbol{A}, \tag{4.31}$$

where the initial condition of $\boldsymbol{P}(0) = \boldsymbol{I}$, where $\boldsymbol{I}$ is an identity matrix and

$$\boldsymbol{A} \equiv \begin{pmatrix} -\lambda_0 & \lambda_0 & 0 & 0 & \dots & 0 & 0 \\ \mu_1 & -(\lambda_1+\mu_1) & \lambda_1 & 0 & \dots & 0 & 0 \\ 0 & \mu_2 & -(\lambda_2+\mu_2) & \lambda_2 & \dots & 0 & 0 \\ \vdots & \vdots & \vdots & \vdots & \dots & \vdots & \vdots \\ 0 & 0 & 0 & 0 & \dots & -(\lambda_{m-1}+\mu_{m-1}) & \lambda_{m-1} \\ 0 & 0 & 0 & 0 & \dots & \mu_m & -\mu_m \end{pmatrix}.$$

Solving (4.31),

$$\boldsymbol{P}(t) = e^{\boldsymbol{A}t} = \boldsymbol{I} + \sum_{n=1}^{\infty} \frac{\boldsymbol{A}^n t^n}{n!}. \tag{4.32}$$

In general, we can compute the transition probabilities from (4.32), however, it is difficult to do so except in simple cases.

4.2.3 Limiting Probabilities

If all states communicate with each other, i.e., the process consists of one recurrent class, then there exists the limiting probabilities

$$P_j \equiv \lim_{t\to\infty} P_{ij}(t) > 0 \qquad (j = 0, 1, 2, \ldots, m),$$

that are independent of the initial State i and $\sum_{j=0}^{m} P_j = 1$, as shown in Sect. 4.1.3. Furthermore,

$$\lim_{t\to\infty} P'_{ij}(t) = 0. \tag{4.33}$$

It can be clearly seen that there exits $\lim_{t\to\infty} P'_{ij}(t)$ from (4.29) because the limiting probabilities exists. Thus, if $P'_{ij}(t) \neq 0$, then there exists $T > 0$ such that

$$P'_{ij}(t) \geq c, \quad \text{or} \quad P'_{ij}(t) \leq -c$$

for $t \geq T$ and $c > 0$. If $P'_{ij}(t) \geq c$, then

$$\int_T^{T_0} P'_{ij}(t)\mathrm{d}t \geq c(T_0 - T)$$

for $T_0 > T$, i.e.,

$$P_{ij}(T_0) - P_{ij}(T) \geq c(T_0 - T),$$

that leads to the mathematical contradiction because the right-hand side goes to infinity as $T_0 \to \infty$. Therefore, (4.33) is proved. It is also proved in the case of $P'_{ij}(t) \leq -c$.

Using the above results, from (4.29) and (4.30), the limiting probabilities satisfy the simultaneous equations,

$$\begin{aligned} -\lambda_0 P_0 + \mu_1 P_1 &= 0, \\ \lambda_{j-1}P_{j-1} - (\lambda_j + \mu_j)P_j + \mu_{j+1}P_{j+1} &= 0 \qquad (j = 1, 2, \ldots, m-1), \\ \lambda_{m-1}P_{m-1} - \mu_m P_m &= 0. \end{aligned} \tag{4.34}$$

Summing for $i = 0, 1, 2, \ldots, j-1$ in (4.34),

$$P_j = \frac{\lambda_{j-1}}{\mu_j} P_{j-1} = \frac{\lambda_{j-1}\lambda_{j-2}}{\mu_j \mu_{j-1}} P_{j-2} = \cdots = \left(\prod_{k=1}^{j} \frac{\lambda_{k-1}}{\mu_k} \right) P_0.$$

Recalling that $\sum_{j=0}^{m} P_j = 1$,

$$P_j = \frac{\prod_{k=1}^{j} (\lambda_{k-1}/\mu_k)}{1 + \sum_{j=1}^{m} \prod_{k=1}^{j} (\lambda_{k-1}/\mu_k)} \qquad (j = 0, 1, \ldots, m), \tag{4.35}$$

where $\prod_{k=1}^{0} \equiv 1$.

Example 4.13 (n-unit parallel system) Consider an n-unit parallel system with one repair person. Then, setting that $\lambda_k = (n-k)\lambda$ and $\mu_{k+1} = \theta$ $(k = 0, 1, \ldots, n)$ in (4.35), the limiting probabilities are

$$P_j = \frac{(\lambda/\theta)^j/(n-j)!}{\sum_{k=0}^{n} (\lambda/\theta)^k/(n-k)!} \qquad (j = 0, 1, 2, \ldots, n).$$

In particular, when $n = 2$, P_j agree with the limiting probabilities P_j of a two-unit parallel system with one repair person in Example 4.11.

4.3 Problems 4

4.1 Prove in Example 4.3 that by the mathematical induction,

$$\boldsymbol{P}^n = \frac{1}{a+b} \begin{pmatrix} b & a \\ b & a \end{pmatrix} + \frac{(1-a-b)^n}{a+b} \begin{pmatrix} a & -a \\ -b & b \end{pmatrix},$$

$$\boldsymbol{\pi}(n) = \left(\frac{b + a(1-a-b)^n}{a+b}, \frac{a[1-(1-a-b)^n]}{a+b} \right).$$

When $a + b = 1$, show that for $n \geq 1$,

$$\boldsymbol{P}^n = \begin{pmatrix} 1-a & a \\ 1-a & a \end{pmatrix}, \quad \boldsymbol{\pi}(n) = (1-a, a).$$

4.2 The unit begins to operate at time 0, and when it fails, it is replaced with a new one [4, p. 113]. In addition, if the unit has not failed until Nth period, it is also replaced with a new one. It is assumed that the unit has a failure probability p_n at time n. Then, the probability that the unit with age n fails at time n is

$$r_n \equiv \frac{p_n}{\sum_{k=n}^{\infty} p_k},$$

that is called a *discrete failure rate* [5, p. 18, 6, p. 9]. Show that the transition probabilities are given by

$$P_{i1} = 1 - P_{ii+1} = r_i \quad (i = 1, 2, \ldots, N-1), \quad P_{N1} = 1,$$

and derive the limiting probabilities π_j $(j = 1, 2, \ldots, N)$.

4.3 In Example 4.9, compute $\boldsymbol{P}$ and $(\boldsymbol{I} - \boldsymbol{P})^{-1}$ by striking out row 5 and column 5 when $\lambda t_0 = 0.9$ and $\lambda L = 0.1$.

4.4 In Example 4.10, prove that by using the binomial distribution with n and $p \equiv 1 - \mathrm{e}^{-\lambda t}$,

$$P_k(t) = \binom{n}{k}(1 - \mathrm{e}^{-\lambda t})^k \mathrm{e}^{-(n-k)\lambda t}.$$

4.5 Derive the transition probabilities of two-unit standby system with one and two repair persons, respectively.

References

1. Osaki S (1992) Applied stochastic system modeling. Springer, Berlin
2. Kemeny JG, Snell JL (1960) Finite Markov chains. Nostrand Co, Princeton
3. Tijms HC (2003) A first course in stochastic models. Wiley, Chichester
4. Ross SM (1983) Stochastic processes. Wiley, New York
5. Barlow RE, Proschan F (1965) Mathematical theory of reliability. Wiley, New York
6. Nakagawa T (2005) Maintenance theory of reliability. Springer, London

Chapter 5
Semi-Markov and Markov Renewal Processes

We consider the same one-unit system in Chap. 4. If $F(t)$ and $G_j(t)$ $(j = 1, 2)$ are not exponential, the process does not have the Markov property at any time t and has its property at only epochal times of transition to states. Let $Z(t)$ denote the state of the process at time t, $N_j(t)$ $(j = 0, 1, 2)$ denote the number of times that the process visits State j in $[0, t]$ when the system begins to operate at time 0, and define that

$$P_j(t) \equiv \Pr\{Z(t) = j\}, \quad F_j(t) \equiv \Pr\{N_j(t) > 0\},$$

where $F_j(t)$ is called a *first-passage distribution* to State j. Then, using the techniques of forming renewal equations in Sect. 3.3,

$$\begin{aligned}
F_0(t) &= \alpha F(t) * G_1(t) + (1 - \alpha)F(t) * G_2(t), \\
F_1(t) &= \alpha F(t) + (1 - \alpha)F(t) * G_2(t) * F_1(t), \\
F_2(t) &= (1 - \alpha)F(t) + \alpha F(t) * G_1(t) * F_2(t),
\end{aligned} \tag{5.1}$$

where the asterisk denotes the pairwise Stieltjes convolution, i.e., $a(t) * b(t) \equiv \int_0^t b(t - u)\mathrm{d}a(u)$.

Furthermore, using $F_j(t)$, the transition probabilities are

$$\begin{aligned}
P_0(t) &= 1 - F(t) + F_0(t) * P_0(t), \\
P_1(t) &= F_1(t) * [1 - G_1(t) + G_1(t) * P_1(t)], \\
P_2(t) &= F_2(t) * [1 - G_2(t) + G_2(t) * P_2(t)].
\end{aligned} \tag{5.2}$$

Thus, using $F_j(t)$ in (5.1), we can obtain the transition probabilities $P_j(t)$. The process $\{Z(t)\}$ is called a *semi-Markov process* because it has the Markov property only at regeneration points at which the state transitions take place. We investigate mainly the properties of the transition probabilities $P_j(t)$ of a semi-Markov process in Sect. 5.2 as an example of a standby redundant system.

T. Nakagawa, *Stochastic Processes*, Springer Series in Reliability Engineering,
DOI: 10.1007/978-0-85729-274-2_5,

Finally, define that renewal functions $M_j(t) \equiv E\{N_j(t)\}$ $(j = 0, 1, 2)$ that represent the expected number of visits to State j in $[0, t]$. In a similar way of obtaining (5.2),

$$\begin{aligned} M_0(t) &= F_0(t) * [1 + M_0(t)], \\ M_1(t) &= F_1(t) * [1 + G_1(t) * M_1(t)], \\ M_2(t) &= F_2(t) * [1 + G_2(t) * M_2(t)]. \end{aligned} \tag{5.3}$$

Thus, using $F_j(t)$ in (5.1), we can obtain $M_j(t)$ from (5.3) (Problem 5.1). The process $\{N_j(t)\}$ is called a *Markov renewal process* that combines a renewal process in Chap. 3 and a Markov chain in Chap. 4. We investigate mainly the properties of $M_j(t)$ for a Markov renewal process in Sect. 5.3. However, there is no clear difference between semi-Markov and Markov renewal processes. Markov chain, Markov process, Semi-Markov process, and Markov renewal process are called *Markov process* in one word, i.e., Markov processes are a general term of the stochastic process with the Markov property.

State space is usually defined by the number of units that are working satisfactorily. As far as the applications to reliability theory is concerned, we consider only a finite number of states, contrast with a queueing theory. We mention only the theory of stationary Markov processes with a finite-state space. It is shown that transition probabilities, first-passage distributions, and renewal functions are given by forming renewal equations. Furthermore, some limiting properties are summarized when all states communicate.

5.1 Markov Process

Consider a stochastic process with a finite state space $\{0, 1, 2, \ldots, m\}$ that makes transitions from state to state in accordance with a Markov chain with stationary transition probabilities. However, in the process, the amount of time spent in each state until the next transition is not always constant and exponential, but has a random variable shown in Definition 5.1.

Consider a stochastic process $\{X(n), S_n, n = 0, 1, 2, \ldots\}$ where S_n denotes the random variable of the nth arrival time at which the process just moves from one state to another, and $X(n)$ denotes the state of the process at time S_n.

Definition 5.1 The stochastic process $\{X(n), S_n\}$ is called a Markov renewal process [1, p. 167, 2, p. 313] if

$$\begin{aligned} &\Pr\{X(n+1) = j, S_{n+1} - S_n \le t | X(0), X(1), \ldots, X(n) = i; S_0, S_1, \ldots, S_n\} \\ &\quad = \Pr\{X(n+1) = j, S_{n+1} - S_n \le t | X(n) = i, S_n\} \equiv Q_{ij}(t) \end{aligned} \tag{5.4}$$

for all $i, j = 0, 1, 2, \ldots, m$ and $t \ge 0$.

The probability $Q_{ij}(t)$ denotes that after making a transition into State i, the next process makes a transition into State j, in an amount of time less than or equal to t. Clearly, we have $Q_{ij}(t) \geq 0$ and $\sum_{j=0}^{m} Q_{ij}(\infty) = 1$, where $Q_{ij}(\infty)$ represents the probability that the next process makes a transition into State j, given that the process goes into State i, i.e., $Q_{ij}(\infty)$ forms a discrete-time Markov chain in Sect. 4.1. We call the probability $Q_{ij}(t)$ a *mass function* or a one-step transition probability [1, p. 167]. Letting

$$G_{ij}(t) = \frac{Q_{ij}(t)}{Q_{ij}(\infty)} \quad \text{for } Q_{ij}(\infty) > 0,$$

then $G_{ij}(t)$ represents the conditional probability that the process makes a transition in an amount of time less than or equal to t, given that the process goes from State i to State j at the next transition. In other words, $G_{ij}(t) = \Pr\{S_{n+1} - S_n \leq t | X(n) = i, X(n+1) = j\}$ is the distribution of the sojourn time that the process takes in State i given that the next visiting state is j.

Because $X(n)$ denotes the state of the process immediately after the nth transition has occurred for $n \geq 1$, the stochastic process $\{X(n), n = 0, 1, 2, \ldots\}$ is called an *embedded Markov chain* with transition probabilities $Q_{ij}(\infty)$. If the process makes a transition from one state to another with one unit of time, i.e., $G_{ij}(t) = 0$ for $t < 1$, and 1 for $t \geq 1$, then an embedded Markov chain becomes a *discrete-time Markov chain* in Sect. 4.1. Furthermore, if an amount of time spent in State i depends only on State i and is distributed exponentially independent of the next state; $G_{ij}(t) = 1 - \mathrm{e}^{-\lambda_i t}$ for constant $\lambda_i > 0$, the process is said to be a *continuous-time Markov chain* in Sect. 4.2. In addition, the process becomes a *renewal process* in Chap. 3 if it has only one state and becomes an alternating renewal process in Sect. 3.4 if it has two states.

If we let $Z(t)$ denote the state of the process at time t, then the stochastic process $\{Z(t), t \geq 0\}$ is called a *semi-Markov process*. Let $N_j(t)$ denote the number of times that the process visits State j in $[0, t]$. It follows from the renewal theory that with probability 1, $N_j(t) < \infty$ for $t \geq 0$. The stochastic process $\{N_0(t), N_1(t), N_2(t), \ldots, N_m(t)\}$ is called a *Markov renewal process*, and $E\{N_j(t)\}$ are called *Markov renewal functions*. The event $\{N_j(t) = k\}$ represents the number of k visits to State j in $[0, t]$. An embedded Markov chain records the state of the process at each transition point, a semi-Markov process records the state of the process at each time point t, and a Markov renewal process records the total number of times that each state has been visited in $[0, t]$.

Letting $H_i(t)$ denote the distribution of an amount of time spent in State i until the process makes a transition to the next state,

$$H_i(t) \equiv \sum_{j=0}^{m} Q_{ij}(t),$$

that is called the *unconditional distribution* for State i. We suppose that $H_i(0) < 1$ for all i. Denoting

$$\eta_i \equiv \int_0^\infty t\mathrm{d}H_i(t) \quad \text{and} \quad \mu_{ij} \equiv \int_0^\infty t\mathrm{d}G_{ij}(t),$$

it is easily seen that

$$\eta_i = \sum_{j=0}^{m} Q_{ij}(\infty)\mu_{ij},$$

which represents the mean time spent in State i.

We denote transition probabilities, first-passage distributions, and Markov renewal functions as, respectively,

$$\begin{aligned} P_{ij}(t) &\equiv \Pr\{Z(t) = j | Z(0) = i\}, \\ F_{ij}(t) &\equiv \Pr\{N_j(t) > 0 | Z(0) = i\}, \\ M_{ij}(t) &\equiv E\{N_j(t) | Z(0) = i\}. \end{aligned}$$

We have the following renewal equations that show the relationships for $P_{ij}(t)$, $F_{ij}(t)$ and $M_{ij}(t)$ in terms of the mass functions $Q_{ij}(t)$:

$$\begin{aligned} P_{ii}(t) &= 1 - H_i(t) + \sum_{k=0}^{m} \int_0^t P_{ki}(t-u)\mathrm{d}Q_{ik}(u), \\ P_{ij}(t) &= \sum_{k=0}^{m} \int_0^t P_{kj}(t-u)\mathrm{d}Q_{ik}(u) \quad \text{for } i \neq j, \end{aligned} \tag{5.5}$$

$$F_{ij}(t) = Q_{ij}(t) + \sum_{k=0,k\neq j}^{m} \int_0^t F_{kj}(t-u)\mathrm{d}Q_{ik}(u), \tag{5.6}$$

$$M_{ij}(t) = Q_{ij}(t) + \sum_{k=0}^{m} \int_0^t M_{kj}(t-u)\mathrm{d}Q_{ik}(u). \tag{5.7}$$

Therefore, the mass functions $Q_{ij}(t)$ determine $P_{ij}(t)$, $F_{ij}(t)$ and $M_{ij}(t)$ uniquely. Furthermore,

$$\begin{aligned} P_{ii}(t) &= 1 - H_i(t) + \int_0^t P_{ii}(t-u)\mathrm{d}F_{ii}(u), \\ P_{ij}(t) &= \int_0^t P_{jj}(t-u)\mathrm{d}F_{ij}(u) \quad \text{for } i \neq j, \end{aligned} \tag{5.8}$$

$$M_{ij}(t) = F_{ij}(t) + \int_0^t M_{jj}(t-u)\mathrm{d}F_{ij}(u). \tag{5.9}$$

Taking the LS transforms of the above equations,

$$P_{ii}^*(s) = \frac{1 - H_i^*(s)}{1 - F_{ii}^*(s)}, \quad P_{ij}^*(s) = F_{ij}^*(s)P_{jj}^*(s) \quad \text{for } i \neq j, \tag{5.10}$$

$$M_{ij}^*(s) = F_{ij}^*(s)[1 + M_{jj}^*(s)], \tag{5.11}$$

where the asterisk denotes the LS transform of the function with itself. Thus, if the first-passage distributions $F_{ij}(t)$ are obtained, then the transition probabilities $P_{ij}(t)$ and renewal functions $M_{ij}(t)$ are given by (5.10) and (5.11), respectively.

Consider the process in which all states communicate, i.e., $G_{ii}(\infty) = 1$, $\mu_{ii} < \infty$ for all i, and each $G_{ii}(t)$ is a nonlattice distribution. It is said that the process consists of one positive recurrent class. Then,

$$G_{ij}(\infty) = 1, \quad \mu_{ij} < \infty,$$

$$\mu_{ij} = \eta_i + \sum_{k=0,k\neq j}^{n} Q_{ik}(\infty)\mu_{kj}.$$

Furthermore,

$$\lim_{t\to\infty} M_{ij}(t) = \int_0^\infty P_{ij}(t)\mathrm{d}t,$$

$$\lim_{t\to\infty} \frac{M_{ij}(t)}{t} = \frac{1}{\mu_{jj}} < \infty, \quad \lim_{t\to\infty} P_{ij}(t) = \frac{\eta_j}{\mu_{jj}} < \infty.$$

In this case, there exist $\lim_{t\to\infty} M_{ij}(t)/t$ and $\lim_{t\to\infty} P_{ij}(t)$. From a Tauberian theorem in Appendix A.3 that if for some nonnegative integer n, $\lim_{s\to 0} s^n\Phi^*(s) = C$, then $\lim_{t\to\infty} \Phi(t)/t^n = C/n!$,

$$\lim_{t\to\infty} \frac{M_{ij}(t)}{t} = \lim_{s\to 0} sM_{ij}^*(s),$$

$$\lim_{t\to\infty} P_{ij}(t) = \lim_{t\to\infty} \frac{1}{t}\int_0^t P_{ij}(u)\mathrm{d}u = \lim_{s\to 0} P_{ij}^*(s).$$

Example 5.1 (Alternating renewal process) An alternating renewal process in Sect. 3.4.1 has the mass functions $Q_{01}(t) = F(t)$ and $Q_{10}(t) = G(t)$. Then, from (5.5),

$$P_{00}(t) = 1 - Q_{01}(t) + \int_0^t P_{10}(t-u)\mathrm{d}Q_{01}(u)$$
$$= \overline{F}(t) + \int_0^t P_{10}(t-u)\mathrm{d}F(u),$$
$$P_{11}(t) = 1 - Q_{10}(t) + \int_0^t P_{01}(t-u)\mathrm{d}Q_{10}(u)$$
$$= \overline{G}(t) + \int_0^t P_{01}(t-u)\mathrm{d}G(u),$$
$$P_{01}(t) = \int_0^t P_{11}(t-u)\mathrm{d}Q_{01}(u) = \int_0^t P_{11}(t-u)\mathrm{d}F(u),$$
$$P_{10}(t) = \int_0^t P_{00}(t-u)\mathrm{d}Q_{10}(u) = \int_0^t P_{00}(t-u)\mathrm{d}G(u),$$

that agree with (3.54). Furthermore, from (5.7),

$$M_{00}(t) = \int_0^t M_{10}(t-u)\mathrm{d}Q_{01}(u) = \int_0^t M_{10}(t-u)\mathrm{d}F(u),$$
$$M_{11}(t) = \int_0^t M_{01}(t-u)\mathrm{d}Q_{10}(u) = \int_0^t M_{01}(t-u)\mathrm{d}G(u),$$
$$M_{01}(t) = Q_{01}(t) + \int_0^t M_{11}(t-u)\mathrm{d}Q_{01}(u) = F(t) + \int_0^t M_{11}(t-u)\mathrm{d}F(u),$$
$$M_{10}(t) = Q_{10}(t) + \int_0^t M_{00}(t-u)\mathrm{d}Q_{10}(u) = G(t) + \int_0^t M_{00}(t-u)\mathrm{d}G(u),$$

that agrees with (3.52).

Example 5.2 (Intermittent fault) [3, p. 113] Faults occur in a communication system according to an exponential distribution $(1 - \mathrm{e}^{-\lambda t})$ and are hidden. When the duration time X in hidden fault exceeds an upper limit time Y, faults become permanent failures eternally, and otherwise, they get out of a hidden state. That is, if the event $\{X \leq Y\}$ occurs, then hidden faults disappear, and conversely, if the event $\{X > Y\}$ occurs, then they become permanent failure. It is assumed that both random variables X and Y are independent and have exponential distributions $\Pr\{X \le t\} = 1 - \mathrm{e}^{-\mu t}$ and $\Pr\{Y \le t\} = 1 - \mathrm{e}^{-\theta t}$, respectively.

We define the following states of intermittent faults:

State 0: No fault occurs and the system is in a normal condition.
State 1: Hidden fault occurs.
State 2: Permanent failure occurs.

We have the following mass functions $Q_{ij}(t)$ $(i = 0, 1,\ j = 0, 1, 2)$ from State i to State j in time t from Fig. 5.1:

$$\begin{aligned} &Q_{01}(t) = 1 - \mathrm{e}^{-\lambda t}, \\ &Q_{10}(t) = \Pr\{X \le Y, X \le t\} = \int_0^t \mathrm{e}^{-\theta u} \mu\, \mathrm{e}^{-\mu u} \mathrm{d}u = \frac{\mu}{\mu + \theta}\left[1 - \mathrm{e}^{-(\mu+\theta)t}\right], \\ &Q_{12}(t) = \Pr\{X > Y, Y \le t\} = \int_0^t \mathrm{e}^{-\mu u} \theta\, \mathrm{e}^{-\theta u} \mathrm{d}u = \frac{\theta}{\mu + \theta}\left[1 - \mathrm{e}^{-(\mu+\theta)t}\right]. \end{aligned} \tag{5.12}$$

Let $P_{ij}(t)$ denote the transition probabilities from State i at time 0 to State j at time t. Then, we have the following renewal equations:

$$\begin{aligned} P_{00}(t) &= 1 - Q_{01}(t) + Q_{01}(t) * P_{10}(t), \\ P_{01}(t) &= Q_{01}(t) * P_{11}(t), \\ P_{10}(t) &= Q_{10}(t) * P_{00}(t), \\ P_{11}(t) &= 1 - Q_{10}(t) - Q_{12}(t) + Q_{10}(t) * P_{01}(t), \\ P_{02}(t) &= Q_{01}(t) * P_{12}(t), \\ P_{12}(t) &= Q_{12}(t) + Q_{10}(t) * P_{02}(t). \end{aligned} \tag{5.13}$$

Taking the LS transforms of (5.13) and rearranging them,

$$\begin{aligned} P^*_{00}(s) &= \frac{1 - Q^*_{01}(s)}{1 - Q^*_{01}(s) Q^*_{10}(s)}, \\ P^*_{01}(s) &= \frac{Q^*_{01}(s)\left[1 - Q^*_{10}(s) - Q^*_{12}(s)\right]}{1 - Q^*_{01}(s) Q^*_{10}(s)}, \\ P^*_{10}(s) &= \frac{Q^*_{10}(s)\left[1 - Q^*_{01}(s)\right]}{1 - Q^*_{01}(s) Q^*_{10}(s)}, \\ P^*_{11}(s) &= \frac{1 - Q^*_{10}(s) - Q^*_{12}(s)}{1 - Q^*_{01}(s) Q^*_{10}(s)}. \end{aligned} \tag{5.14}$$

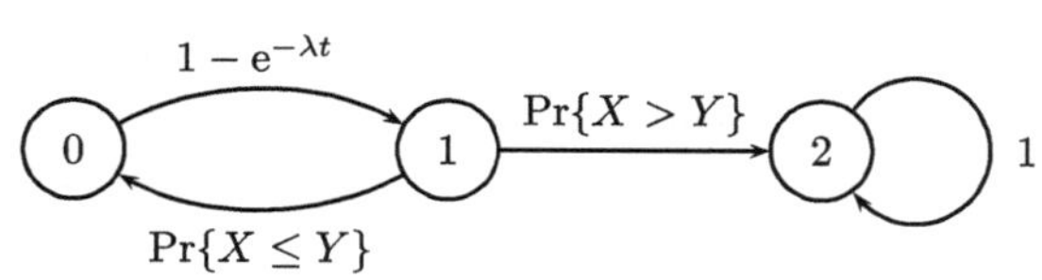

Fig. 5.1 Transition diagram of intermittent fault

Thus, substituting (5.12) for (5.14) and taking the inverse LS transforms, the transition probabilities from State i to State j $(i,j = 0,1,2)$ are

$$\begin{aligned}
P_{00}(t) &= \frac{1}{\omega_1 - \omega_2}[(\mu + \theta - \omega_2)\mathrm{e}^{-\omega_2 t} - (\mu + \theta - \omega_1)\mathrm{e}^{-\omega_1 t}], \\
P_{01}(t) &= \frac{\lambda}{\omega_1 - \omega_2}(\mathrm{e}^{-\omega_2 t} - \mathrm{e}^{-\omega_1 t}), \quad P_{10}(t) = \frac{\mu}{\omega_1 - \omega_2}(\mathrm{e}^{-\omega_2 t} - \mathrm{e}^{-\omega_1 t}), \\
P_{11}(t) &= \frac{1}{\omega_1 - \omega_2}[(\lambda - \omega_2)\mathrm{e}^{-\omega_2 t} - (\lambda - \omega_1)\mathrm{e}^{-\omega_1 t}], \\
P_{02}(t) &= 1 - P_{00}(t) - P_{01}(t), \quad P_{12}(t) = 1 - P_{10}(t) - P_{11}(t),
\end{aligned} \tag{5.15}$$

where

$$\omega_1 \equiv \frac{1}{2}\left[\lambda + \mu + \theta + \sqrt{(\lambda + \mu + \theta)^2 - 4\lambda\theta}\right],$$
$$\omega_2 \equiv \frac{1}{2}\left[\lambda + \mu + \theta - \sqrt{(\lambda + \mu + \theta)^2 - 4\lambda\theta}\right].$$

Next, let $M_{ij}(t)$ $(i,j = 0,1)$ be the expected number of visits to State j from State i before State 2 in time t. Then, we have the following renewal equations:

$$\begin{aligned}
M_{00}(t) &= Q_{01}(t) * M_{10}(t), \quad M_{01}(t) = Q_{01}(t) * [1 + M_{11}(t)], \\
M_{11}(t) &= Q_{10}(t) * M_{01}(t), \quad M_{10}(t) = Q_{10}(t) * [1 + M_{00}(t)].
\end{aligned} \tag{5.16}$$

Taking the LS transforms of (5.16) and rearranging them,

$$M_{00}^*(s) = \frac{Q_{01}^*(s)Q_{10}^*(s)}{1 - Q_{01}^*(s)Q_{10}^*(s)}, \quad M_{01}^*(s) = \frac{Q_{01}^*(s)}{1 - Q_{01}^*(s)Q_{10}^*(s)}. \tag{5.17}$$

Thus, substituting (5.12) for (5.17) and taking the inverse LS transforms,

$$\begin{aligned}
M_{00}(t) &= \frac{\lambda\mu}{\omega_1 - \omega_2}\left[\frac{1}{\omega_2}(1 - \mathrm{e}^{-\omega_2 t}) - \frac{1}{\omega_1}(1 - \mathrm{e}^{-\omega_1 t})\right], \\
M_{01}(t) &= \frac{\lambda}{\omega_1 - \omega_2}\left[\frac{\omega_1 - \mu - \theta}{\omega_1}(1 - \mathrm{e}^{-\omega_1 t}) - \frac{\omega_2 - \mu - \theta}{\omega_2}(1 - \mathrm{e}^{-\omega_2 t})\right].
\end{aligned} \tag{5.18}$$

Example 5.3 (ARQ model) ARQ (Automatic-Repeat-Request) models with intermittent faults that are mainly employed in data transmission to achieve high reliability of communication [3, p. 110]: ARQ strategies are adopted when data transmission due to faults, fail. However, the data throughput decreases significantly if retransmissions are repeated without limitation. To keep the level of data throughput, when all numbers N of transmissions have failed, the system is inspected and maintained. We repeat the above procedure until the data transmission succeeds.

It is assumed in Example 5.2 that when the system is in State 0, the data transmission succeeds with probability 1, when the system is in State 1, it succeeds with probability q and fails with probability $p \equiv 1 - q$, and when the system is in State 2, it fails with probability 1. When the data transmission fails, the retransmission incurs in time T_1. When N retransmissions have failed consecutively including the first one, the system is inspected and preventively maintained, and the same data transmission begins new after time T_2.

Suppose that the system is in State 0 at time 0. Then, the probability that the data transmission succeeds until the Nth number is

$$\overline{P}(N) = 1 - pP_{01}(T_1) - P_{02}(T_1) + pP_{01}(T_1)[1 - pP_{11}(T_1) - P_{12}(T_1)] \sum_{j=0}^{N-2} [pP_{11}(T_1)]^j,$$

and the probability that all N retransmissions have failed is

$$P(N) = P_{02}(T_1) + pP_{01}(T_1) \left\{ P_{12}(T_1) \sum_{j=0}^{N-2} [pP_{11}(T_1)]^j + [pP_{11}(T_1)]^{N-1} \right\},$$

where $\overline{P}(N) + P(N) = 1$.

We call the time from the beginning of transmission to N failures or success as *one period*. Then, the mean time of one period is

$$\begin{aligned} l_1 &= T_1 \left\{ 1 - pP_{01}(T_1) - P_{02}(T_1) + pP_{01}(T_1)[1 - pP_{11}(T_1) - P_{12}(T_1)] \right. \\ &\quad \left. \times \sum_{j=0}^{N-2} (j+2)[pP_{11}(T_1)]^j + NP(N) \right\} \\ &= T_1 \left\{ 1 + (N-1) \left[P_{02}(T_1) + \frac{pP_{01}(T_1)P_{12}(T_1)}{1 - pP_{11}(T_1)} \right] \right. \\ &\quad \left. + pP_{01}(T_1) \frac{1 - pP_{11}(T_1) - P_{12}(T_1)}{1 - pP_{11}(T_1)} \frac{1 - [pP_{11}(T_1)]^{N-1}}{1 - pP_{11}(T_1)} \right\}. \end{aligned}$$

Thus, the mean time in which the data transmission succeeds is

$$l_2(N) = l_1 + P(N)[T_2 + l_2(N)],$$

i.e.,

$$\begin{aligned} l_2(N) &= \frac{l_1 + T_2 P(N)}{\overline{P}(N)} \\ &= \frac{T_1}{1 - pP_{11}(T_1)} - T_2 + \frac{A + (N-1)B}{C + D\left\{1 - [pP_{11}(T_1)]^{N-1}\right\}} \qquad (N = 1, 2, \ldots), \end{aligned} \tag{5.19}$$

where

$$A \equiv T_1 + T_2 - \frac{T_1[P_{00}(T_1) + qP_{01}(T_1)]}{1 - pP_{11}(T_1)},$$

$$B \equiv T_1\left[P_{02}(T_1) + \frac{pP_{01}(T_1)P_{12}(T_1)}{1 - pP_{11}(T_1)}\right],$$

$$C \equiv P_{00}(T_1) + qP_{01}(T_1), \quad D \equiv pP_{01}(T_1)\frac{P_{10}(T_1) + qP_{11}(T_1)}{1 - pP_{11}(T_1)}.$$

From the inequality $l_2(N+1) - l_2(N) \geq 0$,

$$\frac{1 + C/D}{[1 - pP_{11}(T_1)][pP_{11}(T_1)]^{N-1}} - (N-1) \geq \frac{A}{B} + \frac{1}{1 - pP_{11}(T_1)} \quad (N = 1, 2, \ldots). \tag{5.20}$$

Letting $L(N)$ be the left-hand side of (5.20),

$$L(\infty) \equiv \lim_{N\to\infty} L(N) = \infty, \quad L(1) = \frac{1 + C/D}{1 - pP_{11}(T_1)},$$

$$L(N+1) - L(N) = \frac{1 + C/D - [pP_{11}(T_1)]^N}{[P_{11}(T_1)]^N} > 0.$$

Therefore, an optimum N^* $(1 \leq N^* < \infty)$ that minimizes $l_2(N)$ in (5.19) is given by a finite and unique minimum that satisfies (5.20). If $C/D \geq (A/B)[1 - pP_{11}(T_1)]$, then $N^* = 1$ (Problem 5.2).

5.2 Embedded Markov Chain

Consider a standby redundant system with $n+1$ repair persons and one operating unit supported by n identical spare units. Each unit fails according to a general distribution $F(t)$ with finite mean μ and undergoes repair immediately. When failed units are repaired, they become as good as new units and rejoin spare ones. It is assumed that the repair time of each failed unit is an independent random variable with an identical exponential distribution $(1 - \mathrm{e}^{-\theta t})$ with finite mean $1/\theta$.

Let denote by $Z(t)$ the number of units undergoing repair at time t. We say that the system is in State j at time t if $Z(t) = j$ $(j = 0, 1, 2, \ldots, n+1)$. We also say that system failure occurs when the system is in Sate $n+1$. Denote by $\tau_0, \tau_1, \ldots, \tau_k, \ldots$ the failure times of operating units. For simplicity of computations, we restrict ourselves to the case of $\tau_0 \equiv 0$. If we denote $Z_k \equiv Z(\tau_k - 0)$ for $k = 1, 2, \ldots$, then Z_k represents the number of units undergoing repair immediately before the kth failure occurrence, where $Z_0 \equiv 0$. It is noted that the process $\{Z_k, k = 0, 1, 2, \ldots\}$

forms an embedded Markov chain and $\{Z(t), t \geq 0\}$ forms a semi-Markov process with stationary transition probabilities and a finite state space.

5.2.1 Transition Probabilities

We derive the transition probabilities by using binomial moments [4] and the expected number of failures before system failure: Let denote the transition probabilities as

$$P_{ij}^{(k)}(t) = \Pr\{Z_k = j, \tau_k \leq t | Z_0 = i\} \quad (i, j = 0, 1, 2, \ldots, n)$$

for $k = 1, 2, \ldots$. By similar arguments of obtaining the Chapman–Kolmogorov equations of (4.4) and (4.25),

$$P_{ij}^{(k)}(t) = \sum_{l=0}^{n} \int_0^t P_{lj}^{(1)}(t-u) \mathrm{d}P_{il}^{(k-1)}(u). \tag{5.21}$$

Furthermore, because Z_1 has a binomial distribution with parameters $i+1$ and $\mathrm{e}^{-\theta u}$ under the given conditions,

$$\begin{aligned} &\Pr\{Z_1 = j | \tau_1 - \tau_0 = u, Z_0 = i\} \\ &= \begin{cases} \binom{i+1}{j}(1-\mathrm{e}^{-\theta u})^{i+1-j}\mathrm{e}^{-j\theta u} & (i = 0, 1, \ldots, n-1,\ j \leq i+1), \\ 0 & (j > i+1). \end{cases} \end{aligned}$$

Dropping the condition $\tau_1 - \tau_0 = u$, i.e., the operating unit fails in $(u, u + \mathrm{d}u]$,

$$P_{ij}^{(1)}(t) = \begin{cases} \int_0^t \binom{i+1}{j}(1-\mathrm{e}^{-\theta u})^{i+1-j}\mathrm{e}^{-j\theta u}\mathrm{d}F(u) & (i = 0, 1, \ldots, n-1,\ j \leq i+1), \\ 0 & (j > i+1). \end{cases} \tag{5.22}$$

If $Z_0 = n$, then

$$P_{nj}^{(1)}(t) = \int_0^t (n+1)\theta\, \mathrm{e}^{-(n+1)\theta u} P_{n-1j}^{(1)}(t-u)\mathrm{d}u. \tag{5.23}$$

Next, introduce the rth binomial moment of the LS transform of $P_{ij}^{(k)}(t)$ as follows:

$$\Phi_{ir}^{(k)}(s) \equiv \sum_{j=r}^{n} \binom{j}{r} \int_0^\infty e^{-st} dP_{ij}^{(k)}(t) \quad (r = 0, 1, 2, \ldots, n)$$

for $k = 1, 2, \ldots$. Then, taking the LS transforms of (5.21)–(5.23) and arranging them,

$$\begin{aligned}
\Phi_{ir}^{(1)}(s) &= F^*(s + r\theta)\left[\binom{i+1}{r} - \binom{n+1}{r}\frac{(s + r\theta)\delta_{in}}{s + (n+1)\theta}\right], \\
\Phi_{ir}^{(k)}(s) &= F^*(s + r\theta)\left[\Phi_{ir}^{(k-1)}(s) + \Phi_{ir-1}^{(k-1)}(s)\right. \\
&\quad \left. - \binom{n+1}{r}\frac{s + r\theta}{s + (n+1)\theta}\Phi_{in}^{(k-1)}(s)\right] \quad (r = 0, 1, 2, \ldots, n),
\end{aligned} \tag{5.24}$$

where δ_{in} is the Kronecker delta, i.e., $\delta_{in} \equiv 0$ for $i \neq n$ and $\delta_{in} = 1$ for $i = n$. From (5.24), $\Phi_{ir}^{(k)}(s)$ can be determined recursively for every k and r.

It is more convenient to introduce the generating function in Appendix A.4

$$\Phi_{ir}(z, s) \equiv \sum_{k=1}^{\infty} z^k \Phi_{ir}^{(k)}(s) \quad \text{for } |z| \leq 1.$$

Taking the generating function of (5.24),

$$\begin{aligned}
\Phi_{ir}(z, s) &= zF^*(s + r\theta)\left\{\Phi_{ir}(z, s) + \Phi_{ir-1}(z, s) + \binom{i+1}{r}\right. \\
&\quad \left. - \binom{n+1}{r}\frac{s + r\theta}{s + (n+1)\theta}[\Phi_{in}(z, s) + \delta_{in}]\right\}.
\end{aligned}$$

Solving for $\Phi_{ir}(z, s)$,

$$\begin{aligned}
\Phi_{ir}(z, s) &= \frac{zF^*(s + r\theta)}{1 - zF^*(s + r\theta)}\left\{\Phi_{ir-1}(z, s) + \binom{i+1}{r}\right. \\
&\quad \left. - \binom{n+1}{r}\frac{s + r\theta}{s + (n+1)\theta}[\Phi_{in}(z, s) + \delta_{in}]\right\}.
\end{aligned} \tag{5.25}$$

Denote

$$C_r(z, s) \equiv \prod_{l=0}^{r} \frac{zF^*(s + l\theta)}{1 - zF^*(s + l\theta)} \quad (r = 0, 1, \ldots, n), \qquad C_{-1}(z, s) \equiv 1.$$

Then, dividing both side of (5.25) by $C_r(z,s)$,

$$\frac{\Phi_{ir}(z,s)}{C_r(z,s)} = \frac{\Phi_{ir-1}}{C_{r-1}(z,s)} + \binom{i+1}{r}\frac{1}{C_{r-1}(z,s)} - \binom{n+1}{r}\frac{s+r\theta}{C_{r-1}(z,s)}\frac{\Phi_{in}(z,s)+\delta_{in}}{s+(n+1)\theta} \quad (r=1,2,\ldots,n), \tag{5.26}$$

$$\frac{\Phi_{i0}(z,s)}{C_0(z,s)} = 1 - s\frac{\Phi_{in}(z,s)+\delta_{in}}{s+(n+1)\theta}. \tag{5.27}$$

Writing down (5.26) for $r-1, r-2, \ldots, 1$, adding them, and from (5.27),

$$\frac{\Phi_{ir}(z,s)}{C_r(z,s)} = \sum_{l=0}^{r}\binom{i+1}{l}\frac{1}{C_{r-1}(z,s)} - \sum_{l=0}^{r}\binom{n+1}{l}\frac{s+j\theta}{C_{l-1}(z,s)}\frac{\Phi_{in}(z,s)+\delta_{in}}{s+(n+1)\theta} \quad (r=0,1,\ldots,n). \tag{5.28}$$

Setting $r=n$ in (5.28) and rearranging it,

$$\frac{\Phi_{in}(z,s)+\delta_{in}}{s+(n+1)\theta} = \frac{\sum_{l=0}^{i+1}\binom{i+1}{l}/C_{l-1}(z,s)}{\sum_{l=0}^{n+1}\binom{n+1}{l}(s+l\theta)/C_{l-1}(z,s)}. \tag{5.29}$$

Thus, from (5.28),

$$\Phi_{ir}(z,s) = C_r(z,s)\left[\sum_{l=0}^{r}\binom{i+1}{l}\frac{1}{C_{l-1}(z,s)} - \sum_{l=0}^{i+1}\binom{i+1}{l}\frac{1}{C_{l-1}(z,s)}\frac{\sum_{l=0}^{r}\binom{n+1}{l}(s+l\theta)/C_{l-1}(z,s)}{\sum_{l=0}^{n+1}\binom{n+1}{l}(s+l\theta)/C_{l-1}(z,s)}\right], \tag{5.30}$$

that is given in terms of the binomial moment. By inverting the above binomial moment,

$$P_{ij}^*(z,s) \equiv \sum_{k=1}^{\infty} z^k \int_0^{\infty} e^{-st} dP_{ij}^{(k)}(t) = \sum_{r=j}^{n}\binom{r}{j}(-1)^{r-j}\Phi_{ir}(z,s). \tag{5.31}$$

Therefore, the transition probabilities $P_{ij}^{(k)} \equiv \Pr\{Z_k = j | Z_0 = i\}$ for the process $\{Z_k, k = 0, 1, \ldots\}$ are given by the generating function

$$P_{ij}^*(z) \equiv \sum_{k=1}^{\infty} z^k P_{ij}^{(k)} = \lim_{s \to 0} P_{ij}^*(z, s) = \sum_{r=j}^{n} \binom{r}{j} (-1)^{r-j} \Phi_{ir}(z, 0). \tag{5.32}$$

The total expected number of visits to State j in $[0, t]$ immediately before the operating unit fails, initially starting from State i, is $\sum_{k=1}^{\infty} P_{ij}^{(k)}(t)$. In particular, the total expected number of system failures in $[0, t]$ is $\sum_{k=1}^{\infty} P_{in}^{(k)}(t)$.

5.2.2 First-Passage Distributions

We derive the first-passage distributions using the results of transition probabilities in Sect. 5.2.1: Let denote the first-passage distribution

$$F_{ij}^{(k)}(t) \equiv \Pr\{Z_k = j, Z_l \neq j \text{ for } l = 1, 2, \ldots, k-1, \tau_k \leq t | Z_0 = i\}$$
$$(i, j = 0, 1, \ldots, n)$$

for $k = 1, 2, \ldots$. Then, by similar methods of obtaining (4.9),

$$\begin{aligned} P_{ij}^{(1)}(t) &= F_{ij}^{(1)}(t), \\ P_{ij}^{(k)}(t) &= F_{ij}^{(k)}(t) + \sum_{l=1}^{k-1} \int_0^t P_{jj}^{(k-l)}(t-u) \mathrm{d}F_{ij}^{(l)}(u) \end{aligned} \tag{5.33}$$

for $k = 2, 3, \ldots$. The first-passage distribution $F_{ij}^{(k)}(t)$ can be calculated recursively from (5.33) and are uniquely determined.

Introduce the following notation

$$F_{ij}^*(z, s) \equiv \sum_{k=1}^{\infty} z^k \int_0^{\infty} \mathrm{e}^{-st} \mathrm{d}F_{ij}^{(k)}(t) \quad \text{for } |z| \leq 1.$$

Then, taking the generating functions and the LS transforms of (5.33),

$$F_{ij}^*(z, s) = \frac{P_{ij}^*(z, s)}{1 + P_{jj}^*(z, s)}. \tag{5.34}$$

Substituting (5.31) for (5.34),

$$F^*_{ij}(z,s) = \frac{\begin{array}{l}\sum_{r=j}^{n}\binom{r}{j}(-1)^{r-j}C_r(z,s)\\ \times\Big[\sum_{l=0}^{r}\binom{i+1}{l}/C_{l-1}(z,s)\sum_{l=0}^{n+1}\binom{n+1}{l}(s+l\theta)/C_{l-1}(z,s)\\ -\sum_{l=0}^{i+1}\binom{i+1}{l}/C_{l-1}(z,s)\sum_{l=0}^{r}\binom{n+1}{l}(s+l\theta)/C_{l-1}(z,s)\Big]\end{array}}{\begin{array}{l}\sum_{l=0}^{j+1}\binom{j+1}{l}/C_{l-1}(z,s)\\ \times\Big[\sum_{r=j}^{n}\binom{r}{j}(-1)^{r-j}C_r(z,s)\sum_{l=r+1}^{n+1}\binom{n+1}{l}(s+l\theta)/C_{l-1}(z,s)\Big]\end{array}}. \tag{5.35}$$

Therefore, the first-passage distribution

$$F_{ij}^{(k)} \equiv \Pr\{Z_k = j, Z_l \neq k \text{ for } l = 1,2,\ldots,k-1, |Z_0 = i\}$$

is given by the generating function

$$F^*_{ij}(z) \equiv \sum_{k=1}^{\infty} z^k F_{ij}^{(k)}$$

$$= \frac{\begin{array}{l}\sum_{r=j}^{n}\binom{r}{j}(-1)^{r-j}C_r(z)\Big[\sum_{l=0}^{r}\binom{i+1}{l}/C_{l-1}(z)\sum_{l=0}^{n}\binom{n}{j}/C_l(z)\\ -\sum_{l=0}^{i+1}\binom{i+1}{l}/C_{l-1}(z)\sum_{l=0}^{r-1}\binom{n}{l}/C_l(z)\Big]\end{array}}{\sum_{l=0}^{j+1}\binom{j+1}{l}/C_{l-1}(z)\Big[\sum_{r=j}^{n}\binom{r}{j}(-1)^{r-j}C_r(z)\sum_{l=r}^{n}\binom{n}{l}/C_l(z)\Big]}, \tag{5.36}$$

where $C_r(z) \equiv C_r(z,0)(r = -1,0,1,\ldots)$. In particular, for $i<j$,

$$F^*_{ij}(z) = \frac{\sum_{l=0}^{i+1}\binom{i+1}{l}/C_{l-1}(z)}{\sum_{l=0}^{j+1}\binom{i+1}{l}/C_{l-1}(z)}.$$

Next, let l_{ij} be the mean first-passage time from State i to State j. Then, from (5.35), for $i<j$,

$$l_{ij} \equiv \lim_{s\to 0}\frac{1-F^*_{ij}(1,s)}{s} = \mu\sum_{l=1}^{j+1}\left[\binom{j+1}{l} - \binom{i+1}{l}\right]\frac{1}{C_{l-1}}, \tag{5.37}$$

where $C_r \equiv \prod_{l=1}^{r}\{F^*(l\theta)/[1-F^*(l\theta)]\}$ $(r = 1,2,\ldots)$ and $C_0 \equiv 1$.

In particular, when $i = -1$ and $j = n$, l_{-1n} agrees with [5]. Note that State -1 means the initial condition that one unit begins to operate and n units are in standby at time 0.

5.2.3 Expected Numbers of Visits to States

We obtain two reliability quantities from the first-passage distributions; (1) the expected number of visits to State j before system failure and (2) the expected number of failures and repairs before system failure.

First, we derive the probabilities of visits to State j at least once before returning to State i. Let denote $f_{m;ij}^{(k)}$ $(k = 1, 2, \ldots)$ for an embedded Markov chain $\{Z_k, k = 0, 1, 2, \ldots\}$ the probability of first-passage from State i to State k without passing through State m. In general, $f_{m;ij}^{(k)}$ is known as *taboo* probability for a Markov chain. We evidently have, for $j \neq m$,

$$\begin{aligned} F_{ij}^{(1)} &= f_{m;ij}^{(1)}, \\ F_{ij}^{(k)} &= f_{m;ij}^{(k)} + \sum_{l=1}^{k-1} F_{mj}^{(k-l)} f_{j;im}^{(l)} \quad (k = 2, 3, \ldots), \end{aligned} \tag{5.38}$$

because, the first-passage from State i to State j for a Markov chain $\{Z_k, k = 1, 2, \ldots\}$ occurs without passing through State m $(\neq j)$ or by passing through State m. By exchanging j and m in (5.38), (5.38) is hold and uniquely determine $f_{m;ij}^{(k)}$.

Introduce the following generating function

$$F_{m;ij}^{*}(z) \equiv \sum_{k=1}^{\infty} z^k f_{m;ij}^{(k)} \quad \text{for } |z| \leq 1.$$

Then, from (5.38), for $j \neq m$,

$$F_{m;ij}^{*}(z) = \frac{F_{ij}^{*}(z) - F_{im}^{*}(z) F_{mj}^{*}(z)}{1 - F_{jm}^{*}(z) F_{mj}^{*}(z)}. \tag{5.39}$$

Substituting (5.36) for (5.39), we can obtain $F_{m;ij}^{*}(z)$ and the probability

$$f_{m;ij} \equiv \sum_{k=1}^{\infty} f_{m;ij}^{(k)} = \lim_{z \to 1} F_{m;ij}^{*}(z) \tag{5.40}$$

of the first-passage from State i to State j without passage through State m.

Furthermore, because $f_{i;ij} = 1 - f_{j;ii}$ (Problem 5.3), from (5.37) and (5.39),

$$f_{i;ij} = 1 \Bigg/ \left[\sum_{r=i}^{j} (-1)^{r-i} \binom{r}{i} C_r \sum_{l=r+1}^{j+1} \binom{j+1}{l} \frac{1}{C_{l-1}} \right] \quad \text{for } i < j \leq n, \tag{5.41}$$

that is the probability of visits to State j at least once before returning to State i.

Next, we give two reliability quantities from the probability $f_{j;jn}$. Let denote by $M_j^{(k)}$ $(j = 0, 1, \ldots, n-1)$ the number of visits to State j for a Markov chain $\{Z_k, k = 1, 2, \ldots\}$, including the first before system failure. Clearly,

$$\Pr\{M_j^{(k)} = l\} = \sum_{k_1+k_2+\cdots+k_j=k, k_1,k_2,\ldots,k_j > 0} f_{n;jj}^{(k_1)} f_{n;jj}^{(k_2)} \cdots f_{n;jj}^{(k_{j-1})} f_{j;jn}^{(k_j)} \quad (k, l = 1, 2, \ldots). \tag{5.42}$$

By operating the convolution in (5.42),

$$\sum_{k=1}^{\infty} z^k \Pr\{M_j^{(k)} = l\} = [F_{n;jj}^*(z)]^{l-1} F_{j;jn}^*(z), \tag{5.43}$$

and from (5.40) and (5.43),

$$\sum_{k=1}^{\infty} \Pr\{M_j^{(k)} = l\} = (1 - f_{j;jn})^{l-1} f_{j;jn}.$$

Thus, the expected number M_j of visits to State j before system failure is

$$M_j = \sum_{l=1}^{\infty} l \sum_{k=1}^{\infty} \Pr\{M_j^{(k)} = l\} = \frac{1}{f_{j;jn}},$$

and from (5.41),

$$M_j = \sum_{r=j}^{n} (-1)^{r-j} \binom{r}{j} C_r \sum_{l=r+1}^{n+1} \binom{n+1}{l} \frac{1}{C_{l-1}}. \tag{5.44}$$

Let M denote the total expected number of failures before system failures, when one unit starts to operate and n units are in standby as spare ones at time 0. Then,

$$M = 1 + \sum_{j=0}^{n-1} M_j,$$

because each failure causes one visit before system failure and the last failure causes system failure. From (5.44),

$$M = \sum_{j=1}^{n+1} \binom{n+1}{j} \frac{1}{C_{j-1}}.$$

The expected number of repairs before system failure is $M - (n+1)$. Note that μM is also the mean time to system failure that agrees with l_{-1n} in (5.37).

5.2.4 Optimization Problems

Consider the total cost of the operation for the system before system failure. It is assumed that the cost of each unit is c_0, the repair cost incurred for each failed unit is c_1, and the cost incurred for the system failure is c_2. Then, the total expected cost including the cost of system failure is

$$(n+1)c_0 + [M - (n+1)]c_1 + c_2,$$

and hence, the expected cost rate is

$$C(n) = \frac{(n+1)(c_0 - c_1) + Mc_1 + c_2}{\mu M}. \tag{5.45}$$

We can easily determine an optimum number of spare units that minimizes the total expected cost before system failure. If the total cost C before system failure is previously given, we can find an upper bound of the number of spare units such that

$$(n+1)c_0 + [M - (n+1)]c_1 + c_2 \le C. \tag{5.46}$$

We can determine an optimum number of spare units that minimizes (5.45) within this number. In this case, because $C(n)$ in (5.45) decreases to c_0/μ (Problem 5.4), an optimum n^* is given by a unique maximum that satisfies (5.46).

5.3 Markov Renewal Process with Nonregeneration Points

This section explains unique modifications of Markov renewal processes and applies them to redundant repairable systems including some nonregeneration points [6]. It has already been shown that such modifications give powerful plays for analyzing two-unit redundant systems [7] and communication systems [8].

It is assumed that the Markov renewal process under consideration has only one positive recurrent class, because we restrict ourselves to applications of reliability models. Consider the case where epochs at which the process enters some states are not regeneration points. Then, we partition a state space S into $S = S^* \cup S^\dagger$ ($S^* \cap S^\dagger = \emptyset$), where S^* is the portion of the state space such that the epoch entering State i ($i \in S^*$) is not a regeneration point, and $S^\dagger$ is such that the epoch entering State i ($i \in S^\dagger$) is a regeneration point, where S^* and $S^\dagger$ are assumed not to be empty.

Define the mass function $Q_{ij}(t)$ from State i ($i \in S^\dagger$) to State j ($j \in S$) by the probability that after entering State i, the process makes a transition into State j, in an amount of time less than or equal to t. However, it is impossible to define mass functions $Q_{ij}(t)$ for i ($i \in S^*$), because the epoch entering State i is not a regeneration point. We define the new mass function $Q_{ij}^{(k_1,k_2,\ldots,k_l)}(t)$ that is the probability that after entering State i ($i \in S^{\dagger}$), the process next makes transition into State $k_1, k_2, \ldots, k_l$ ($k_1, k_2, \ldots, k_l \in S^*$), and finally, enters State j ($j \in S$), in an amount of time less than or equal to t.

5.3.1 Type 1 Markov Renewal Process

Consider a Markov renewal process with $m+1$ states that consists of $S^{\dagger} = \{0\}$ and $S^* = \{1, 2, \ldots, m\}$ in Fig. 5.2. The process starts in State 0, i.e., $Z(0) = 0$, and makes transitions into State $1, 2, \ldots, m$, and comes back to State 0. Then, from straightforward renewal arguments, the first-passage distributions are

$$\begin{aligned} F_{01}(t) &= Q_{01}(t), \\ F_{0j}(t) &= Q_{0j}^{(1,2,\ldots,j-1)}(t) \quad (j = 2, 3, \ldots, m), \\ F_{00}(t) &= Q_{00}^{(1,2,\ldots,m)}(t), \end{aligned} \tag{5.47}$$

the renewal functions are

$$\begin{aligned} M_{01}(t) &= Q_{01}(t) + Q_{00}^{(1,2,\ldots,m)}(t) * M_{01}(t) \\ &= Q_{01}(t) + F_{00}(t) * M_{01}(t), \\ M_{0j}(t) &= Q_{0j}^{(1,2,\ldots,j-1)}(t) + F_{00}(t) * M_{0j}(t) \quad (j = 2, 3, \ldots, m), \\ M_{00}(t) &= F_{00}(t) + F_{00}(t) * M_{00}(t), \end{aligned} \tag{5.48}$$

and the transition probabilities are

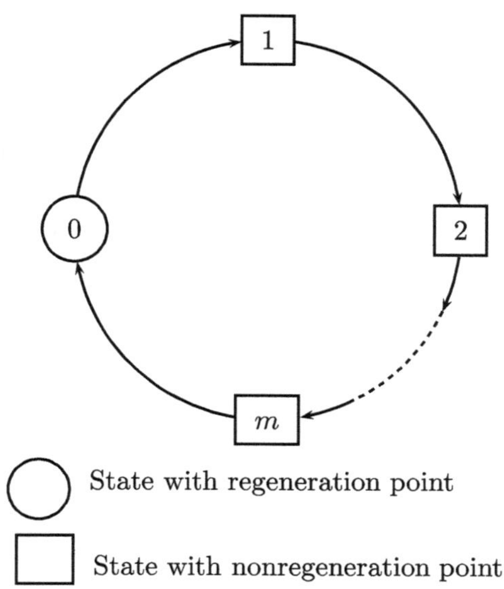

Fig. 5.2 State transition diagram for Type 1 Markov renewal process

$$\begin{aligned} P_{01}(t) &= Q_{01}(t) - Q_{02}^{(1)}(t) + Q_{00}^{(1,2,\ldots,m)}(t) * P_{01}(t) \\ &= Q_{01}(t) - Q_{02}^{(1)}(t) + F_{00}(t) * P_{01}(t), \\ P_{0j}(t) &= Q_{0j}^{(1,2,\ldots,j-1)}(t) - Q_{0j+1}^{(1,2,\ldots,j)}(t) + F_{00}(t) * P_{0j}(t) \quad (j = 2, 3, \ldots, m), \\ P_{00}(t) &= 1 - Q_{01}(t) + F_{00}(t) * P_{00}(t), \end{aligned} \tag{5.49}$$

where $Q_{0m+1}^{(1,2,\ldots,m)}(t) = Q_{00}^{(1,2,\ldots,m)}(t) = F_{00}(t)$.

Taking the LS transforms on both sides of (5.48) and (5.49),

$$\begin{aligned} M_{01}^*(s) &= \frac{Q_{01}^*(s)}{1 - F_{00}^*(s)}, \\ M_{0j}^*(s) &= \frac{Q_{0j}^{*(1,2,\ldots,j-1)}(s)}{1 - F_{00}^*(s)} \quad (j = 2, 3, \ldots, m), \\ M_{00}^*(s) &= \frac{F_{00}^*(s)}{1 - F_{00}^*(s)}, \end{aligned} \tag{5.50}$$

and

$$\begin{aligned} P_{01}^*(s) &= \frac{Q_{01}^*(s) - Q_{02}^{*(1)}(s)}{1 - F_{00}^*(s)}, \\ P_{0j}^*(s) &= \frac{Q_{0j}^{*(1,2,\ldots,j-1)}(s) - Q_{0j+1}^{*(1,2,\ldots,j)}(s)}{1 - F_{00}^*(s)} \quad (j = 2, 3, \ldots, m), \\ P_{00}^*(s) &= \frac{1 - Q_{01}^*(s)}{1 - F_{00}^*(s)}, \end{aligned} \tag{5.51}$$

where note that $\sum_{j=0}^{m} P_{0j}^*(s) = 1$. From (5.50) and (5.51), the renewal functions and the transition probabilities are computed explicitly upon inversion, however, this might not be easy except in simple cases.

Example 5.4 (Parallel system) Consider an n-unit parallel redundant system: When at least one of n units is operating, the system is operating. When all units are down simultaneously, the system fails and will begin to operate again immediately by replacing all failed units with new ones. Each unit operates independently and has an identical failure distribution $F(t)$. The states are denoted by the total number of failed units. When all units begin to operate at time 0, the mass functions are

$$\begin{aligned} Q_{01}(t) &= 1 - [\overline{F}(t)]^n, \\ Q_{0j}^{(1,2,\ldots,j-1)}(t) &= \sum_{i=j}^{n} \binom{n}{i} [F(t)]^i [\overline{F}(t)]^{n-i} \quad (j = 2, 3, \ldots, n). \end{aligned}$$

Because the system begins to operate immediately at system failure,

$$Q_{00}^{(1,2,\ldots,n)}(t) = Q_{0n}^{(1,2,\ldots,n-1)}(t) = F(t)^n.$$

If the system operates again after a maintenance time Y with distribution $G(t) \equiv \Pr\{Y \leq t\}$ at system failure, then States 0 and n are regeneration points and $Q_{n0}(t) = G(t)$, $Q_{00}^{(1,2,\ldots,n)}(t) = F(t)^n * G(t) = F_{00}(t)$. Thus, substituting the above equations for (5.50) and (5.51), we can obtain the renewal functions and the transition probabilities.

5.3.2 Type 2 Markov Renewal Process

Consider a Markov renewal process with $S^\dagger = \{0\}$ and $S^* = \{1, 2, \ldots, m\}$ in Fig. 5.3. The process starts in State 0 and is only permitted to make a transition into one State $j \in S^*$, and then return to State 0.

The LS transforms of the first-passage distributions, the renewal functions, and the transition probabilities are

$$\begin{aligned} F_{00}^*(s) &= \sum_{i=1}^{m} Q_{00}^{*(i)}(s), \\ F_{0j}^*(s) &= Q_{0j}^*(s) \Big/ \left[1 - \sum_{i=1, i \neq j}^{m} Q_{00}^{*(i)}(s)\right] \quad (j = 1, 2, \ldots, m), \end{aligned} \tag{5.52}$$

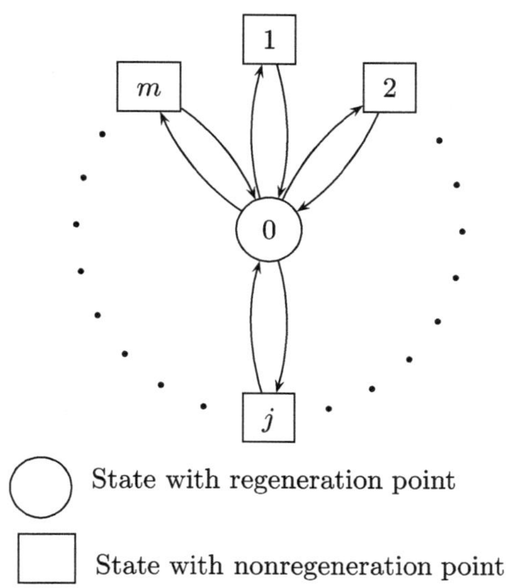

Fig. 5.3 State transition diagram for Type 2 Markov renewal process

$$
\begin{aligned}
M_{00}^*(s) &= \frac{F_{00}^*(s)}{1 - F_{00}^*(s)}, \\
M_{0j}^*(s) &= \frac{Q_{0j}^*(s)}{1 - F_{00}^*(s)} \quad (j = 1, 2, \ldots, m),
\end{aligned} \tag{5.53}
$$

$$
\begin{aligned}
P_{00}^*(s) &= \frac{1 - \sum_{j=1}^{m} Q_{0j}^*(s)}{1 - F_{00}^*(s)}, \\
P_{0j}^*(s) &= \frac{Q_{0j}^*(s) - Q_{00}^{*(j)}(s)}{1 - F_{00}^*(s)} \quad (j = 1, 2, \ldots, m).
\end{aligned} \tag{5.54}
$$

The process corresponds to a special one of Type 1 when $m = 1$. That is, it is the simplest state space with one nonregeneration point. The process takes two alternate States 0 and 1. When the epoch entering State 1 is also a regeneration point, the process becomes an alternating renewal process described in Sect. 3.4.

Example 5.5 (Inventory model) We give the inventory model of spare units [6] as one example of Markov renewal processes with nonregeneration points: A unit begins to operate at time 0. If the operating unit fails before time T, the spare unit is ordered immediately and is placed into service as soon as it is delivered after a lead time L. If the operating unit does not fail up to time T, the spare one is ordered at time T and is delivered after the lead time L. In this case, the spare unit is put into service when the original unit has failed or is put into inventory when it does not fail.

Define the following states:

State 0: Unit is operating and spare one is not delivered.
State 1: Unit is operating and spare one is ordered.
State 2: Unit is operating and spare one is in inventory.
State 3: Operating unit has failed during the order of spare unit.
State 4: Spare unit is ordered because of the failure of operating unit before time T.

The system with the above five states forms a Markov renewal process with regeneration point $\{0\}$ and nonregeneration points $\{1, 2, 3, 4\}$ (Fig. 5.4).

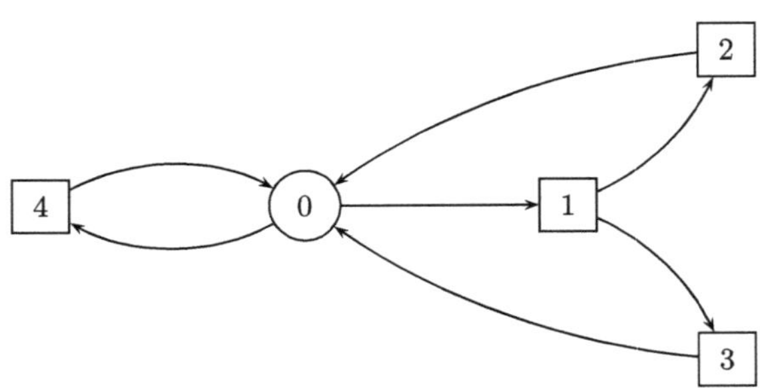

Fig. 5.4 State transition diagram of spare part inventory

Assume that the failure times of each operating unit are identical and have a general distribution $F(t)$ with mean time μ. In addition, assume that $\Pr\{T \leq t\} \equiv A(t)$ and $\Pr\{L \leq t\} = B(t)$ are the degenerate distributions placing unit masses at T and L, respectively. Then, the mass functions of the above process are

$$Q_{01}(t) = \int_0^t \overline{F}(u)\mathrm{d}A(u),$$

$$Q_{02}^{(1)}(t) = \int_0^t \overline{F}(u)\mathrm{d}[A(u) * B(u)],$$

$$Q_{00}^{(1,2)}(t) = \int_0^t [A(u) * B(u)]\mathrm{d}F(u),$$

$$Q_{03}^{(1)}(t) = \int_0^t [A(u) * \overline{B}(u)]\mathrm{d}F(u),$$

$$Q_{00}^{(1,3)}(t) = \int_0^t \left\{ \int_0^u [F(u) - F(v)]\mathrm{d}A(v) \right\} \mathrm{d}B(u - v),$$

$$Q_{04}(t) = \int_0^t \overline{A}(u)\mathrm{d}F(u), \quad Q_{40}(t) = B(t).$$

Thus, the LS transforms of the above mass functions are

$$Q_{01}^*(s) = \mathrm{e}^{-sT}\overline{F}(T),$$

$$Q_{02}^{*(1)}(s) = \mathrm{e}^{-s(T+L)}\overline{F}(T + L),$$

$$Q_{00}^{*(1,2)}(s) = \int_{T+L}^{\infty} \mathrm{e}^{-st}\mathrm{d}F(t),$$

$$Q_{03}^{*(1)}(s) = \int_T^{\infty} \mathrm{e}^{-st}\mathrm{d}F(t) - \int_{T+L}^{\infty} \mathrm{e}^{-st}\mathrm{d}F(t),$$

$$Q_{00}^{*(1,3)}(s) = \mathrm{e}^{-s(T+L)}[F(T + L) - F(T)],$$

$$Q_{04}^*(s) = \int_0^T \mathrm{e}^{-st}\mathrm{d}F(t),$$

$$Q_{40}^*(s) = \mathrm{e}^{-sL}.$$

The LS transform of the recurrence time distribution to State 0 is

$$H_{00}^{*}(s) = Q_{00}^{*(1,2)}(s) + Q_{00}^{*(1,3)}(s) + Q_{04}^{*}(s)Q_{40}^{*}(s)$$
$$= \int_{T+L}^{\infty} \mathrm{e}^{-st}\mathrm{d}F(t) + \mathrm{e}^{-s(T+L)}[F(T+L) - F(T)] + \mathrm{e}^{-sL}\int_{0}^{T} \mathrm{e}^{-st}\mathrm{d}F(t),$$

where note that $H_{00}^{*}(0) = 1$. Thus, the mean recurrence time to State 0 is

$$l_{00} \equiv \int_{0}^{\infty} t\mathrm{d}H_{00}(t) = \mu + \int_{T}^{T+L} F(t)\mathrm{d}t.$$

Using the above results, the LS transforms of the transition probabilities are

$$P_{00}^{*}(s) = \frac{1 - Q_{01}^{*}(s) - Q_{04}^{*}(s)}{1 - H_{00}^{*}(s)},$$
$$P_{01}^{*}(s) = \frac{Q_{01}^{*}(s) - Q_{02}^{*(1)}(s) - Q_{03}^{*(1)}(s)}{1 - H_{00}^{*}(s)},$$
$$P_{02}^{*}(s) = \frac{Q_{02}^{*(1)}(s) - Q_{00}^{*(1,2)}(s)}{1 - H_{00}^{*}(s)},$$
$$P_{03}^{*}(s) = \frac{Q_{03}^{*(1)}(s) - Q_{00}^{*(1,3)}(s)}{1 - H_{00}^{*}(s)},$$
$$P_{04}^{*}(s) = \frac{Q_{04}^{*}(s)[1 - Q_{40}^{*}(s)]}{1 - H_{00}^{*}(s)},$$

where note that $P_{00}^{*}(s) + P_{01}^{*}(s) + P_{02}^{*}(s) + P_{03}^{*}(s) + P_{04}^{*}(s) = 1$. Thus, the limiting probabilities $P_j \equiv \lim_{t\to\infty} P_{0j}(t) = \lim_{s\to 0} P_{0j}^{*}(s)$ are

$$P_0 = \frac{1}{l_{00}}\int_{0}^{T} \overline{F}(t)\mathrm{d}t, \quad P_1 = \frac{1}{l_{00}}\int_{T}^{T+L} \overline{F}(t)\mathrm{d}t,$$
$$P_2 = \frac{1}{l_{00}}\int_{T+L}^{\infty} \overline{F}(t)\mathrm{d}t, \quad P_3 = \frac{1}{l_{00}}\left[\int_{T}^{T+L} [\overline{F}(T) - \overline{F}(t)]\mathrm{d}t\right],$$
$$P_4 = \frac{LF(T)}{l_{00}},$$

where also note that $P_0 + P_1 + P_2 + P_3 + P_4 = 1$ (Problem 5.5).

Furthermore, the LS transforms of the expected number of visits to State j are

$$M_{00}^*(s) = \frac{H_{00}^*(s)}{1 - H_{00}^*(s)}, \quad M_{01}^*(s) = \frac{Q_{01}^*(s)}{1 - H_{00}^*(s)},$$
$$M_{02}^*(s) = \frac{Q_{02}^{*(1)}(s)}{1 - H_{00}^*(s)}, \quad M_{03}^*(s) = \frac{Q_{03}^{*(1)}(s)}{1 - H_{00}^*(s)},$$
$$M_{04}^*(s) = \frac{Q_{04}^*(s)}{1 - H_{00}^*(s)},$$

and the limiting expected number of visits to State j per unit of time, i.e., $M_j \equiv \lim_{t\to\infty} M_{0j}(t)/t = \lim_{s\to 0} sM_{00}^*(s)$ are

$$M_0 = \frac{1}{l_{00}}, \quad M_1 = \frac{\overline{F}(T)}{l_{00}}, \quad M_2 = \frac{\overline{F}(T+L)}{l_{00}},$$
$$M_3 = \frac{F(T+L) - F(T)}{l_{00}}, \quad M_4 = \frac{F(T)}{l_{00}},$$

where note that $M_0 = M_2 + M_3 + M_4$ and $M_1 = M_2 + M_3$ (Problem 5.6).

5.4 Problems 5

5.1 Derive $F_j^*(s)$, $P_j^*(s)$ and $M_j(s) (j = 0, 1, 2)$ by taking the LS transforms of (5.1)–(5.3).

5.2 It is assumed in Example 5.3 that when the system is in State i ($i = 0, 1, 2$), the data transmission succeeds with probability q_i and fails with probability $p_i \equiv 1 - q_i$, where $0 \le p_0 \le p_1 \le p_2 \le 1$ [3, p. 116]. Then, obtain the mean time $l_3(N)$ in which the data transmission succeeds and derive analytically an optimum number N^* that minimizes $l_3(N)$.

5.3 Prove that $f_{i;ij} = 1 - f_{j;ii}$.

5.4 Prove in (5.45) that $C(n)$ decrease to c_0/μ for $c_0 \ge c_1$.

5.5 Derive P_j and M_j ($j = 0, 1, 2, 3, 4$) in Example 5.4.

5.6 It is assumed in Example 5.5 that c_1 is the expedited order cost of a spare unit when the original unit fails before time T, c_2 ($<c_1$) is the ordinary order cost at time T, c_4 is the inventory cost per unit of time in State 2, and c_3 ($>c_4$) is the system failure cost per unit of time in States 3 and 4. Then, when $F(t) = 1 - e^{-\lambda t}$, derive an optimum time T^* that minimizes the expected cost rate

$$C(T) = c_1 M_4 + c_2 M_1 + c_3(P_3 + P_4) + c_4 P_2.$$

References

1. Osaki S (1992) Applied stochastic system modeling. Springer, Berlin
2. Çinlar E (1975) Introduction to stochastic processes. Prentice-Hall, Englewood Cliffs
3. Nakagawa T (2008) Advanced reliability models and maintenance policies. Springer, London
4. Takács L (1962) Introduction to the theory of queues. Oxford University Press, New York
5. Srinivasan VS (1968) First emptiness in the spare parts problem for repairable components. Oper Res 16:407–415
6. Nakagawa T, Osaki S (1976) Markov renewal process with some nonregeneration points and their applications to reliability theory. Microelectron Reliab 15:633–636
7. Nakagawa T (2002) Two-unit redundant models. In: Osaki S (ed) Stochastic models in reliability and maintenance. Springer, Berlin, pp 165–185
8. Yasui K, Nakagawa T, Sandoh H (2002) Reliability models in data communication systems. In: Osaki S (ed) Stochastic models in reliability and maintenance. Springer, New York, pp 281–306

Chapter 6
Cumulative Processes

Failures of units or systems are generally classified into two failure modes: Catastrophic failure in which units fail by some sudden shock, and degradation failure in which units fail by physical deterioration suffered from some damage. In the latter case, units fail when the total damage due to shocks has exceeded a critical failure level. This is called a *cumulative damage model* or *shock model* with additive damage and can be described theoretically by a cumulative process [1] in stochastic processes. Damage models can be applied to actual units that are working in architecture, industry, service, information, and computers, and were summarized [2].

Let $\{N(t)\}$ be a counting process that represents the total numbers of events up to time t defined in Chaps. 2 and 3. Next, let W_k be some amount of increments due to the kth event. Then, define

$$Z(t) \equiv \sum_{k=0}^{N(t)} W_k \quad (N(t) = 0, 1, 2, \ldots), \tag{6.1}$$

that is the total amount of increments at time t. The stochastic process $\{Z(t)\}$ is a cumulative process [1] or is called a jump process [3] or doubly stochastic process [4] (Problem 6.1).

In Sects. 6.1 and 6.2, we investigate the properties of the process $\{Z(t)\}$ when increments due to each event are additive or not, respectively. We apply such results to damage models and obtain their reliabilities, mean times to failure, and failure rates. Furthermore, Sect. 6.3 proposes four modified damage models. In Sect. 6.4, we summarize compactly replacement policies for damage models where a unit is replaced at a planned time, a shock number or a damage level. The continuous damage model where the total damage increases with time will be dealt in Chap. 7 by applying a Brownian motion and a Lévy process. In this way, cumulative processes can be applied to not only reliability models but also stochastic models such as queueing, insurance, finance, shot noise, and stochastic duel [2, p. 170].

T. Nakagawa, *Stochastic Processes*, Springer Series in Reliability Engineering, DOI: 10.1007/978-0-85729-274-2_6, © Springer-Verlag London Limited 2011

6.1 Standard Cumulative Process

Consider a cumulative process in the context of Cox [1]: Let random variables X_k $(k = 1, 2, \ldots)$ denote a sequence of interarrival times between successive events that form Poisson or renewal processes introduced in Chaps. 2 and 3. Next, suppose that an increment W_k $(k = 1, 2, \ldots)$ such as damage, wear, fatigue, cost, and risk is associated with the kth event, where $W_0 \equiv 0$. Two random variables X_k and W_k might be possibly dependent. It is assumed that the sequence of $\{W_k\}$ is nonnegative, independently, and identically distributed, and is independent of X_j $(k \neq j)$. Then, consider the cumulative process $\{Z(t)\}$ defined in (6.1).

Example 6.1 (Replacement cost) Suppose that each failed unit is replaced immediately at failures with a new one, i.e., X_k is the failure time between replacements. If W_k is the replacement cost for the kth failure, then $Z(t)$ is the total replacement cost up to time t. Furthermore, if X_k is less than a planned replacement time T, W_k is the replacement cost c_1 after failure, and conversely, if X_k is greater than T, W_k is the replacement cost c_2 before failure, then $Z(t)$ is the total replacement cost for an age replacement policy in Sect. 3.3 and $Z(t) = \widehat{C}(t)$ in (3.42). On the other hand, suppose that an operating unit is replaced only at periodic times kT and each failed unit undergoes some repair, i.e., X_k is always equal to T. If W_k is the sum of repair and replacement costs in the kth replacement interval, $Z(t)$ is the total cost up to time t for periodic replacement policies in Sect. 3.3.2.

Example 6.2 (Cumulative damage model) A unit is subjected to shocks and suffers some damage, and each damage is additive. It is assumed that the unit fails when the total damage has exceeded a failure level K $(0 < K < \infty)$ for the first time (Fig. 6.1).

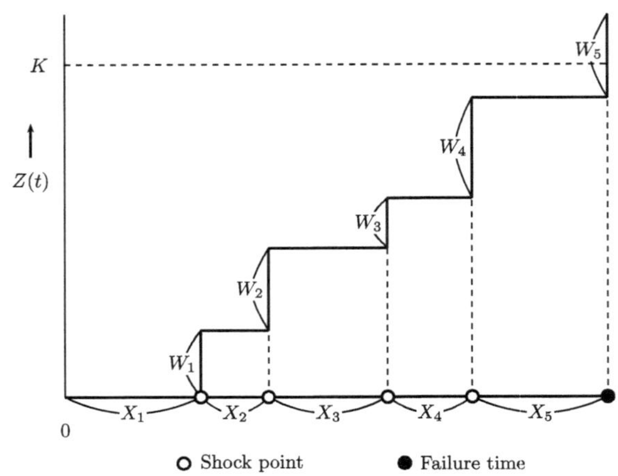

Fig. 6.1 Process for a standard cumulative damage model

We are interested in the following quantities of the cumulative process $\{Z(t)\}$:

(i) $\Pr\{Z(t) \leq x\}$; the distribution of the total increment at time t.
(ii) $E\{Z(t)\}$; the total expected increment at time t.

In addition, let K be the threshold level of the total increment. Defining $Y \equiv \min_t\{Z(t) > K\}$,

(iii) $\Pr\{Y \leq t\}$; the first-passage distribution to a critical point.
(iv) $E\{Y\}$; the mean time to a critical point.

Suppose that the sequences of $\{X_k\}$ and $\{W_k\}$ are non-negative, independently, and identically distributed, and both of X_k and W_k are independent of each other. It is assumed that $F(t) \equiv \Pr\{X_k \leq t\}$ with finite mean $E\{X_k\} \equiv \mu$ and $G(x) \equiv \Pr\{W_k \leq x\}$ with finite mean $E\{W_k\} \equiv \omega$ for any $t, x \geq 0$. Then, from (3.3), the number $N(t)$ of events in $[0, t]$ has the following probability

$$\Pr\{N(t) = n\} = F^{(n)}(t) - F^{(n+1)}(t) \quad (n = 0, 1, 2, \ldots),$$

where $\varphi^{(n)}(t)$ denotes the n-fold Stieltjes convolution of any function $\varphi(t)$ with itself, and $\varphi^{(0)}(t) \equiv 1$ for $t \geq 0$. Thus,

$$\begin{aligned}\Pr\left\{\sum_{k=0}^{N(t)} W_k \leq x, N(t) = n\right\} &= \Pr\left\{\sum_{k=0}^{n} W_k \leq x | N(t) = n\right\} \Pr\{N(t) = n\} \\ &= G^{(n)}(x)[F^{(n)}(t) - F^{(n+1)}(t)].\end{aligned}$$

Therefore, the distribution of $Z(t)$ defined in (6.1) is

$$\begin{aligned}\Pr\{Z(t) \leq x\} &= \Pr\left\{\sum_{k=0}^{N(t)} W_k \leq x\right\} \\ &= \sum_{n=0}^{\infty} \Pr\left\{\sum_{k=0}^{n} W_k \leq x | N(t) = n\right\} \Pr\{N(t) = n\} \\ &= \sum_{n=0}^{\infty} G^{(n)}(x)[F^{(n)}(t) - F^{(n+1)}(t)],\end{aligned} \tag{6.2}$$

and the survival probability is

$$\Pr\{Z(t) > x\} = \sum_{n=0}^{\infty}[G^{(n)}(x) - G^{(n+1)}(x)]F^{(n+1)}(t). \tag{6.3}$$

The total expected increment at time t is

$$\begin{aligned}E\{Z(t)\} &= \int_0^{\infty} x \mathrm{d} \Pr\{Z(t) \leq x\} \\ &= \omega \sum_{n=1}^{\infty} F^{(n)}(t) = \omega M_F(t),\end{aligned} \tag{6.4}$$

where $M_F(t) \equiv E\{N(t)\} = \sum_{n=1}^{\infty} F^{(n)}(t)$ is a renewal function with distribution $F(t)$ and represents the expected number of events occurred in $[0, t]$, and its properties have been investigated in Sect. 3.2. It can be intuitively known that $E\{Z(t)\}$ is given by the product of the average amount of increment suffered from each event and the expected number of events in time t. This is useful for estimating the total expected increment at time t.

Furthermore, from Theorem 3.1, if $\mu_i \equiv \int_0^\infty t^i \,\mathrm{d}F(t) < \infty (i = 2, 3)$ and $\sigma_F^2 \equiv \mu_2 - \mu^2$, then as $t \to \infty$,

$$M_F(t) = \frac{t}{\mu} + \left(\frac{\sigma_F^2}{2\mu^2} - \frac{1}{2}\right) + o(1),$$
$$V\{N(t)\} = \frac{\sigma_F^2 t}{\mu^3} + \left(\frac{5\sigma_F^4}{4\mu^4} + \frac{2\sigma_F^2}{\mu^2} + \frac{3}{4} - \frac{2\mu_3}{3\mu^3}\right) + o(1),$$

where $o(h)/h \to 0$ as $h \to 0$. Thus, when $F(G)$ has finite mean $\mu(w)$ and variance $\sigma_F^2(\sigma_G^2)$, approximately, for large t (Problem 6.2),

$$\begin{aligned}
E\{Z(t)\} &= E\left\{E\left\{\sum_{k=1}^{N(t)} W_k | N(t)\right\}\right\} = E\{N(t)\}E\{W_k\} \\
&\approx \omega\left(\frac{t}{\mu} + \frac{\sigma_F^2 - \mu^2}{2\mu^2}\right), \\
V\{Z(t)\} &= E\{Z^2(t)\} - (E\{Z(t)\})^2 \\
&= E\left\{E\left\{\sum_{k=1}^{N(t)} W_k \sum_{j=1}^{N(t)} W_j | N(t)\right\}\right\} - (E\{Z(t)\})^2 \\
&= V\{N(t)\}(E\{W_k\})^2 + E\{N(t)\}V\{W_k\} \\
&\approx \omega^2\left[\frac{t}{\mu}\left(\frac{\sigma_F^2}{\mu^2} + \frac{\sigma_G^2}{\omega^2}\right) + \frac{5\sigma_F^4}{4\mu^4} + \frac{2\sigma_F^2}{\mu^2} + \frac{3}{4} - \frac{2\mu_3}{3\mu^3}\right] + \frac{\sigma_G^2}{2}\left(\frac{\sigma_F^2}{\mu^2} - 1\right).
\end{aligned} \tag{6.5}$$

Moreover, because

$$\lim_{t\to\infty} \frac{E\{Z(t)\}}{t} = \frac{\omega}{\mu}, \quad \lim_{t\to\infty} \frac{V\{Z(t)\}}{t} = \frac{\omega^2}{\mu}\left(\frac{\sigma_F^2}{\mu^2} + \frac{\sigma_G^2}{\omega^2}\right), \tag{6.6}$$

by applying Theorem 3.5 to the cumulative process $\{Z(t)\}$, we have the following asymptotic distribution of $Z(t)$ given by a normal distribution:

$$\lim_{t\to\infty} \Pr\left\{\frac{Z(t) - \omega t/\mu}{(\omega^2 t/\mu)(\sigma_F^2/\mu^2 + \sigma_G^2/\omega^2)} \le x\right\} = \frac{1}{\sqrt{2\pi}} \int_{-\infty}^{x} \mathrm{e}^{-u^2/2} \mathrm{d}u. \tag{6.7}$$

Example 6.3 (Cumulative damage model) In the cumulative damage model given in Example 6.2, we wish to estimate the total damage when the probability that is more than z in $t = 30$ days of operation is given by 0.90. The distributions $F(t)$ and $G(x)$ of shock times and the amount of damage are unknown, but from sample data, the following estimations of means and variances are made:

$$\mu = 2\,\text{days}, \quad \sigma_F^2 = 5\,(\text{days})^2, \quad \omega = 1, \quad \sigma_G^2 = 0.5.$$

In this case, from (6.5), $E\{Z(30)\} \approx 15.125$. Then, from (6.7), when $t = 30$,

$$\frac{Z(t) - \omega t/\mu}{\sqrt{(\omega^2 t/\mu)(\sigma_F^2/\mu^2 + \sigma_G^2/\omega^2)}} = \frac{Z(30) - 15}{5.12}$$

is approximately normally distributed with mean 0 and variance 1. Hence,

$$\Pr\{Z(30) > z\} = \Pr\left\{\frac{Z(30) - 15}{5.12} > \frac{z - 15}{5.12}\right\}$$
$$\approx \frac{1}{\sqrt{2\pi}} \int_{(z-15)/5.12}^{\infty} \mathrm{e}^{-u^2/2} \mathrm{d}u = 0.90.$$

Because $u_0 = -1.28$ satisfies $(1/\sqrt{2\pi}) \int_{u_0}^{\infty} \mathrm{e}^{-u^2/2} \mathrm{d}u = 0.90, z = 15 - 5.12 \times 1.28 \approx 8.45$. Thus, the total damage is more than 8.45 in 30 days with probability 0.90.

Next, when a failure level is known as $K = 10$,

$$\Pr\{Z(t) > 10\} = \Pr\left\{\frac{Z(30) - 15}{5.12} > \frac{10 - 15}{5.12}\right\}$$
$$\approx \frac{1}{\sqrt{2\pi}} \int_{-0.98}^{\infty} \mathrm{e}^{-u^2/2} \mathrm{d}u \approx 0.836.$$

Thus, the probability that the unit with a failure level $K = 10$ fails in 30 days is about 0.836 (Problem 6.3).

Next, suppose that the process ends when the total increment exceeds a threshold level K that is called a critical point. In this case, the probability that the total increment exceeds a threshold level K at the $(n+1)$th event is $p_{n+1} = G^{(n)}(K) - G^{(n+1)}(K)$ $(n = 0, 1, 2, \ldots)$. Define $Y \equiv \min_t\{Z(t) > K\}$. Because the event of $\{Y \le t\}$ and $\{Z(t) > K\}$ are equivalent, the first-passage distribution to a critical point is, from (6.3),

$$\begin{aligned} \Phi(t) &\equiv \Pr\{Y \le t\} = \Pr\{Z(t) > K\} \\ &= \sum_{n=0}^{\infty} [G^{(n)}(K) - G^{(n+1)}(K)] F^{(n+1)}(t), \end{aligned} \tag{6.8}$$

and its Laplace-Stieltjes (LS) transform from Appendix A is

$$\Phi^*(s) \equiv \int_0^\infty e^{-st} d\Phi(t) = \sum_{n=0}^\infty [G^{(n)}(K) - G^{(n+1)}(K)][F^*(s)]^{n+1}, \tag{6.9}$$

where $\varphi^*(s)$ denotes the LS transform of any function $\varphi(t)$, i.e., $\varphi^*(s) \equiv \int_0^\infty e^{-st} d\varphi(t)$. Thus, the mean time to a critical point is

$$\begin{aligned} E\{Y\} &= \int_0^\infty t d\Pr\{Y \le t\} = -\left.\frac{d\Phi^*(s)}{ds}\right|_{s=0} \\ &= \mu \sum_{n=0}^\infty G^{(n)}(K) = \mu[1 + M_G(K)], \end{aligned} \tag{6.10}$$

where $M_G(K) \equiv \sum_{n=1}^\infty G^{(n)}(K)$ represents the expected number of events before a critical point.

Similarly, when $G(x)$ has finite mean ω and variance σ_G^2, from (3.17), approximately,

$$E\{Y\} \approx \mu\left(\frac{K}{\omega} + \frac{\sigma_G^2 + \omega^2}{2\omega^2}\right). \tag{6.11}$$

In addition, when $G(x)$ has an IFR property defined in Sect. 3.2.2 (Example 6.4), it has been shown that

$$\frac{x}{\omega} - 1 < M_G(x) \le \frac{x}{\omega},$$

from (a) of Sect. 3.2.2.3. Thus,

$$\frac{\mu K}{\omega} < E\{Y\} \le \mu\left(\frac{K}{\omega} + 1\right).$$

In Example 6.3, $E\{Y\}$ is approximately 21.5 days and $20 < E\{Y\} \le 22$.

Next, suppose that events occur in a nonhomogeneous Poisson process with an intensity function $h(t)$ and a mean value function $H(t) \equiv \int_0^t h(u)du$ in Sect. 2.3. Then, from (2.43),

$$\Pr\{N(t) = n\} = \frac{[H(t)]^n}{n!} e^{-H(t)} \quad (n = 0, 1, 2, \ldots).$$

Thus, by replacing $F^{(n)}(t)$ with $\sum_{j=n}^\infty \{[H(t)]^j / j!\} e^{-H(t)}$ formally, we can obtain the following quantities:

$$\Pr\{Z(t) \le x\} = \sum_{n=0}^{\infty} G^{(n)}(x) \frac{[H(t)]^n}{n!} e^{-H(t)}, \quad E\{Z(t)\} = \omega H(t),$$
$$E\{Y\} = \sum_{n=0}^{\infty} G^{(n)}(K) \int_0^{\infty} \frac{[H(t)]^n}{n!} e^{-H(t)} dt. \tag{6.12}$$

If each event occurs at a constant time $T\ (0<T<\infty)$, i.e., $F(t)$ is the degenerate distribution placing unit mass at time T, and $F(t) \equiv 0$ for $t<T$ and 1 for $t \ge T$, then

$$\Pr\{Y \le t\} = 1 - G^{([t/T])}(K), \quad E\{Y\} = \int_0^{\infty} G^{([t/T])}(K)dt,$$

where $[t/T]$ denotes the greatest integer less than or equal to t/T.

Finally, when $G(x) \equiv 0$ for $x<1$ and 1 for $x \ge 1$, and $K = n$,

$$\Pr\{Y \le t\} = F^{(n+1)}(t), \quad E\{Y\} = (n+1)\mu.$$

6.2 Independent Process

Consider the independent process where the total increment is not additive, and the process ends when one amount of increment exceeds a threshold level K. In the damage model, when the damage due to some shock has exceeded for the first time a failure level K, then the unit fails (Fig. 6.2). The practical examples of such models are the fracture of brittle materials and semiconductor parts that have failed by some over-current or fault voltage (Problem 6.4). The generalized model with three types of shocks where shocks with a small level of damage cause no damage to the unit, shocks with a large level of damage result in failure, and shocks with an intermediate level result in failure only with some probability are considered [5].

In this case, the probability that the increment exceeds firstly a threshold level K at the $(n+1)$th event is $P_{n+1} = [G(K)]^n - [G(K)]^{n+1}$. Thus, the first-passage distribution to a critical point is

$$\Pr\{Y \le t\} = \sum_{n=0}^{\infty} \{[G(K)]^n - [G(K)]^{n+1}\} F^{(n+1)}(t), \tag{6.13}$$

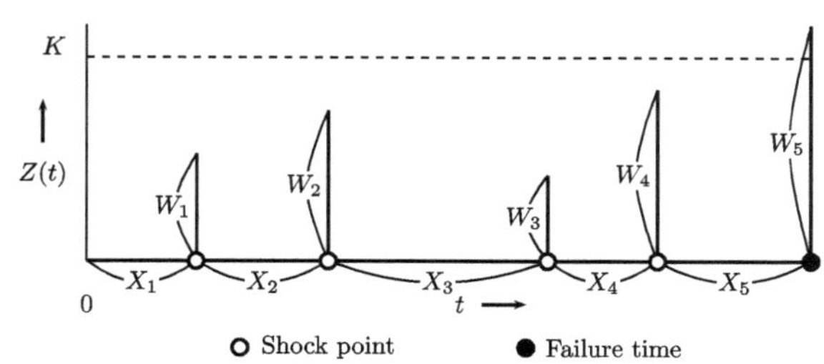

Fig. 6.2 Process for a standard cumulative damage model

and its LS transform is

$$\int_0^\infty \mathrm{e}^{-st}\,\mathrm{d}\Pr\{Y \le t\} = \frac{[1 - G(K)]F^*(s)}{1 - G(K)F^*(s)}. \tag{6.14}$$

Hence, the mean time to a critical point is

$$E\{Y\} = \frac{\mu}{1 - G(K)}. \tag{6.15}$$

If shocks occur in a nonhomogeneous Poisson process with a mean value function $H(t)$, then

$$\begin{aligned} \Pr\{Y \le t\} &= \sum_{n=0}^{\infty}\{1 - [G(K)]^n\}\frac{[H(t)]^n}{n!}\mathrm{e}^{-H(t)} \\ &= 1 - \mathrm{e}^{-[1-G(K)]H(t)}, \\ E\{Y\} &= \int_0^\infty \mathrm{e}^{-[1-G(K)]H(t)}\,\mathrm{d}t. \end{aligned} \tag{6.16}$$

If events occur at a constant time T,

$$\Pr\{Y \le t\} = 1 - [G(K)]^{[t/T]},$$

$$E\{Y\} = \int_0^\infty [G(K)]^{[t/T]}\mathrm{d}t.$$

Example 6.4 (Failure rate) It is important to investigate the properties of failure rates practically and theoretically. Let $\overline{P}_n$ denote the probability of surviving the first n shocks $(n = 0, 1, 2, \ldots)$, where $\overline{P}_0 \equiv 1$, and $F_n(t)$ be the probability that n shocks occur in time t, where $F_0(t) \equiv 1$. Then, the survival function that the unit does not fail in time t is written in the following general form: From (6.2),

$$\overline{\Phi}(t) = \Pr\{Y > t\} = \sum_{n=0}^{\infty}\overline{P}_n \Pr\{N(t) = n\} = \sum_{n=0}^{\infty}\overline{P}_n[F_n(t) - F_{n+1}(t)].$$

Let $f_n(t)$ be a density function of $F_n(t)$. The failure rate is

$$r(t) \equiv \frac{\Phi'(t)}{\overline{\Phi}(t)} = \frac{\sum_{n=0}^{\infty}(\overline{P}_n - \overline{P}_{n+1})f_{n+1}(t)}{\sum_{n=0}^{\infty}\overline{P}_n[F_n(t) - F_{n+1}(t)]}. \tag{6.17}$$

In particular, when shocks occur in a Poisson process with rate λ, i.e., $F_n(t) = \sum_{j=n}^{\infty}[(\lambda t)^j/j!]\mathrm{e}^{-\lambda t}$,

$$\overline{\Phi}(t) = \sum_{n=0}^{\infty} \overline{P}_n \frac{(\lambda t)^n}{n!}\mathrm{e}^{-\lambda t}. \tag{6.18}$$

The probabilistic properties of $\overline{\Phi}(t)$ were extensively investigated [6, 7]. The failure rate is, from (6.17),

$$r(t) = \lambda\left\{1 - \frac{\sum_{n=0}^{\infty}\overline{P}_{n+1}[(\lambda t)^n/n!]}{\sum_{n=0}^{\infty}\overline{P}_n[(\lambda t)^n/n!]}\right\} \le \lambda.$$

When $\overline{P}_n = q^n$, i.e., $q = G(K)$ and the total damage is not additive, from (6.18),

$$\overline{\Phi}(t) = \mathrm{e}^{-\lambda(1-q)t}, \quad r(t) = \lambda(1-q),$$

that is constant.

Any distribution $F(t)$ is said to have the property of IFR (Increasing Failure Rate) if and only if $[F(t+x) - F(t)]/\overline{F}(t)$ increases with t for $x > 0$ and $F(t) < 1$ [8, p. 23], where $\overline{F}(t) \equiv 1 - F(t)$. Furthermore, it has been proved that $F(t)$ is IFR if and only if $h(t) \equiv F'(t)/\overline{F}(t)$ increases with t. The failure rate has been discussed in Sect. 2.3. In this model, the following properties (i) and (ii) were proved [9]:

(i) The failure rate $r(t)$ in (6.17) increases with t if $(\overline{P}_n - \overline{P}_{n+1})/\overline{P}_n$ increases with n.

When $\overline{P}_{n+1}/\overline{P}_n$ decreases with n and $f_{n+1}(t) = [(\lambda t)^n/n!]\mathrm{e}^{-\lambda t}$, we prove that $r(t)$ increases with t. Differentiating $r(t)$ in (6.17) with respect to t,

$$\frac{\lambda^2}{[\overline{\Phi}(t)]^2}\left[\sum_{n=0}^{\infty}\frac{(\lambda t)^n}{n!}\sum_{k=0}^{\infty}(\overline{P}_{n+1}\overline{P}_{k+1} - \overline{P}_{n+2}\overline{P}_k)\right].$$

The bracket is

$$\sum_{n=0}^{\infty}\frac{(\lambda t)^n}{n!}\overline{P}_{n+1}\sum_{k=0}^{\infty}\frac{(\lambda t)^k}{k!}\overline{P}_k\left(\frac{\overline{P}_{k+1}}{\overline{P}_k} - \frac{\overline{P}_{n+1}}{\overline{P}_{n+1}}\right)$$

$$= \sum_{n=0}^{\infty}\frac{(\lambda t)^n}{n!}\overline{P}_{n+1}\left[\sum_{k=0}^{n+1}\frac{(\lambda t)^k}{k!}\overline{P}_k\left(\frac{\overline{P}_{k+1}}{\overline{P}_k} - \frac{\overline{P}_{n+2}}{\overline{P}_{n+1}}\right) + \sum_{k=n+1}^{\infty}\frac{(\lambda t)^k}{k!}\overline{P}_k\left(\frac{\overline{P}_{k+1}}{\overline{P}_k} - \frac{\overline{P}_{n+2}}{n+1}\right)\right]$$

$$= \sum_{n=0}^{\infty}\frac{(\lambda t)^n}{n!}\overline{P}_{n+1}\sum_{k=0}^{n+1}\frac{(\lambda t)^k}{k!}\overline{P}_k\left(\frac{\overline{P}_{k+1}}{\overline{P}_k} - \frac{\overline{P}_{n+2}}{\overline{P}_{n+1}}\right)$$

$$+ \sum_{n=0}^{\infty}\frac{(\lambda t)^{n+1}}{(n+1)!}\overline{P}_{n+1}\sum_{k=0}^{n+1}k\frac{(\lambda t)^{k-1}}{k!}\overline{P}_k\left(\frac{\overline{P}_{n+2}}{\overline{P}_{n+1}} - \frac{\overline{P}_{k+1}}{\overline{P}_k}\right)$$

$$= \sum_{n=0}^{\infty}\frac{(\lambda t)^n}{(n+1)!}\overline{P}_{n+1}\sum_{k=0}^{n+1}\frac{(\lambda t)^k}{k!}\overline{P}_k\left(\frac{\overline{P}_{k+1}}{\overline{P}_k} - \frac{\overline{P}_{n+2}}{\overline{P}_{n+1}}\right)(n+1-k) \ge 0,$$

that completes the proof.

In addition, when the total damage is additive, from (6.8),

$$\overline{\Phi}(t) = \sum_{n=0}^{\infty} G^{(n)}(K)\frac{(\lambda t)^n}{n!}\mathrm{e}^{-\lambda t}. \tag{6.19}$$

(ii) The failure rate average $\int_0^t r(u)\mathrm{d}u/t$ increases with t because $[G^{(n)}(x)]^{1/n}$ decreases with n. Note that if $r(t)$ increases, then $\int_0^t r(u)\mathrm{d}u/t$ also increases. In particular, when $\overline{P}_n = G^{(n)}(K) = \sum_{j=n}^{\infty}((\theta K)^j/j!)\mathrm{e}^{-\theta K}, \overline{P}_{n+1}/\overline{P}_n$ decreases strictly with n from $(1-\mathrm{e}^{-\theta K})$ to 0. Thus, the failure rate $r(t)$ increases strictly from $\lambda \mathrm{e}^{-\theta K}$ to λ (Example 6.5).

In general, for the cumulative damage model with a failure level K, the failure rate is, setting that $\overline{P}_n = G^{(n)}(K)$ and $F_n(t) = F^{(n)}(t)$ in (6.17),

$$r(t) \equiv \frac{\sum_{n=0}^{\infty}[G^{(n)}(K) - G^{(n+1)}(K)]f^{(n+1)}(t)}{\sum_{n=0}^{\infty} G^{(n)}(K)[F^{(n)}(t) - F^{(n+1)}(t)]}. \tag{6.20}$$

When shocks occur in a nonhomogeneous Poisson process with an intensity function $h(t)$ and a mean value function $H(t)$ [7], from (6.12),

$$\overline{\Phi}(t) = \sum_{n=0}^{\infty} \overline{P}_n \frac{[H(t)]^n}{n!}\mathrm{e}^{-H(t)}. \tag{6.21}$$

(iii) The failure rate $r(t)$ increases if both $h(t)$ and $(\overline{P}_n - \overline{P}_{n+1})/\overline{P}_n$ increase.
(iv) The failure rate average $\int_0^t r(u)\mathrm{d}u/t$ increases if both $r(t)/t$ and $(\overline{P}_n - \overline{P}_{n+1})/\overline{P}_n$ increase.

When the total damage is additive, (6.19) is

$$\overline{\Phi}(t) = \sum_{n=0}^{\infty} G^{(n)}(K)\frac{[H(t)]^n}{n!}\mathrm{e}^{-H(t)}. \tag{6.22}$$

Then, properties (iii) and (iv) are rewritten as (Problem 6.5):

(v) The failure rate $r(t)$ increases if both $h(t)$ and $[G^{(n)}(K) - G^{(n+1)}(K)]/G^{(n)}(K)$ increase.
(vi) The failure rate average $\int_0^t r(u)\mathrm{d}u/t$ increases if both $H(t)/t$ and $[G^{(n)}(K) - G^{(n+1)}(K)]/G^{(n)}(K)$ increase.

Such results were compactly summarized [10]. Moreover, when shocks occur in the birth process [11], in the counting process [12], and in the Lévy process [13, 14], similar results were obtained.

Example 6.5 (Exponential damage and shock times) Suppose that $F(t) = 1 - \mathrm{e}^{-\lambda t}$ and $G(x) = 1 - \mathrm{e}^{-\theta x}$, i.e., shocks occur in a Poisson process with rate λ and each damage due to shocks is exponential with mean $1/\theta$. In this case, both nonhomogeneous Poisson and renewal processes form the same Poisson process, i.e.,

$$F_n(t) = F^{(n)}(t) = \sum_{k=n}^{\infty} \frac{[H(t)]^k}{k!} e^{-H(t)} = \sum_{k=n}^{\infty} \frac{(\lambda t)^k}{k!} e^{-\lambda t} \quad (n = 0, 1, 2, \ldots).$$

In the cumulative damage model, from (6.2),

$$\int_0^{\infty} e^{-sx} \, d\Pr\{Z(t) \le x\} = e^{-\lambda t s/(s+\theta)}.$$

By inversion [8, p. 80],

$$\Pr\{Z(t) \le x\} = e^{-\lambda t} \left[1 + \sqrt{\lambda \theta t} \int_0^x e^{-\theta u} u^{-1/2} I_1(2\sqrt{\lambda \theta t u}) du \right],$$

where $I_i(x)$ is the Bessel function of order i for the imaginary argument defined by

$$I_i(x) \equiv \sum_{j=0}^{\infty} \left(\frac{x}{2}\right)^{2j+i} \frac{1}{j!(j+i)!}.$$

Thus, from (6.8),

$$\Phi(t) = \Pr\{Y \le t\} = 1 - e^{-\lambda t} \left[1 + \sqrt{\lambda \theta t} \int_0^K e^{-\theta u} u^{-1/2} I_1(2\sqrt{\lambda \theta t u}) du \right].$$

Furthermore, from (6.4), (6.6), and (6.10),

$$E\{Z(t)\} = \frac{\lambda t}{\theta}, \quad V\{Z(t)\} = \frac{2\lambda t}{\theta^2}, \quad E\{Y\} = \frac{\theta K + 1}{\lambda}.$$

The failure rate is, from (6.17),

$$r(t) = \frac{\lambda e^{-\lambda t - \theta K} I_0(2\sqrt{\lambda \theta t K})}{1 + \sqrt{\lambda \theta t} \int_0^K e^{-\theta u} u^{-1/2} \, I_1(2\sqrt{\lambda \theta t u}) du}.$$

Because

$$\frac{G^{(n)}(K) - G^{(n+1)}(K)}{G^{(n)}(K)} = \frac{(\theta K)^n / n!}{\sum_{j=n}^{\infty} (\theta K)^j / j!}$$

increases with n from $e^{-\theta K}$ to 1 (Problem 6.6), $r(t)$ increases from $\lambda e^{-\theta K}$ to λ.

In the independent damage model, from (6.13),

$$\Pr\{Y \le t\} = 1 - \exp\left(-\lambda t e^{-\theta K}\right),$$

and from (6.15),

$$E\{Y\} = \frac{1}{r(t)} = \frac{1}{\lambda}e^{\theta K}.$$

Example 6.6 (Replacement policy) Consider a replacement policy with two kinds of damage: Shocks occur in a renewal process according to $F(t)$ with finite mean μ, and the amount W_k of damage 1 due to the kth shock has an identical distribution $G(x) \equiv \Pr\{W_k \leq x\}$ with finite mean ω. Each damage 1 is additive, and the unit fails when the total damage exceeds a failure level K. Suppose that the unit is replaced at time $T\,(0<T\leq\infty)$ or at failure, whichever occurs first. Then, the probability that the unit is replaced before failure at time T is, from (6.8),

$$\overline{\Phi}(T) = \sum_{n=0}^{\infty} G^{(n)}(K)[F^{(n)}(T) - F^{(n+1)}(T)],$$

and the probability that it is replaced at failure is

$$\Phi(T) = \sum_{n=0}^{\infty}[G^{(n)}(K) - G^{(n+1)}(K)]F^{(n+1)}(T).$$

Thus, the mean time to replacement is

$$T\overline{\Phi}(T) + \int_0^T t\mathrm{d}\Phi(t) = \sum_{n=0}^{\infty} G^{(n)}(K)\int_0^T [F^{(n)}(t) - F^{(n+1)}(t)]\mathrm{d}t.$$

Introduce the replacement costs: c_K is the replacement cost at failure and c_T is the replacement cost at time T, where $c_K > c_T$. Then, from (3.19), the expected cost rate is

$$\widetilde{C}(T) = \frac{c_K - (c_K - c_T)\sum_{n=0}^{\infty} G^{(n)}(K)[F^{(n)}(T) - F^{(n+1)}(T)]}{\sum_{n=0}^{\infty} G^{(n)}(K)\int_0^T [F^{(n)}(t) - F^{(n+1)}(t)]\mathrm{d}t}. \tag{6.23}$$

Next, that another damage 2 occurs in a non homogeneous Poisson process with an intensity function $h(t)$ and a mean value function $H(t) \equiv \int_0^t h(u)\mathrm{d}u$, i.e., the probability of n occurrences of damage 2 in $[0, t]$ is given in (2.43). It is assumed that damage 2 occurs independently for damage 1 and its damage is not additive. When damage 2 occurs, the unit undergoes only minimal repair and its amount of damage 2 becomes zero. Thus, the expected number of occurrences of damage 2, i.e., minimal repair, before the replacement is, from (2.45),

$$H(T)\sum_{n=0}^{\infty}G^{(n)}(K)[F^{(n)}(T)-F^{(n+1)}(T)]+\sum_{n=0}^{\infty}[G^{(n)}(K)-G^{(n+1)}(K)]\int_0^T H(t)\mathrm{d}F^{(n+1)}(t)$$

$$=\sum_{n=0}^{\infty}G^{(n)}(K)\int_0^T[F^{(n)}(t)-F^{(n+1)}(t)]h(t)\mathrm{d}t.$$

Therefore, letting c_M be the minimal repair cost and adding the total minimal repair cost before replacement to $\widetilde{C}(T)$ in (6.23),

$$C(T)=\frac{\begin{array}{c}c_K-(c_K-c_T)\sum_{n=0}^{\infty}G^{(n)}(K)[F^{(n)}(T)-F^{(n+1)}(T)]\\ +c_M\sum_{n=0}^{\infty}G^{(n)}(K)\int_0^T[F^{(n)}(t)-F^{(n+1)}(t)]h(t)\mathrm{d}t\end{array}}{\sum_{n=0}^{\infty}G^{(n)}(K)\int_0^T[F^{(n)}(t)-F^{(n+1)}(t)]\mathrm{d}t}. \tag{6.24}$$

Clearly,

$$C(0)\equiv\lim_{T\to 0}C(T)=\infty,$$

$$C(\infty)\equiv\lim_{T\to\infty}C(T)=\frac{c_K+c_M\sum_{n=0}^{\infty}G^{(n)}(K)\int_0^{\infty}[F^{(n)}(t)-F^{(n+1)}(t)]h(t)\mathrm{d}t}{\mu[1+M_G(K)]},$$

whose denominator represents the mean time to failure in (6.10). Thus, there exists a positive T^* $(0<T^*\leq\infty)$ that minimizes $C(T)$ in (6.24).

We find an optimum time T^* that minimizes $C(T)$. Differentiating $C(T)$ with respect to T and setting it equal to zero,

$$\begin{aligned}(c_K-c_T)&\left\{r(T)\sum_{n=0}^{\infty}G^{(n)}(K)\int_0^T[F^{(n)}(t)-F^{(n+1)}(t)]\mathrm{d}t\right.\\ &\left.-\sum_{n=0}^{\infty}[G^{(n)}(K)-G^{(n+1)}(K)]F^{(n+1)}(T)\right\}\\ &+c_M\sum_{n=0}^{\infty}G^{(n)}(K)\int_0^T[F^{(n)}(t)-F^{(n+1)}(t)][h(T)-h(t)]\mathrm{d}t=c_T,\end{aligned} \tag{6.25}$$

where $r(T)$ is given in (6.20). It can be clearly seen that if $r(t)$ increases strictly and $h(t)$ increases, or $r(t)$ increases and $h(t)$ increases strictly (Problem 6.7), then the left-hand side of (6.25) increases strictly from 0 to

$$(c_K - c_T)\{\mu r(\infty)[1 + M_G(K)] - 1\}$$
$$+ c_M\left\{\mu h(\infty)[1 + M_G(K)] - \sum_{n=0}^{\infty} G^{(n)}(K) \int_0^{\infty} [F^{(n)}(t) - F^{(n+1)}(t)]h(t)\mathrm{d}t\right\}. \tag{6.26}$$

Thus, if (6.26) is greater than c_T, then there exists a finite and unique T^* $(0 < T^* < \infty)$ that satisfies (6.25).

When $H(t) = \alpha t$, i.e., $h(t) = \alpha$ $(\alpha > 0)$, from (6.26), if $\mu r(\infty)[1 + M_G(K)] > c_K/(c_K - c_T)$ then there exists a finite T^* that satisfies (6.25). In addition, when $F(t) = 1 - \mathrm{e}^{-\lambda t}$ and $G^{(n)}(K) = \sum_{j=n}^{\infty}[(\theta K)^j/j!]\mathrm{e}^{-\theta K}$,

$$r(T) = \frac{\lambda \sum_{n=0}^{\infty}[G^{(n)}(K) - G^{(n+1)}(K)][(\lambda T)^n/n!]}{\sum_{n=0}^{\infty} G^{(n)}(K)[(\lambda T)^n/n!]}$$

increases strictly from $\lambda \mathrm{e}^{-\theta K}$ to λ, as shown in Example 6.5. Thus, if $\theta K > c_T/(c_K - c_T)$, then there exists a finite and unique T^* $(0 < T^* < \infty)$ that satisfies (6.25).

6.3 Modified Damage Models

We give the following four modified damage models [2, p. 28]: Damage model with imperfect shock where some shock may produce no damage, a failure level K is a random variable with a general distribution $L(x)$, the total damage decreases exponentially with time, and the total damage increases with time. A variety of shock models were widely summarized [14].

6.3.1 Imperfect Shock

It has been assumed that shocks occur in a renewal process with $F(t)$ and each damage due to shocks is distributed generally with $G(x)$. However, it may be considered the case where some shocks do not produce any damage.

Suppose that each shock does some damage with probability p $(0 < p \le 1)$ and no damage with probability $q \equiv 1 - p$. Then, the distribution of time where some shocks does the damage in time t is

$$F_1(t) \equiv [1 + qF(t) + qF(t) * qF(t) + \cdots] * pF(t),$$

and its LS transform is

$$F_1^*(s) = \frac{pF^*(s)}{1 - qF^*(s)}. \tag{6.27}$$

Hence, the mean time to the first damage due to shocks is

$$E\{Y_1\} \equiv \int_0^\infty t \mathrm{d}F_1(t) = \frac{\mu}{p}.$$

Thus, by replacing $F(t)$ with $F_1(t)$, we can get the results in the cases where shocks are imperfect. In particular, when $F(t) = 1 - \mathrm{e}^{-\lambda t}, F_1(t) = 1 - \mathrm{e}^{-p\lambda t}$ and $E\{Y_1\} = 1/(p\lambda)$, and

$$\Pr\{N(t) = n\} = \frac{(p\lambda t)^n}{n!}\mathrm{e}^{-p\lambda t} \quad (n = 0, 1, 2, \ldots).$$

Similarly, when shocks occur in a non homogeneous Poisson process with a mean value function $H(t)$,

$$\Pr\{N(t) = n\} = \frac{[pH(t)]^n}{n!}\mathrm{e}^{-pH(t)} \quad (n = 0, 1, 2, \ldots),$$
$$E\{N(t)\} = V\{N(t)\} = pH(t).$$

Therefore, substituting $F_1(t)$ in (6.27) for $F(t)$ in (6.2), (6.4), (6.8), (6.10) and (6.20), $\Pr\{Z(t) \le x\}, E\{Z(t)\}, \Pr\{Y \le t\}, E\{Y\}$, and $r(t)$ are given. In particular, from (6.9) and (6.10),

$$\int_0^\infty \mathrm{e}^{-st}\,\mathrm{d}\Pr\{Y \le t\} = \sum_{n=0}^\infty [G^{(n)}(K) - G^{(n+1)}(K)]\left[\frac{pF^*(s)}{1 - qF^*(s)}\right]^{n+1},$$
$$E\{Y\} = \frac{\mu}{p}[1 + M_G(K)].$$

The corresponding results for the independent damage model are, from (6.14) and (6.15),

$$\int_0^\infty \mathrm{e}^{-st}\,\mathrm{d}\Pr\{Y \le t\} = \frac{p[1 - G(K)]F^*(s)}{1 - [q + pG(K)]F^*(s)},$$
$$E\{Y\} = \frac{\mu}{p[1 - G(K)]}.$$

Example 6.7 (Parallel system) Consider a parallel redundant system with n identical units, each of which fails at shocks with probability p, where $q \equiv 1 - p$, and shocks occur in a renewal process with mean interval μ (Problem 6.8). Let W_k be the total number of units that fail at the kth ($k = 1, 2, \ldots$) shock. Then, because the probability that one unit fails until the kth shock is $\sum_{j=1}^k pq^{j-1} = 1 - q^k$, the mean time to system failure is [2, p. 62]

$$\sum_{k=1}^{\infty} k\mu \Pr\{W_1 + W_2 + \cdots + W_{k-1} \le n-1 \text{ and } W_1 + \cdots + W_k = n\}$$
$$= \sum_{k=1}^{\infty} k\mu[(1-q^k)^n - (1-q^{k-1})^n]$$
$$= \mu \sum_{k=0}^{\infty}[1-(1-q^k)^n] = \mu \sum_{k=1}^{n} \binom{n}{k} (-1)^{k+1} \frac{1}{1-q^k}.$$

6.3.2 Random Failure Level

Most units have individual variations in their ability to withstand shocks and are operating in different environments. In such cases, a failure level K is not constant and would be random. Consider the case where a failure level K is a random variable with a general distribution $L(x)$ such that $L(0) = 0$. Then, for the cumulative damage model, the distribution of time to failure is, from (6.8),

$$\Pr\{Y \le t\} = \sum_{n=0}^{\infty} F^{(n+1)}(t) \int_0^{\infty} [G^{(n)}(x) - G^{(n+1)}(x)]\mathrm{d}L(x), \tag{6.28}$$

and its mean time is, from (6.10),

$$E\{Y\} = \mu \int_0^{\infty} [1 + M_G(x)]\mathrm{d}L(x). \tag{6.29}$$

The failure rate is, from (6.20),

$$r(t) = \frac{\sum_{n=0}^{\infty} f^{(n+1)}(t) \int_0^{\infty} [G^{(n)}(x) - G^{(n+1)}(x)]\mathrm{d}L(x)}{\sum_{n=0}^{\infty} [F^{(n)}(t) - F^{(n+1)}(t)] \int_0^{\infty} G^{(n)}(x)\mathrm{d}L(x)}. \tag{6.30}$$

For the independent damage model,

$$\begin{aligned} \Pr\{Y \le t\} &= \sum_{n=0}^{\infty} F^{(n+1)}(t) \int_0^{\infty} \{[G(x)]^n - [G(x)]^{n+1}\}\mathrm{d}L(x), \\ E\{Y\} &= \mu \sum_{n=0}^{\infty} \int_0^{\infty} [G(x)]^n \mathrm{d}L(x). \end{aligned} \tag{6.31}$$

For the cumulative model with imperfect shock,

$$\int_0^\infty \mathrm{e}^{-st}\,\mathrm{d}\Pr\{Y\le t\} = \sum_{n=0}^\infty \left[\frac{pF^*(s)}{1-qF^*(s)}\right]^{n+1} \int_0^\infty [G^{(n)}(x) - G^{(n+1)}(x)]\mathrm{d}L(x). \quad (6.32)$$

Example 6.8 (Random failure level) Suppose that all random variables are exponential. When $F(t) = 1 - \mathrm{e}^{-\lambda t}, F_1^*(s) = p\lambda/(s + p\lambda)$, i.e., $F_1(t) = 1 - \mathrm{e}^{-p\lambda t}$ by inversion. Thus, substituting λ for $p\lambda$, we can obtain the corresponding results.

When $G(x) = 1 - \mathrm{e}^{-\theta x}$ and a failure level K also has an exponential distribution $L(x) = 1 - \mathrm{e}^{-\beta x}$,

$$\int_0^\infty [G^{(n)}(x) - G^{(n+1)}(x)]\mathrm{d}L(x) = \frac{\beta\theta^n}{(\beta+\theta)^{n+1}}.$$

Thus, from (6.28),

$$\int_0^\infty \mathrm{e}^{-st}\mathrm{d}\Pr\{Y\le t\} = \sum_{n=0}^\infty \left(\frac{\lambda}{s+\lambda}\right)^{n+1} \frac{\beta\theta^n}{(\beta+\theta)^{n+1}} = \frac{\lambda\beta}{s(\beta+\theta)+\lambda\beta}.$$

By inversion,

$$\Pr\{Y\le t\} = 1 - \exp\left(-\frac{\lambda\beta t}{\beta+\theta}\right),$$

$$E\{Y\} = \frac{1}{r(t)} = \frac{1}{\lambda}\left(\frac{\theta}{\beta}+1\right).$$

For the independent damage model,

$$\Pr\{Y > t\} = \int_0^\infty \exp(-\lambda t\mathrm{e}^{-\theta x})\beta\mathrm{e}^{-\beta x}\mathrm{d}x$$

$$= \sum_{n=0}^\infty \frac{(-\lambda t)^n}{n!}\int_0^\infty \beta\mathrm{e}^{-(\beta+n\theta)x}\mathrm{d}x = \sum_{n=0}^\infty \frac{(-\lambda t)^n}{n!}\frac{\beta}{\beta+n\theta},$$

$$E\{Y\} = \frac{1}{r(t)} = \frac{1}{\lambda}\int_0^\infty \mathrm{e}^{\theta x}\beta\mathrm{e}^{-\beta x}\mathrm{d}x = \begin{cases} \frac{\beta}{\lambda(\beta-\theta)} & (\beta>\theta), \\ \infty & (\beta\le\theta). \end{cases}$$

6.3.3 Damage with Annealing

The total damage in the usual reliability models is additive and does not decrease. In some materials, annealing, i.e., lessening the damage, can take place in materials such as rubber, fiber reinforced plastics, and polyurethane.

Takaćs [15] considered the following damage model: If a unit suffers some damage W due to shock, then its damage after time duration t is reduced to $We^{-\alpha t}$ $(0<\alpha<\infty)$. Define

$$Z(t) = \sum_{n=1}^{N(t)} W_n \exp[-\alpha(t - S_n)], \tag{6.33}$$

where $S_n \equiv \sum_{k=1}^{n} X_k$ (Fig. 6.3).

Suppose that shocks occur in a Poisson process with rate λ. Then, $\Phi(t,x) \equiv \Pr\{Z(t) \le x\}$ forms the following renewal equation [15, p. 105]

$$\frac{\mathrm{d}\Phi(t,x)}{\mathrm{d}t} = -\lambda\left\{\Phi(t,x) - \int_0^x G[(x-y)e^{-\alpha t}]\mathrm{d}_y\Phi(t,y)\right\},$$

and its LS transform is

$$\frac{\mathrm{d}\Phi^*(t,s)}{\mathrm{d}t} = -\lambda[1 - G^*(se^{-\alpha t})]\Phi^*(t,s),$$

where $\Phi^*(t,s) \equiv \int_0^\infty e^{-sx}\mathrm{d}\Phi(t,x)$ and $G^*(s) \equiv \int_0^\infty e^{-sx}\mathrm{d}G(x)$. Solving the differential equation,

$$\begin{aligned} \Phi^*(t,s) &= \exp\left\{-\lambda\int_0^t [1 - G^*(se^{-\alpha u})]\mathrm{d}u\right\}, \\ E\{Z(t)\} &= -\left.\frac{\mathrm{d}\Phi^*(t,s)}{\mathrm{d}s}\right|_{s=0} = \frac{\lambda\omega(1-e^{-\alpha t})}{\alpha}. \end{aligned} \tag{6.34}$$

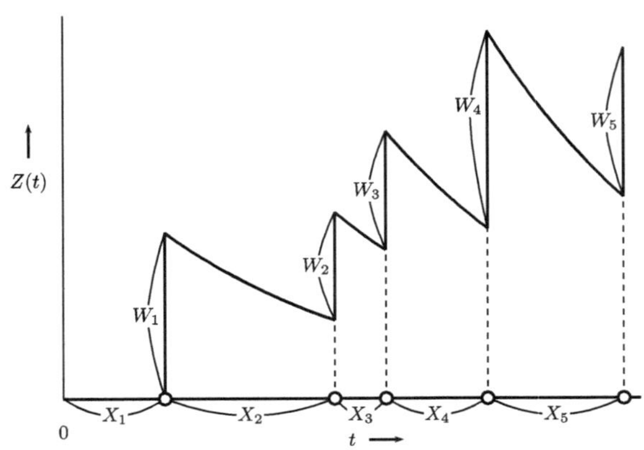

Fig. 6.3 Process for a cumulative damage model with annealing

In addition, when $\omega < \infty$, $\lim_{t\to\infty} \Pr\{Z(t) \le x\}$ exits and its LS transform is

$$\Phi^*(\infty, s) = \exp\left\{ -\frac{\lambda}{\alpha} \int_0^1 \frac{1 - G^*(su)}{u} \mathrm{d}u \right\}. \tag{6.35}$$

We derive (6.34) by another method: When shocks occur in a Poisson process, from (2.29) and (2.37), unordered random variables of n shock times $S_1, S_2, \ldots, S_n$, given that $N(t) = n$, are independent and identically distributed uniformly over the interval $[0, t]$, i.e., $t - S_k$ $(k = 1, 2, \ldots)$ are also distributed uniformly over $[0, t]$. Thus, when $N(t) = 1, Z(t) = W_1 \exp[-\alpha(t - S_1)]$ and its LS transform is

$$\int_0^\infty \mathrm{e}^{-sx} \mathrm{d}\Pr\{Z(t) \le x\} = E\{G^*[s \exp(-\alpha S_1)]\} = \frac{1}{t} \int_0^t G^*(s\mathrm{e}^{-\alpha u}) \mathrm{d}u.$$

Generally, when $N(t) = n$,

$$\int_0^\infty \mathrm{e}^{-sx} \mathrm{d}\Pr\{Z(t) \le x\} = (E\{G^*[s \exp(-\alpha S_1)]\})^n = \left[\frac{1}{t} \int_0^t G^*(s\mathrm{e}^{-\alpha u}) \mathrm{d}u \right]^n.$$

So that, using the law of total probability,

$$\begin{aligned}
\int_0^\infty \mathrm{e}^{-st} \mathrm{d}\Pr\{Z(t) \le x\} &= \sum_{n=0}^\infty \frac{(\lambda t)^n}{n!} \mathrm{e}^{-\lambda t} \left[\frac{1}{t} \int_0^t G^*(s\mathrm{e}^{-\alpha u}) \mathrm{d}u \right]^n \\
&= \exp\left\{ -\lambda \int_0^t [1 - G^*(s\mathrm{e}^{-\alpha u})] \mathrm{d}u \right\}.
\end{aligned}$$

Furthermore, (6.34) is also derived directly as follows [16, p. 40]:

$$\begin{aligned}
E\{Z(t)|N(t) = n\} &= E\left\{ \sum_{k=1}^{N(t)} W_k \exp[-\alpha(t - S_k)] \middle| N(t) = n \right\} \\
&= \sum_{k=1}^n E\{W_k \exp[-\alpha(t - S_k)] | N(t) = n\} \\
&= \sum_{k=1}^n E\{W_k | N(t) = n\} E\{\exp[-\alpha(t - S_k)] | N(t) = n\} \\
&= E\{W_k\} \mathrm{e}^{-\alpha t} \sum_{k=1}^n E\{\exp(\alpha S_k) | N(t) = n\}.
\end{aligned}$$

Recalling that S_k is distributed uniformly over $[0, t]$,

$$E\{Z(t)|N(t) = n\} = \omega e^{-\alpha t} E\left\{\sum_{k=1}^{n} \exp(\alpha S_k)\right\}$$
$$= \omega e^{-\alpha t}\frac{n}{t}\int_0^t e^{\alpha x}dx = \frac{\omega n}{\alpha t}(1 - e^{-\alpha t}).$$

So that, because $E\{N(t)\} = \lambda t$,

$$E\{Z(t)\} = E\left\{\frac{\omega N(t)}{\alpha t}(1 - e^{-\alpha t})\right\} = \frac{\lambda\omega}{\alpha}(1 - e^{-\alpha t}).$$

Example 6.9

(1) When $G(x) = 1 - e^{-\theta x}$,

$$\Phi^*(t, s) = \left(\frac{s + \theta e^{\alpha t}}{s + \theta}\right)^{\nu} e^{-\lambda t},$$

where $\nu \equiv \lambda/\alpha$. Thus, by inversion [2, p. 32],

$$\Pr\{Z(t) \le x\} = e^{-\lambda t}\sum_{j=0}^{\infty}\binom{\nu + j - 1}{j}(1 - e^{-\alpha t})^j \sum_{i=j}^{\infty}\frac{(\theta x e^{\alpha t})^i}{i!}\exp(-\theta x e^{\alpha t}).$$

In a similar way,

$$\Phi^*(\infty, s) = \left(\frac{\theta}{s + \theta}\right)^{\nu}, \quad \lim_{t\to\infty}\Pr\{Z(t) \le x\} = \int_0^x \frac{\theta(\theta u)^{\nu-1}}{\Gamma(\nu)}e^{-\theta u}du,$$

that is a gamma distribution with mean ν/θ.

(2) When $G(x) \equiv 0$ for $x < \omega$ and 1 for $x \ge \omega$, i.e., the damage due to each shock is constant and its amount is ω. From the results [15, p. 129],

$$\Phi^*(\infty, s) = \left(\frac{1}{s\omega\gamma}\right)^{\nu}\exp\left(-\nu\int_{\omega}^{\infty}\frac{e^{-su}}{u}du\right),$$

where $\gamma \equiv e^C = 1.781072\ldots$ and $C \equiv 0.577215\ldots$ that is Euler's constant. By inversion,

$$\lim_{t\to\infty}\Pr\{Z(t) \le x\} = \frac{x^{\nu} + \sum_{j=1}^{\infty}\left[(-1)^j(\nu)^j/j!\right]\int_{j\omega}^{x}(x - u)^{\nu}I^{(j)}(u)du}{(\omega\gamma)^{\nu}\Gamma(1 + \nu)},$$

where $I(u)$ is uniform over $[0, \omega]$.

6.3.4 Increasing Damage with Time

Consider the cumulative damage model with two kinds of damage (Fig. 6.4). One of them is caused by shock and is additive, and the other increases proportionately with time. That is, the total damage is accumulated subject to shocks and time at the rate of constant α $(\alpha > 0)$, independent of shocks. A unit fails whether the total damage exceeds with time or has exceeded a failure level K at some shocks, and its failure is detected only at the time of shocks. Such models would be the life of dry and storage batteries. The replacement policy where the unit is replaced before failure will be discussed in Example 7.4 of Chap. 7.

Suppose that $S_n \equiv \sum_{k=1}^{n} X_k$ and $Z_n \equiv \sum_{k=1}^{n} W_k$, where $S_0 = W_0 = 0$. Because $\Pr\{S_n \le t\} = F^{(n)}(t)$ and $\Pr\{Z_n \le x\} = G^{(n)}(x)$, the distribution of time to detect a failure at some shock is [2, p. 35],

$$\Pr\{Y \le t\} = \sum_{n=0}^{\infty} \Pr\{Z_n + \alpha S_n < K \le Z_{n+1} + \alpha S_{n+1}, S_{n+1} \le t\}$$
$$= \sum_{n=0}^{\infty} \int_0^t \left\{ \int_0^{t-u} [G^{(n)}(K - \alpha u) - G^{(n+1)}(K - \alpha(u + x))] \mathrm{d}F(x) \right\} \mathrm{d}F^{(n)}(u), \tag{6.36}$$

where note that $G^{(n)}(x) = 0$ for $x < 0$. Thus, the mean time to detect a failure at some shock is

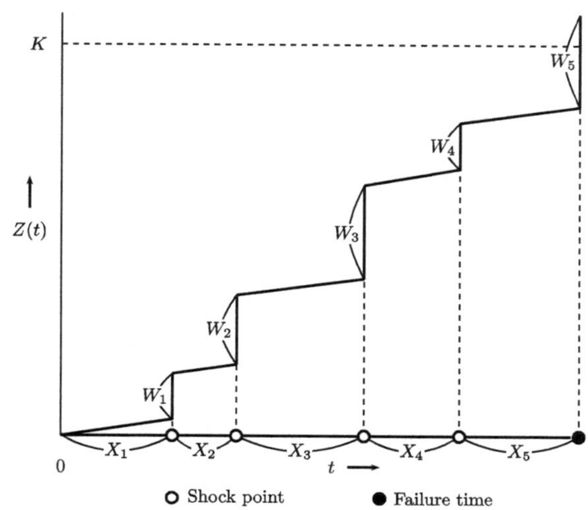

Fig. 6.4 Process for a cumulative damage model with two kinds of damages

$$E\{Y\} = \sum_{n=0}^{\infty} \int_0^{\infty} \left\{ \int_0^{\infty} (t+x) \left[G^{(n)}(K - \alpha t) - G^{(n+1)}(K - \alpha(t+x)) \right] \mathrm{d}F(x) \right\} \mathrm{d}F^{(n)}(t)$$
$$= \mu \sum_{n=0}^{\infty} \int_0^{K/\alpha} G^{(n)}(K - \alpha t) \mathrm{d}F^{(n)}(t). \tag{6.37}$$

Finally, if the total damage increases exponentially, i.e.,

$$Z(t) = \sum_{n=1}^{N(t)} W_n \exp[\alpha(t - S_n)], \tag{6.38}$$

then by arguments similar to those of Sect. 6.3.3, when $F(t) = 1 - \mathrm{e}^{-\lambda t}$,

$$\Phi^*(t,s) = \exp\left\{ -\lambda \int_0^t [1 - G^*(s\mathrm{e}^{\alpha u})] \mathrm{d}u \right\},$$
$$E\{Z(t)\} = \frac{\lambda \omega (\mathrm{e}^{\alpha t} - 1)}{\alpha}, \tag{6.39}$$
$$\Phi^*(\infty, s) = \exp\left[-\frac{\lambda}{\alpha} \int_1^{\infty} \frac{1 - G^*(su)}{u} \mathrm{d}u \right].$$

6.4 Replacement Models

Suppose that shocks occur in a renewal process with distribution $F(t)$ and each causes a random amount of damage with distribution $G(x)$ to a unit. These damages are additive, and the unit fails when the total damage has exceeded a failure level K. When the failure during actual operation may be costly or dangerous, it is of great importance to avoid such terrible situation. It would be wise to exchange the unit before failure at a lower cost. If we have no information on the condition of the unit, its maintenance should be done at a planned time. On the other hand, if we could get the number of shocks up to now and the amount of damage at shock times, its maintenance should be done at a pre-specified number of shocks or at a damage level. A variety of replacement models subjected to shocks were summarized [2].

This section summarizes the replacement policies for cumulative damage models using the theories of cumulative processes: In Sect. 6.4.1, the unit is replaced before failure at a planned time T, at a shock number N, or a damage level Z, whichever occurs first. Introducing the respective replacement costs for T, N and Z, we obtain the expected cost rate $C(T, N, Z)$. In Sect. 6.4.2, we derive analytically optimum T^*, N^* and Z^* that minimize the expected cost rates for three policies. Furthermore, we discuss optimum N^* and Z^* that minimize the expected cost rate $C(N,Z)$ and give a numerical example. Finally, we consider the periodic replacement with minimal repair at failures for a cumulative damage model and obtain the expected cost rate.

6.4.1 Three Replacement Policies

As per the preventive replacement policies, the unit is replaced before failure at time T $(0<T\leq\infty)$, at shock number N $(N=1,2,\ldots)$, or at damage level Z $(0\leq Z\leq K)$, whichever occurs first. In addition, it is assumed that the unit is replaced at K or Z without considering to replace it at N when the total damage has exceeded K or Z at shock N.

The probability that the unit is replaced at time T is

$$P_T=\sum_{n=0}^{N-1}\left[F^{(n)}(T)-F^{(n+1)}(T)\right]G^{(n)}(Z),$$

the probability that it is replaced at shock N is

$$P_N=F^{(N)}(T)G^{(N)}(Z),$$

the probability that it is replaced at damage Z is

$$P_Z=\sum_{n=0}^{N-1}F^{(n+1)}(T)\int_0^Z[G(K-x)-G(Z-x)]\mathrm{d}G^{(n)}(x),$$

and the probability that it is replaced at a failure level K is

$$P_K=\sum_{n=0}^{N-1}F^{(n+1)}(T)\int_0^Z\overline{G}(K-x)\mathrm{d}G^{(n)}(x),$$

where note that $P_T+P_N+P_Z+P_K=1$ (Problem 6.9). Thus, the mean time to replacement is

$$\begin{aligned}E\{Y\}&=T\sum_{n=0}^{N-1}\left[F^{(n)}(T)-F^{(n+1)}(T)\right]G^{(n)}(Z)+G^{(N)}(Z)\int_0^T t\,\mathrm{d}F^{(N)}(t)\\&\quad+\sum_{n=0}^{N-1}\int_0^T t\,\mathrm{d}F^{(n+1)}(t)\int_0^Z[G(K-x)-G(Z-x)]\mathrm{d}G^{(n)}(x)\\&\quad+\sum_{n=0}^{N-1}\int_0^T t\,\mathrm{d}F^{(n+1)}(t)\int_0^Z\overline{G}(K-x)\mathrm{d}G^{(n)}(x)\\&=\sum_{n=0}^{N-1}G^{(n)}(Z)\int_0^T\left[F^{(n)}(t)-F^{(n+1)}(t)\right]\mathrm{d}t.\end{aligned}$$

For the above replacement model, we introduce the following replacement costs: Costs c_T, c_N, c_Z, and c_K are the respective replacement costs at time T, at shock N, at damage Z, and at level K, where c_K is higher than the three costs c_T, c_N and c_Z. Then, the expected cost rate is, from (3.19) in Sect. 3.2,

$$C(T,N,Z) = \frac{c_T P_T + c_N P_N + c_Z P_Z + c_K P_K}{E\{Y\}}$$

$$= \frac{\begin{array}{l} c_K - (c_K - c_T)\sum_{n=0}^{N-1}\left[F^{(n)}(T) - F^{(n+1)}(T)\right]G^{(n)}(Z) \\ -(c_K - c_N)F^{(n)}(T)G^{(N)}(Z) \\ -(c_K - c_Z)\sum_{n=0}^{N-1} F^{(n+1)}(T)\int_0^Z [G(K-x) - G(Z-x)]\mathrm{d}G^{(n)}(x) \end{array}}{\sum_{n=0}^{N-1} G^{(n)}(Z)\int_0^T [F^{(n)}(t) - F^{(n+1)}(t)]\mathrm{d}t}. \tag{6.40}$$

When the unit is replaced only after failure, the expected cost rate is, from (6.10),

$$C \equiv \lim_{\substack{T\to\infty \\ N\to\infty \\ Z\to K}} C(T,N,Z) = \frac{c_K}{\mu[1 + M_G(K)]}. \tag{6.41}$$

The unit might be working as long as possible in the case where the replacement cost after failure is not so high. In this case, the unit should be replaced before failure at time T, shock number N, or damage level Z, whichever occurs *last*. Then, the probability that the unit is replaced at time T is

$$P_T = \sum_{n=N}^{\infty}\left[F^{(n)}(T) - F^{(n+1)}(T)\right]\left[G^{(n)}(K) - G^{(n)}(Z)\right],$$

the probability that it is replaced at shock N is

$$P_N = \overline{F}^{(N)}(T)\left[G^{(N)}(K) - G^{(N)}(Z)\right],$$

the probability that it is replaced at damage Z is

$$P_Z = \sum_{n=N}^{\infty}\overline{F}^{(n+1)}(T)\int_0^Z [G(K-x) - G(Z-x)]\mathrm{d}G^{(n)}(x),$$

and the probability that it is replaced at a failure level K is

$$P_K = \sum_{n=N}^{\infty}\left[F^{(n)}(T) - F^{(n+1)}(T)\right]\overline{G}^{(n)}(K) + \overline{F}^{(N)}(T)\overline{G}^{(N)}(K)$$
$$+ \sum_{n=N}^{\infty}\overline{F}^{(n+1)}(T)\int_0^Z \overline{G}(K-x)\mathrm{d}G^{(n)}(x),$$

where note that $P_T + P_N + P_Z + P_K = 1$ (Problem 6.9). Thus, the mean time to replacement is

$$
\begin{aligned}
&T\sum_{n=N}^{\infty}\left[F^{(n)}(T)-F^{(n+1)}(T)\right]\left[G^{(n)}(K)-G^{(n)}(Z)\right] \\
&+\int_T^{\infty} t\,\mathrm{d}F^{(N)}(t)\left[G^{(N)}(K)-G^{(N)}(Z)\right] \\
&+\sum_{n=N}^{\infty}\int_T^{\infty} t\,\mathrm{d}F^{(n+1)}(t)\int_0^Z [G(K-x)-G(Z-x)]\mathrm{d}G^{(n)}(x) \\
&+\sum_{n=0}^{\infty}\int_0^{T} t\,\mathrm{d}F^{(n+1)}(t)\left[G^{(n)}(K)-G^{(n+1)}(K)\right] \\
&+\sum_{n=0}^{N-1}\int_T^{\infty} t\,\mathrm{d}F^{(n+1)}(t)\left[G^{(n)}(K)-G^{(n+1)}(K)\right] \\
&+\sum_{n=N}^{\infty}\int_T^{\infty} t\,\mathrm{d}F^{(n+1)}(t)\int_0^Z \overline{G}(K-x)\mathrm{d}G^{(n)}(x) \\
&=\mu[1+M_G(K)]-\sum_{n=N}^{\infty}\left[G^{(n)}(K)-G^{(n)}(Z)\right]\int_T^{\infty}\left[F^{(n)}(t)-F^{(n+1)}(t)\right]\mathrm{d}t.
\end{aligned}
$$

Therefore, by the similar method of obtaining (6.40), the expected cost rate is

$$
C(T,N,Z)=\frac{\begin{array}{l} c_K-(c_K-c_T)\sum_{n=N}^{\infty}\left[F^{(n)}(T)-F^{(n+1)}(T)\right]\left[G^{(n)}(K)-G^{(n)}(Z)\right] \\ -(c_K-c_N)\overline{F}^{(N)}(T)\left[G^{(N)}(K)-G^{(N)}(Z)\right] \\ -(c_K-c_Z)\sum_{n=N}^{\infty}\overline{F}^{(n+1)}(T)\int_0^Z[G(K-x)-G(Z-x)]\mathrm{d}G^{(n)}(x)\end{array}}{\mu[1+M_G(K)]-\sum_{n=N}^{\infty}\left[G^{(n)}(K)-G^{(n)}(Z)\right]\int_T^{\infty}\left[F^{(n)}(t)-F^{(n+1)}(t)\right]\mathrm{d}t}. \tag{6.42}
$$

Clearly, when $T\to\infty, N\to\infty$, or $Z\to K$, the expected cost rate agrees with (6.41).

6.4.2 Optimum Policies

We derive analytically the optimum planned time T^*, shock number N^*, and damage level Z^* that minimize the expected cost rates.

6.4.2.1 Optimum T^*

Suppose that the unit is replaced only at time T or at failure, whichever occurs first. Then, the expected cost rate is, from (6.40),

$$C_1(T) \equiv \lim_{\substack{N\to\infty \\ Z\to K}} C(T,N,Z)$$
$$= \frac{c_K - (c_K - c_T)\sum_{n=0}^{\infty} G^{(n)}(K)\left[F^{(n)}(T) - F^{(n+1)}(T)\right]}{\sum_{n=0}^{\infty} G^{(n)}(K)\int_0^T [F^{(n)}(t) - F^{(n+1)}(t)]\mathrm{d}t}. \quad (6.43)$$

Note that $\lim_{T\to 0} C_1(T) = \infty$ and $\lim_{T\to\infty} C_1(T)$ is given in (6.41).

We seek an optimum time T^* that minimizes $C_1(T)$ in (6.43) for $c_K > c_T$. Then, differentiating $C_1(T)$ with respect to T and setting it equal to zero,

$$r(T)\sum_{n=0}^{\infty} G^{(n)}(K)\int_0^T \left[F^{(n)}(t) - F^{(n+1)}(t)\right]\mathrm{d}t - \sum_{n=0}^{\infty}\left[G^{(n)}(K) - G^{(n+1)}(K)\right]F^{(n+1)}(T)$$
$$= \frac{c_T}{c_K - c_T}, \quad (6.44)$$

where $r(T)$ is given in (6.20). It can be clearly seen that if $r(T)$ increases strictly with T, then the left-hand side of (6.44) also increases strictly from 0 to $\mu r(\infty)[1 + M_G(K)] - 1$ (Problem 6.10). Thus, if

$$r(\infty)[1 + M_G(K)] > \frac{c_K}{\mu(c_K - c_T)},$$

then there exists a finite and unique T^* that satisfies (6.44), and the resulting cost rate is

$$C_1(T^*) = (c_K - c_T)r(T^*).$$

Conversely, if

$$r(\infty)[1 + M_G(K)] \le \frac{c_K}{\mu(c_K - c_T)},$$

then $T^* = \infty$, i.e., the unit is replaced only at failure, and the expected cost rate is given in (6.41).

If a failure level K is distributed according to $L(x)$, as shown in Sect. 6.3.2, the expected cost rate becomes

$$C_1(T) = \frac{c_K - (c_K - c_T)\sum_{n=0}^{\infty}\left[F^{(n)}(T) - F^{(n+1)}(T)\right]\int_0^{\infty} G^{(n)}(x)\mathrm{d}L(x)}{\sum_{n=0}^{\infty}\int_0^T [F^{(n)}(t) - F^{(n+1)}(t)]\mathrm{d}t \int_0^{\infty} G^{(n)}(x)\mathrm{d}L(x)}. \quad (6.45)$$

In particular, suppose that shocks occur in a nonhomogeneous Poisson process and K is distributed exponentially, i.e., $L(x) = 1 - \mathrm{e}^{-\beta x}$. Then, replacing $F^{(n)}(t) - F^{(n+1)}(t)$ with $\{[H(t)]^n/n!\}\mathrm{e}^{-H(t)}$ and setting that $\int_0^\infty G^{(n)}(x)\mathrm{d}L(x) = [G^*(\beta)]^n$,

$$C_1(T) = \frac{c_K - (c_K - c_T)\mathrm{e}^{-[1-G^*(\beta)]H(T)}}{\int_0^T \mathrm{e}^{-[1-G^*(\beta)]H(t)}\mathrm{d}t}, \tag{6.46}$$

that corresponds to an age replacement policy by setting that $\overline{F}(t) = \mathrm{e}^{-[1-G^*(\beta)]H(t)}$ in (3.42).

6.4.2.2 Optimum N*

Suppose that the unit is replaced only at shock $N\,(N = 1, 2, \ldots)$ or at failure, whichever occurs first. Then, the expected cost rate is, from (6.40),

$$\begin{aligned} C_2(N) &\equiv \lim_{\substack{T\to\infty \\ Z\to K}} C(T, N, Z) \\ &= \frac{c_K - (c_K - c_N)G^{(N)}(K)}{\mu \sum_{n=0}^{N-1} G^{(n)}(K)} \qquad (N = 1, 2, \ldots). \end{aligned} \tag{6.47}$$

Forming the inequality $C_2(N+1) - C_2(N) \geq 0$ to seek an optimum N^* that minimizes $C_2(N)$ for $c_K > c_N$,

$$r_{N+1} \sum_{n=0}^{N-1} G^{(n)}(K) - [1 - G^{(N)}(K)] \geq \frac{c_N}{c_K - c_N} \qquad (N = 1, 2, \ldots), \tag{6.48}$$

where

$$r_{N+1} \equiv \frac{G^{(N)}(K) - G^{(N+1)}(K)}{G^{(N)}(K)}.$$

If r_N increases strictly with N, i.e., $G^{(N+1)}(x)/G^{(N)}(x)$ decreases strictly with N, then the left-hand side of (6.48) also increases strictly with N to $r_\infty[1 + M_G(K)] - 1$, where $r_\infty \equiv \lim_{N\to\infty} r_N \leq 1$ (Problem 6.10). Thus, if $r_\infty[1 + M_G(K)] > c_K/(c_K - c_N)$, then there exists a finite and unique minimum N^* $(1 \leq N^* < \infty)$ that satisfies (6.48).

In particular, when $G(x) = 1 - \mathrm{e}^{-\theta x}$, r_{N+1} increases to 1. In this case, if $\theta K > c_N/(c_K - c_N)$ then there exists a finite and unique minimum N^* $(1 \leq N^* < \infty)$ that satisfies (6.48).

6.4.2.3 Optimum Z*

Suppose that the unit is replaced only at damage $Z\,(0 \leq Z \leq K)$ or at failure, whichever occurs first. Then, the expected cost rate is, from (6.40),

$$C_3(Z) \equiv \lim_{\substack{T\to\infty \\ N\to\infty}} C(T,N,Z)$$
$$= \frac{c_K - (c_K - c_Z)\left[G(K) - \int_0^Z \overline{G}(K-x)\mathrm{d}M_G(x)\right]}{\mu[1+M_G(Z)]}. \tag{6.49}$$

We seek an optimum level Z^* that minimizes $C_3(Z)$ for $c_K > c_Z$. Differentiating $C_3(Z)$ with respect to Z and setting it equal to zero,

$$\int_{K-Z}^{K} [1 + M_G(K-x)]\mathrm{d}G(x) = \frac{c_Z}{c_K - c_Z}, \tag{6.50}$$

whose left-hand side increases strictly from 0 to $M_G(K)$. Thus, if $M_G(K) > c_Z/(c_K - c_Z)$, then there exists a finite and unique Z^* $(0 < Z^* < K)$ that satisfies (6.50), and its resulting cost rate is

$$C_3(Z^*) = \frac{(c_K - c_Z)\overline{G}(K - Z^*)}{\mu}.$$

In particular, when $G(x) = 1 - \mathrm{e}^{-\theta x}$, (6.50) is simplified as

$$\theta Z\,\mathrm{e}^{-\theta(K-Z)} = \frac{c_Z}{c_K - c_Z}.$$

6.4.2.4 Optimum N* and Z*

We can consider optimum policies (N^*, Z^*) and (T^*, Z^*) that minimize the expected cost rates $C_4(N,Z)$ and $C_5(T,Z)$ (Problem 6.11) from $C(T,N,Z)$ in (6.40), respectively. In particular, we derive optimum policies N^* and Z^* that minimize $C_4(N,Z)$: The expected cost rate is, from (6.40),

$$C_4(N,Z) \equiv \lim_{T\to\infty} C(T,N,Z)$$
$$= \frac{c_Z + (c_N - c_Z)G^{(N)}(Z) + (c_K - c_Z)\sum_{n=0}^{N-1}\int_0^Z [1 - G(K-x)]\mathrm{d}G^{(n)}(x)}{\mu\sum_{n=0}^{N-1} G^{(n)}(Z)}. \tag{6.51}$$

Differentiating $C_4(N,Z)$ with respect to Z and setting it equal to zero,

$$(c_N - c_Z)\left[\frac{g^{(N)}(Z)}{\sum_{n=1}^{N-1} g^{(n)}(Z)}\sum_{n=0}^{N-1} G^{(n)}(Z) - G^{(N)}(Z)\right]$$
$$+ (c_K - c_Z)\sum_{n=0}^{N-1}\int_0^Z [G(K-x) - G(K-Z)]\mathrm{d}G^{(n)}(x) = c_Z, \tag{6.52}$$

where $g^{(n)}(x) \equiv \mathrm{d}G^{(n)}(x)/\mathrm{d}x$. From the inequality $C_4(N+1,Z) - C_4(N,Z) \geq 0$,

$$- (c_N - c_Z)\left[\frac{G^{(N)}(Z) - G^{(N+1)}(Z)}{G^{(N)}(Z)}\sum_{n=0}^{N-1} G^{(n)}(Z) + G^{(N)}(Z)\right]$$
$$+ (c_K - c_Z)\left[\frac{\sum_{n=0}^{N-1} G^{(n)}(Z)}{G^{(N)}(Z)}\int_0^Z [1 - G(K-x)]\mathrm{d}G^{(N)}(x)\right.$$
$$\left. - \sum_{n=0}^{N-1}\int_0^Z [1 - G(K-x)]\mathrm{d}G^{(n)}(x)\right] \geq c_Z. \tag{6.53}$$

Substituting (6.52) for (6.53),

$$- (c_N - c_Z)\left[\frac{G^{(N)}(Z) - G^{(N+1)}(Z)}{G^{(N)}(Z)} + \frac{g^{(N)}(Z)}{\sum_{n=1}^{N-1} g^{(n)}(Z)}\right]$$
$$- (c_K - c_Z)\frac{1}{G^{(N)}(Z)}\int_0^Z [G(K-x) - G(K-Z)]\mathrm{d}G^{(N)}(x) \geq 0. \tag{6.54}$$

Thus, if $c_N \geq c_Z$ then there exists no finite N^* that satisfies (6.54) for any Z.

It is assumed that $c_K > c_Z > c_N$ and $G(x) = 1 - \mathrm{e}^{-\theta x}$. Then, (6.52) and (6.54) are simplified as, respectively,

$$- (c_Z - c_N)\left\{\frac{(\theta Z)^{N-1}/(N-1)!}{\sum_{j=0}^{N-2}(\theta Z)^j/j!}\sum_{n=0}^{N-1}\sum_{j=n}^{\infty}\frac{(\theta Z)^j}{j!}\mathrm{e}^{-\theta Z} - \sum_{j=N}^{\infty}\frac{(\theta Z)^j}{j!}\mathrm{e}^{-\theta Z}\right\}$$
$$+ (c_K - c_Z)\sum_{n=0}^{N-1}\sum_{j=n+1}^{\infty}\frac{(\theta Z)^j}{j!}\mathrm{e}^{-\theta K} = c_Z, \tag{6.55}$$

and

$$(c_Z - c_N)\left\{\frac{(\theta Z)^N/N!}{\sum_{j=N}^{\infty}(\theta Z)^j/j!} + \frac{(\theta Z)^{N-1}/(N-1)!}{\sum_{j=0}^{N-2}(\theta Z)^j/j!}\right\}$$
$$- (c_K - c_Z)\frac{\sum_{j=N+1}^{\infty}(\theta Z)^j/j!\mathrm{e}^{-\theta K}}{\sum_{j=N}^{\infty}(\theta Z)^j/j!\mathrm{e}^{-\theta Z}} \geq 0. \tag{6.56}$$

Because $[(\theta Z)^n/n!]/\sum_{j=n}^{\infty}[(\theta Z)^j/j!]$ increases from $\mathrm{e}^{-\theta Z}$ to 1, the left-hand side of (6.56) goes to $c_Z - c_N$ as $N \to \infty$. Thus, there exists a finite $N^*(1 \leq N^* < \infty)$ that satisfies (6.56).

Table 6.1 presents the optimum shock number N^* and damage level Z^*, and expected cost rate $C_4(N^*, Z^*)$ for $K = 10, 20, 30, 40, 50$ when $\theta = 0.5, \mu = 1, c_N = 1, c_Z = 10$ and $c_K = 100$. In case of $N = 1$, $C_4(1, Z)$ decreases with Z and

Table 6.1 Optimum N^* and Z^*, and expected cost rate $C_4(N^*, Z^*)$ when $\theta = 0.5, \mu = 1$, $c_N = 1, c_Z = 10$ and $c_K = 100$

K	10	20	30	40	50
N^*	1	3	6	9	12
Z^*	∞	18.2	26.7	36.2	45.8
$C_4(N^*, Z^*)$	1.606	0.421	0.204	0.128	0.091

$C_4(1, \infty) \approx 1.606$. Both optimum N^* and Z^* increase with K for $K \geq 20$. Note that the ratio of Z^* to K are approximately 90%, i.e., $Z^*/K \approx 0.90$, and $N^*(1/\theta)/K$ increase slowly from 0.20 to 0.48 with K for $K \geq 10$. Furthermore, by the regression analysis, we have the regression line $N^* = 0.28K - 2.2$ for $K \geq 10$ (Problem 6.12).

6.4.2.5 Periodic Replacement

It is assumed in Sect. 6.4.1 that the total damage due to shocks is additive when it has not exceeded a failure level K, however, it is not additive at any shock after it has exceeded K (Fig. 6.5). In this case, the minimal maintenance is done at each shock and the damage level remains in K.

Suppose that the unit is replaced at periodic times kT $(k = 1, 2, \ldots)$ for a planned time T $(0 < T \leq \infty)$ (Problem 6.14). Then, the expected number of minimal maintenances, i.e., the expected number of shocks in the case where the total damage remains in K when it has reached K, is

$$\sum_{n=0}^{\infty}\left[G^{(n)}(K) - G^{(n+1)}(K)\right]\sum_{k=0}^{\infty}(k+1)\int_0^T\left[F^{(k)}(T-t) - F^{(k+1)}(T-t)\right]\mathrm{d}F^{(n+1)}(t)$$

$$= \sum_{n=1}^{\infty}[1 - G^{(n)}(K)]F^{(n)}(T).$$

Let c_M be the cost of minimal maintenance. Then, by the similar method of obtaining (6.40), the expected cost rate is

$$C(T) = \frac{1}{T}\left\{c_T + c_M\sum_{n=1}^{\infty}F^{(n)}(T)[1 - G^{(n)}(K)]\right\}, \tag{6.57}$$

where c_T is the replacement cost at time T. We seek an optimum time T^* that minimizes $C(T)$. Differentiating $C(T)$ with respect to T and setting it equal to zero,

$$\sum_{n=1}^{\infty}[1 - G^{(n)}(K)][Tf^{(n)}(T) - F^{(n)}(T)] = \frac{c_T}{c_M},$$

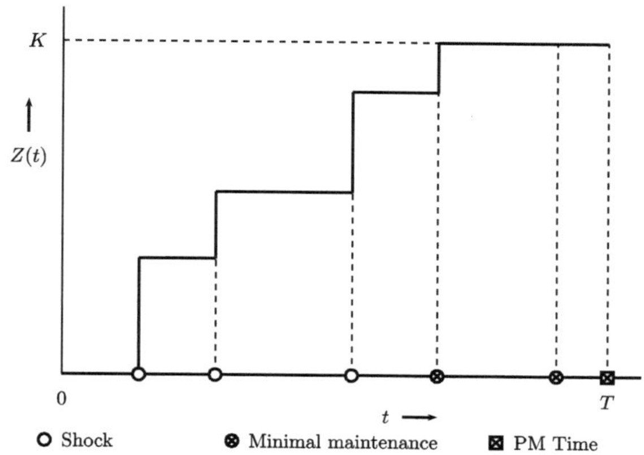

Fig. 6.5 Process for PM at time T

i.e.,

$$\sum_{n=1}^{\infty}\left[1-G^{(n)}(K)\right]\int_0^T t\,\mathrm{d}f^{(n)}(t)=\frac{c_T}{c_M}, \tag{6.58}$$

where $f^{(n)}(t)\equiv \mathrm{d}F^{(n)}(t)/\mathrm{d}t$. Thus, if

$$\sum_{n=1}^{\infty}[1-G^{(n)}(K)]T[f^{(n)}(T)]'>0,$$

and

$$\sum_{n=1}^{\infty}[1-G^{(n)}(K)]\int_0^{\infty} t\,\mathrm{d}f^{(n)}(t)>\frac{c_T}{c_M},$$

then there exists a finite and unique T^* $(0<T^*<\infty)$ that satisfies (6.58).

In particular, when $F(t)=1-\mathrm{e}^{-\lambda t}$, (6.58) is

$$\sum_{n=1}^{\infty}\frac{(\lambda T)^n}{n!}\mathrm{e}^{-\lambda T}\sum_{k=1}^{n}[G^{(k)}(K)-G^{(n)}(K)]=\frac{c_T}{c_M}, \tag{6.59}$$

whose left-hand side increases strictly from 0 to $M_G(K)$. Therefore, if $M_G(K)>c_T/c_M$ then there exists a finite and unique $T^*(0<T^*<\infty)$ that satisfies (6.59).

6.5 Problems 6

6.1 Search for examples of cumulative processes on your daily life and investigate some examples [2].

6.2 Derive (6.5).

6.3 In Example 6.3, compute the total damage when $\alpha = 0.95$, 0.99 and the probability that the unit with $K = 15$, 20 fails in 30 days.

6.4 Give some examples of independent damage models [2, p. 21].

6.5 Prove (v) and (vi) in Example 6.4.

6.6 Prove in Example 6.5 that $[(\theta K)^n/n!]/\sum_{k=n}^{\infty}[(\theta K)^k/k!]$ increases with n from $e^{-\theta K}$ to 1.

6.7 Prove in (6.25) that if $r(t)$ increases strictly and $h(t)$ increases, or $r(t)$ increases and $h(t)$ increases strictly, the left-hand side of (6.25) increases strictly from 0 to (6.26).

6.8 In Example 6.7, consider the optimization problem in which the system is replaced when $N\,(N = 1, 2, \ldots, n)$ units have failed as the preventive replacement policy [2, p. 62].

6.9 Prove that $P_T + P_N + P_Z + P_K = 1$ in Sect. 6.4.1

6.10 Prove in Sect. 6.4.2 that if $r(T)$ and r_N increase strictly with T and N, respectively, then (6.44) and (6.48) increase strictly.

6.11 Derive in Sect. 6.4.2.4 optimum T^* and Z^* that minimize $C_5(T, Z) \equiv \lim_{N\to\infty} C(T, N, Z)$ when $c_T = c_Z$ and shocks occur in a Poisson process [2, p. 53].

6.12 In Table 6.1, derive the regression line

$$N^* = 0.28K - 2.2.$$

6.13 In Sect. 6.4.2.5, obtain the expected cost $C(T, N)$ where the unit is replaced at time T or shock number N, whichever occurs first.

References

1. Cox DR (1962) Renewal theory. Methuen, London
2. Nakagawa T (2007) Shock and damage models in reliability theory. Springer, London
3. Abdel-Hameed M (1984) Life distribution properties of devices subject to a pure jump damage process. J Appl Probab 21:816–825
4. Grandell J (1976) Doubly stochastic poisson process. Lecture notes in mathematics 529. Springer, New York
5. Finkelstein MS, Zarudnij VI (2001) A shock process with a non-cumulative damage. Relib Eng Syst Saf 71:103–107
6. Gaver DP Jr (1963) Random hazard in reliability problems. Technometrics 5:211–226
7. Barlow RE, Proschan F (1975) Statistical theory of reliability and life testing. Halt, Rinehart & Winston, New York
8. Barlow RE, Proschan F (1965) Mathematical theory of reliability. Wiley, New York
9. Esary JD, Marshall AW, Proschan F (1973) Shock models and wear processes. Ann Probab 1:627–649
10. Shaked M (1984) Wear and damage processes from shock models in reliability theory. In: Abdel-Hameed M, Çinlar E, Quinn J (eds) Reliability theory and models. Academic, Orlando
11. Abdel-Hameed M, Proschan F (1975) Shock models with underlying birth process. J Appl Probab 12:18–28
12. Block HW, Savits TH (1978) Shock models with NBUE survival. J Appl Probab 15:621–628

13. Abdel-Hameed M (1984) Life distribution properties of devices subject to a Lévy wear process. Math Oper Res 9:606–614
14. Nikulim MS, Limnios N, Balakrishnan N, Kahle W, Huber-Carol C (2010) Advances in degradation modeling. Birhäuser, Boston
15. Takács L (1960) Stochastic processes. Wiley, New York
16. Ross SM (1983) Stochastic processes. Wiley, New York

Chapter 7
Brownian Motion and Lévy Processes

Wiener (1923) and Lévy (1939) mathematically gave the theoretical foundation and construction of the Brownian Motion based on the phenomenon of small particles observed by Brown (1827) and Einstein (1905). This is called the *Brownian Motion* or *Wiener process*, and is an example of a Markov process with continuous time and state space. This is now one of the most useful stochastic processes in applied sciences such as physics, economics, communication theory, biology, management science, and mathematical statistics, and recently, it is also used to make several valuable models in finance.

Section 7.1 describes a brief outline of a Brownian Motion process and considers its application to a continuous damage level. The total damage might be estimated only at periodic times or with time roughly. Section 7.2 takes up three damage models and compares numerically their optimum replacement policies. Section 7.3 introduces a Lévy process and gives two damage models to show how to apply it to reliability models.

7.1 Brownian Motion and Wiener Processes

We define a Brownian Motion process or a Wiener process [1, p. 184, 2, p. 340]:

Definition 7.1 A stochastic process $\{X(t)\}$ is said to be a Brownian Motion process if

(i) $X(0) = 0$,
(ii) the process $\{X(t)\}$ has stationary and independent increment,
(iii) the process $\{X(t)\}$ is normally distributed with mean 0 and variance $\sigma^2 t$ for any $t > 0$.

For any $t, u > 0$, the increments $X(t+u) - X(u)$ and $X(t) - X(u)$ for $t > u$ have a normal density with mean 0, and variance $\sigma^2 t$ and $\sigma^2(t-u)$, respectively, i.e., the density $f(x,t)$ of $\Pr\{X(t+u) - X(u) \le x\} = \Pr\{X(t) \le x\}$ is

T. Nakagawa, *Stochastic Processes*, Springer Series in Reliability Engineering,
DOI: 10.1007/978-0-85729-274-2_7,

$$f(x,t) = \frac{1}{\sqrt{2\pi t}\sigma} e^{-x^2/(2\sigma^2 t)}, \tag{7.1}$$

and its Laplace transform is (Problem 7.1)

$$\int_{-\infty}^{\infty} e^{-sx} f(x,t) dx = e^{s^2\sigma^2 t/2}. \tag{7.2}$$

When $\sigma = 1$, $B(t) \equiv X(t)/\sigma$ is called a *standard Brownian Motion* because $V\{X(t)/\sigma\} = t$. Thus, any Brownian Motion can convert to the standard process.

For any $0 \le t_0 < t_1 < \cdots < t_n < t$, from the properties of independent and stationary increments,

$$\begin{aligned} \Pr\{X(t) \le x | X(t_0) = x_0, \ldots, X(t_n) = x_n\} &= \Pr\{X(t) \le x | X(t_n) = x_n\} \\ &= \Pr\{X(t) - X(t_n) \le x - x_n\}, \end{aligned}$$

that states that the process has a Markov property, and its distribution function is, from (7.1),

$$\Pr\{X(t) - X(t_n) \le x - x_n\} = \frac{1}{\sqrt{2\pi(t - t_n)}\sigma} \int_{-\infty}^{x - x_n} e^{-u^2/[2\sigma^2(t - t_n)]} du, \tag{7.3}$$

i.e., $X(t) - X(t_n)$ has a normal density with mean 0 and variance $\sigma^2(t - t_n)$ for any $t \ge t_n$, that does not depend on t_n.

When $Z(t) = \mu t + \sigma B(t)$, where $B(t)$ is a standard Brownian Motion, $Z(t)$ is called a Brownian Motion with drift parameter μ and variance σ^2 [1, 2].

Example 7.1 (Continuous damage model) The total damage or wear $Z(t)$ usually increases swayingly with time t from $Z(0) = 0$, and the unit fails when $Z(t)$ has exceeded a failure level K (Fig. 7.1). It is assumed that $Z(t) = A(t)t + \sigma B(t)$ for $A(t) \ge 0$ and $B(t)$ is a standard Brownian Motion. Then, the reliability $R(t)$ at time t, in other words, the probability that the total damage has not exceeded K in time t, is

$$R(t) = \Pr\{Z(t) \le K\} = \Pr\{A(t)t + \sigma B(t) \le K\}, \tag{7.4}$$

and the MTTF is given by $\int_0^\infty R(t) dt$.

(1) When $A(t) \equiv \mu$ (constant), $K = k$ (constant), and $B(t)$ is normally distributed with mean 0 and variance t,

$$R(t) = \Pr\{\sigma B(t) \le k - \mu t\} = \Phi\left(\frac{k - \mu t}{\sigma\sqrt{t}}\right), \tag{7.5}$$

where $\Phi(x)$ is the standard normal distribution with mean 0 and variance 1, i.e., $\Phi(x) \equiv (1/\sqrt{2\pi}) \int_{-\infty}^{x} e^{-u^2/2} du$ (Problem 7.2).

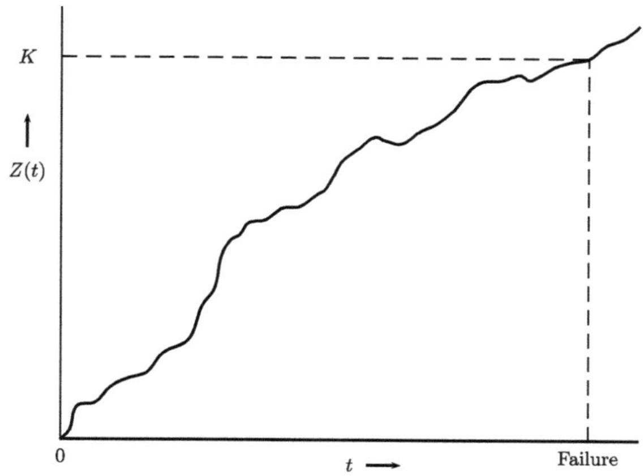

Fig. 7.1 Process for a continuous damage model

(2) When $B(t) \equiv 0$, $K = k$ and $A(t)$ is normally distributed with mean μ and variance σ^2/t,

$$R(t) = \Pr\{A(t) \leq k/t\} = \Phi\left(\frac{k - \mu t}{\sigma\sqrt{t}}\right), \tag{7.6}$$

that becomes equal to (7.5).

(3) When $A(t) \equiv \mu$, $B(t) \equiv 0$ and K is normally distributed with mean k and variance σ^2,

$$R(t) = \Pr\{\mu t \leq K\} = \Phi\left(\frac{k - \mu t}{\sigma}\right). \tag{7.7}$$

When K is normally distributed with mean k and variance $\sigma^2 t$, $R(t)$ is equal to (7.5) and (7.6).

Replacing $\alpha \equiv \sigma/\sqrt{\mu k}$ and $\beta \equiv k/\mu$ in (7.5) or (7.6),

$$R(t) = \Phi\left[\frac{1}{\alpha}\left(\sqrt{\frac{\beta}{t}} - \sqrt{\frac{t}{\beta}}\right)\right],$$

which is called the Birnbaum–Saunders distribution [3]. This is widely applied to fatigue failure for material strength subject to stresses [4, 5].

When $Z(t) = A(t)t + \sigma B(t)$, if $A(t), B(t)$, and K are deterministic, i.e., $A(t) \equiv \mu$, $B(t) = 1$, and $K = k$, then the unit fails at time $(k - \sigma)/\mu$. By fitting appropriate probability distributions to $A(t), B(t)$, and K, and estimating their parameters for actual units, the function $Z(t)$ can be used as a continuous damage or wear function in cumulative damage models [6, p. 26].

Furthermore, the process defined by $X(t) \equiv e^{Z(t)}$ for $t \geq 0$ is called a *geometric Brownian Motion*, and its state space is the interval $[0, \infty)$ [1, p. 193]. The process is applied to models of share prices, interest rates, and stock options in finance. When $Z(t) = \mu t + B(t)$ defined in (7.1), from (7.2),

$$E\left\{e^{sZ(t)}\right\} = \exp\left[st\left(\mu + \frac{1}{2}s\sigma^2\right)\right],$$

and hence, setting that $s = 1$,

$$E\{X(t)\} = E\left\{e^{Z(t)}\right\} = e^{(\mu+\sigma^2/2)t}. \tag{7.8}$$

Example 7.2 [7, p. 379]. Suppose that a unit begins to operate at time 0 and a failed unit undergoes repair, where the repair time is negligible. It is assumed that $Y(t) \equiv c_1[X(t) - 1]$ is the total maintenance cost due to failures over the time interval $[0, t]$ and is given by

$$c(t) \equiv E\{Y(t)\} = c_1 E\left\{e^{Z(t)} - 1\right\} = c_1\left[e^{(\mu+\sigma^2/2)t} - 1\right].$$

Suppose that the unit is replaced at time T. The expected cost rate is, from Sect. 3.3.2.2,

$$C(T) \equiv \frac{c(T) + c_2}{T} = \frac{c_1\left[e^{(\mu+\sigma^2/2)T} - 1\right] + c_2}{T}, \tag{7.9}$$

where c_2 is the replacement cost at time T. Clearly,

$$\lim_{T\to 0} C(T) = \lim_{T\to\infty} C_1(T) = \infty.$$

Thus, there exists an optimum T^* that minimizes (7.9). Differentiating $C(T)$ with respect to T and setting it equal to zero,

$$\left(\mu + \frac{\sigma^2}{2}\right)Te^{(\mu+\sigma^2/2)T} - \left[e^{(\mu+\sigma^2/2)T} - 1\right] = \frac{c_2}{c_1}, \tag{7.10}$$

whose left-hand side increases from 0 to ∞. Therefore, there exists a unique optimum T^* $(0 < T^* < \infty)$ that satisfies (7.10), and the resulting cost rate is (Problems 7.3 and 7.4)

$$C(T^*) = c_1\left(\mu + \frac{\sigma^2}{2}\right)e^{(\mu+\sigma^2/2)T^*}. \tag{7.11}$$

7.2 Three Replacements of Cumulative Damage Models

We consider three cumulative damage models: A unit is subjected to shocks and suffers some damage due to it (see Sect. 7.2.1.1); the amount of damage due to shocks is measured only at periodic times (see Sect. 7.2.1.2), and the total damage increases linearly with time (see Sect. 7.2.1.3). The total damage is additive and the unit fails when the total damage has exceeded a failure level K $(0<K<\infty)$ for three models. Models 2 and 3 would be practically used as the approximation ones of Model 1 and Models 1 and 2, respectively. As per the replacement policy, the unit is replaced at a planned time T. The expected cost rates of each model are obtained, and optimum policies that minimize them are derived analytically. Furthermore, optimum replacement times of each model are compared numerically when each model has the same mean amount of damage at any periodic times.

7.2.1 Three Damage Models

We consider the following three damage models for an operating unit [8].

7.2.1.1 Standard Model

The unit is subjected to shocks and suffers some damage at each shock. Let random variables X_k $(k=1,2,\ldots)$ denote a sequence of interarrival times between successive shocks, and random variables W_k $(k=1,2,\ldots)$ denote the damage produced by the kth shock, where $W_0=0$ (Fig. 6.1). In addition, let $N(t)$ denote the random variable that is the total number of shocks up to time t. Then, define a random variable

$$Z(t)=\sum_{k=0}^{N(t)} W_k \quad (N(t)=0,1,2,\ldots), \tag{7.12}$$

that represents the total damage at time t.

It is assumed that $F(t)\equiv \Pr\{X_k\le t\}$ with finite mean μ and variance σ_F^2 and $G(x)\equiv \Pr\{W_k\le x\}$ with finite mean ω and variance σ_G^2 for $t,x\ge 0$. Then, from (6.3), (6.5) and (6.8),

$$\begin{aligned}\Pr\{Z(t)\le x\} &= \sum_{k=0}^{\infty} G^{(k)}(x)[F^{(k)}(t)-F^{(k+1)}(t)],\\ E\{Z(t)\} &= \omega M_F(t), \quad V\{Z(t)\}\approx \frac{\omega^2 t}{\mu}\left(\frac{\sigma_F^2}{\mu^2}+\frac{\sigma_G^2}{\omega^2}\right),\end{aligned} \tag{7.13}$$

where $\phi^{(k)}(t)$ $(k = 1, 2, \ldots)$ denotes the k-fold Stieltjes convolution of any function $\phi(t)$ with itself and $\phi^{(0)}(t) \equiv 1$ for $t \geq 0$, and $M_\phi(t) \equiv \sum_{k=1}^{\infty} \phi^{(k)}(t)$.

Suppose that the unit fails when the total damage has exceeded a failure level K. As per the replacement policy, the unit is replaced at time T or at failure, whichever occurs first. Letting c_K be the replacement cost at failure and c_T $(c_T < c_K)$ be the replacement cost at time T, the expected cost rate is, from (6.23) and (6.43),

$$C_1(T) = \frac{c_K - (c_K - c_T)\sum_{k=0}^{\infty} G^{(k)}(K)[F^{(k)}(T) - F^{(k+1)}(T)]}{\sum_{k=0}^{\infty} G^{(k)}(K) \int_0^T [F^{(k)}(t) - F^{(k+1)}(t)]\mathrm{d}t}. \tag{7.14}$$

We find an optimum time T_1^* that minimizes $C_1(T)$: Let $f(t)$ be a density function of $F(t)$ and

$$Q_1(T) \equiv \frac{\sum_{k=0}^{\infty} f^{(k+1)}(T)[G^{(k)}(K) - G^{(k+1)}(K)]}{\sum_{k=0}^{\infty} G^{(k)}(K)[F^{(k)}(T) - F^{(k+1)}(T)]}.$$

By the method similar to Example 6.6, if $Q_1(T)$ increases strictly with T and $Q_1(\infty)[1 + M_G(K)] > c_K/[\mu(c_K - c_T)]$, then there exists a finite and unique T_1^* $(0 < T_1^* < \infty)$ that satisfies (Problem 7.5)

$$Q_1(T) \sum_{k=0}^{\infty} G^{(k)}(K) \int_0^T [F^{(k)}(t) - F^{(k+1)}(t)]\mathrm{d}t - \sum_{k=0}^{\infty} F^{(k+1)}(T)[G^{(k)}(K) - G^{(k+1)}(K)] = \frac{c_T}{c_K - c_T}. \tag{7.15}$$

7.2.1.2 Periodic Model

Each amount W_n $(n = 1, 2, \ldots)$ of damage is measured only at periodic times nT_0 $(n = 1, 2, \ldots)$ for given T_0 $(0 < T_0 < \infty)$ and has an identical distribution $G_T(x) \equiv \Pr\{W_n \leq x\}$ with mean ω_T and variance σ_T^2. Then,

$$\begin{aligned} \Pr\{Z(nT_0) \leq x\} &= G_T^{(n)}(x), \\ E\{Z(nT_0)\} = n\omega_T, &\quad V\{Z(nT_0)\} = n\sigma_T^2. \end{aligned} \tag{7.16}$$

Suppose that the unit is replaced at time NT_0 or at failure, whichever occurs first. Then, the expected cost rate is, from (6.47),

$$C_2(N) = \frac{c_K - (c_K - c_T)G_T^{(N)}(K)}{T_0 \sum_{n=0}^{N-1} G_T^{(n)}(K)} \quad (N = 1, 2, \ldots). \tag{7.17}$$

We find an optimum number N^* that minimizes $C_2(N)$. From the inequality $C_2(N+1) - C_2(N) \geq 0$,

$$Q_2(N+1)\sum_{n=0}^{N-1} G_T^{(n)}(K) - [1 - G_T^{(N)}(K)] \geq \frac{c_T}{c_K - c_T}, \tag{7.18}$$

where

$$Q_2(N) \equiv \frac{G_T^{(N-1)}(K) - G_T^{(N)}(K)}{G_T^{(N-1)}(K)}.$$

Thus, if $Q_2(N)$ increases strictly with N and $Q_2(\infty)[1 + M_{G_T}(K)] > c_K/(c_K - c_T)$, then there exists a finite and unique minimum N^* $(1 \leq N^* < \infty)$ that satisfies (7.18) (Problem 7.6).

7.2.1.3 Continuous Model

The total damage $Z(t)$ usually increases with time t. Suppose that $Z(t) = A(t)t$ where $A(t)$ has a distribution $L_A(t)$ with mean a and variance σ_A^2/t. Then,

$$\begin{aligned} \Pr\{Z(t) \leq x\} &= \Pr\{A(t) \leq x/t\} = L_A(x/t), \\ E\{Z(t)\} &= at, \quad V\{Z(t)\} = \sigma_A^2 t. \end{aligned} \tag{7.19}$$

Suppose that the unit is replaced at time T or at failure, whichever occurs first. Then, the expected cost rate is

$$C_3(T) = \frac{c_K - (c_K - c_T)L_A(K/T)}{\int_0^T L_A(K/t)\mathrm{d}t}. \tag{7.20}$$

We find an optimum time T_3^* that minimizes $C_3(T)$: Let $l_A(t)$ be a density function of $L_A(t)$ and $r_A(t)$ be the failure rate of $L_A(t)$, i.e., $r_A(t) \equiv l_A(t)/L_A(t)$. Then, if $r_A(t)$ increases strictly, then there exists a finite and unique T_3^* $(0 < T_3^* < \infty)$ that satisfies (Problem 7.7)

$$r_A(K/T)\int_0^T L_A(K/t)\mathrm{d}t - [1 - L_A(K/T)] = \frac{c_T}{c_K - c_T}. \tag{7.21}$$

7.2.2 Numerical Examples

We compute optimum policies numerically for the three models: It is assumed that $F(t) = 1 - \mathrm{e}^{-\lambda t}$ and $G(x) = 1 - \mathrm{e}^{-x/\omega}$ for Model 1, $G_T(x) = 1 - \mathrm{e}^{-x/\omega_T}$ for Model 2, and $A(t)$ has a normal distribution $N(\omega_A, \sigma_A^2/t)$ for Model 3.

In addition, we set the mean and variance of the total damage at any time nT_0 as equal for all models, i.e.,

$$E\{Z(nT_0)\} = \lambda\omega nT_0 = n\omega_T = \omega_A nT_0, \quad V\{Z(nT_0)\} = 2\lambda\omega^2 nT_0 = n\omega_T^2 = \sigma_A^2 nT_0.$$

Hence,

$$\lambda = \frac{2}{T_0}, \quad \omega = \frac{\omega_T}{2}, \quad \omega_A = \frac{\omega_T}{T_0}, \quad \sigma_A^2 = \frac{\omega_T^2}{T_0}.$$

When $T_0 = 1$ and $\omega_T = 1, \lambda = 2, \omega = 1/2, \omega_A = 1$ and $\sigma_A^2 = 1$. Optimum policies are, for Model 1, from (7.15),

$$\frac{2\sum_{j=0}^{\infty}[(2T)^j(2K)^j/j!j!]}{\sum_{j=0}^{\infty}[(2K)^j/j!]\sum_{k=0}^{j}[(2T)^k/k!]}\sum_{j=0}^{\infty}\frac{(2K)^j}{j!}e^{-2K}\sum_{k=0}^{j}\int_0^T\frac{(2t)^k}{k!}e^{-2t}dt$$
$$-\sum_{j=1}^{\infty}\frac{(2T)^j}{j!}e^{-2T}\sum_{k=0}^{j-1}\frac{(2K)^k}{k!}e^{-2K} = \frac{1}{c_K/c_T - 1}, \tag{7.22}$$

for Model 2, from (7.18),

$$\frac{K^N/N!}{\sum_{k=N}^{\infty}(K^k/k!)}\sum_{k=0}^{N-1}\sum_{j=k}^{\infty}\frac{K^j}{j!}e^{-K} - \sum_{k=0}^{N-1}\frac{N^k}{k!}e^{-K} \geq \frac{1}{c_K/c_T - 1}, \tag{7.23}$$

and for Model 3, from (7.21),

$$\frac{(K/T+1)\phi(K/\sqrt{T}-\sqrt{T})}{2\sqrt{T}\Phi(K/\sqrt{T}-\sqrt{T})}\int_0^T\Phi\left(\frac{K}{\sqrt{t}}-\sqrt{t}\right)dt - \left[1-\Phi\left(\frac{K}{\sqrt{T}}-\sqrt{T}\right)\right]$$
$$= \frac{1}{c_K/c_T - 1}, \tag{7.24}$$

where $\phi(x) \equiv (1/\sqrt{2\pi})e^{-x^2/2}$ and $\Phi(x) \equiv (1/\sqrt{2\pi})\int_{-\infty}^{x} e^{-u^2/2}du$.

Table 7.1 presents optimum T_1^* and $C_1(T_1^*)$ for $c_K/c_T = 5$, 10, 20, 50 and $K = 5$, 10, 15, 20. This indicates that T_1^* decreases when c_K/c_T and $1/K$ increase. It is of interest that $C_1(T_1^*)$ have their minimum values at $K = 15$ for $c_K/c_T \geq 10$. Table 7.2 presents optimum N^*T_0 and $C_2(N^*T_0)$ for the same parameters c_K/c_T and K as Table 7.1. This shows the similar tendencies to Table 7.1, however, $C_2(N^*)$ decreases with K and c_T/c_K. Table 7.3 presents optimum T_3^* and $C_3(T_3^*)$. This shows the same tendencies as Table 7.2 for both T_3^* and $C_3(T_3^*)$. These tables indicate that T_1^* is equal to N^*T_0 and $N^*T_0 > T_3^*$ for large c_K/c_T and K, and T_1^*/T_3^* changes from 1.0 to 2.2 in this case. It is natural that T_1^*, N^*T_0 and T_3^* decrease when c_K/c_T and $1/K$ increase. The values of T_3^* are too small compared with N^*T_0. If T_0 and $1/\lambda$ would be small, T_3^* would approach N^*T_0.

Table 7.1 Optimum T_1^* and $C_1(T_1^*)$ when $T_0 = 1$ and $\mu_{T_0} = 1$

c_K/c_T	K							
	5		10		15		20	
	T_1^*	$C(T_1^*)$	T_1^*	$C(T_1^*)$	T_1^*	$C(T_1^*)$	T_1^*	$C(T_1^*)$
5	2.9	0.520	5.9	0.212	9.3	0.126	13.3	0.110
10	2.2	0.668	5.0	0.247	8.2	0.141	12.2	0.145
20	1.7	0.840	4.3	0.284	7.3	0.157	11.5	0.207
50	1.3	1.127	3.6	0.336	6.4	0.178	11.0	0.385

Table 7.2 Optimum N^*T_0 and $C_2(N^*)$ when $T_0 = 1$ and $\mu_{T_0} = 1$

c_K/c_T	K							
	5		10		15		20	
	N^*T_0	$C_2(N^*)$	N^*T_0	$C_2(N^*)$	N^*T_0	$C_2(N^*)$	N^*T_0	$C_2(N^*)$
5	3.0	0.508	6.0	0.213	9.0	0.128	13.0	0.089
10	2.0	0.684	5.0	0.253	8.0	0.146	12.0	0.100
20	2.0	0.887	4.0	0.299	7.0	0.164	10.0	0.110
50	1.0	1.330	4.0	0.377	6.0	0.190	9.0	0.122

Table 7.3 Optimum T_3^* and $C_3(T_3^*)$ when $T_0 = 1$ and $\mu_{T_0} = 1$

c_K/c_T	K							
	5		10		15		20	
	T_3^*	$C(T_3^*)$	T_3^*	$C(T_3^*)$	T_3^*	$C(T_3^*)$	T_3^*	$C(T_3^*)$
5	2.5	0.483	6.0	0.198	9.2	0.119	13.1	0.084
10	2.0	0.559	5.1	0.219	8.7	0.130	12.0	0.090
20	1.8	0.626	4.6	0.237	7.8	0.139	11.3	0.095
50	1.6	0.704	4.1	0.260	7.1	0.149	10.6	0.102

7.3 Lévy Process

We briefly explain a Lévy process appearing in stochastic processes [9].

Definition 7.2 A stochastic process $\{X(t)\}$ is said to be a Lévy process if

(i) $X(0) = 0$,
(ii) the increments $X(t_1) - X(t_0), X(t_2) - X(t_1), \ldots, X(t_n) - X(t_{n-1})$ are independent for all $0 \le t_0 < t_1 < \cdots < t_n$,
(iii) the distribution of increment $X(t_k) - X(t_{k-1})$ depends only the difference $t_k - t_{k-1}$.

A Lévy process is a stochastic process with independent and stationary increments. For the strict definition of a Lévy process, we have to add two regularity conditions, however, we omit them to exceed mathematically the limit of the book

[10, 11]. From this definition, the increment $X(t) - X(u)$ is independent of the process $\{X(x), 0 \le x \le u\}$ and has the same distribution of $X(t-u)$ for $t \ge u \ge 0$. Special cases of Lévy processes are the Brownian Motion with continuous increment and the Poisson process with jump increment. In other words, the Poisson process is the jump Lévy process and the Brownian Motion is the continuous Lévy process.

The Lévy process has been applied to various fields such as physics, finance, and statistics [11]. However, there have been few papers treated with reliability models using Lévy processes: When damage and wear models form a Lévy process, the damage models have been discussed analytically [12–14].

Example 7.3 (Continuous damage model) It is assumed in Sect. 7.2.1.3 that $Z(t) = \omega t + B_t$, where B_t has an exponential distribution $1 - \mathrm{e}^{-x/\sigma\sqrt{t}}$. That is, the total damage usually increases linearly with time t, however, it undergoes positively some change according to an exponential distribution. Then,

$$\begin{aligned} \Pr\{Z(t) \le x\} &= \Pr\{B_t \le x - \omega t\} = 1 - \exp[-(x - \omega t)/\sigma\sqrt{t}], \\ E\{Z(t)\} &= \omega t + \sigma\sqrt{t}, \\ V\{Z(t)\} &= \sigma^2 t. \end{aligned} \tag{7.25}$$

Thus, when the unit is replaced at time T, the expected cost rate is, from (7.20),

$$C_1(T) = \frac{c_K - (c_K - c_T)\{1 - \exp[-(K - \omega T)/\sigma\sqrt{T}]\}}{\int_0^T \{1 - \exp[-(K - \omega t)/\sigma\sqrt{t}]\}\mathrm{d}t}. \tag{7.26}$$

Differentiating $C_1(T)$ with respect to T and setting it equal to zero,

$$\begin{aligned} &\frac{(K/T + \omega)\exp[-(K - \omega T)/\sigma\sqrt{T}]}{2\sigma\sqrt{T}\{1 - \exp[-(K - \omega T)/\sigma\sqrt{T}]\}} \int_0^T \{1 - \exp[-(K - \omega t)/\sigma\sqrt{t}]\}\mathrm{d}t \\ &\quad - \exp[-(K - \omega T)/\sigma\sqrt{T}] = \frac{1}{c_K/c_T - 1}. \end{aligned} \tag{7.27}$$

Table 7.4 presents optimum T^* and expected cost rate $C_1(T^*)/c_T$ for $c_K/c_T = 5, 10, 20, 50$ and $K = 5, 10, 15, 20$ when $\sigma^2 = 1$ and $\omega = 1$. Compared to Table 7.3, optimum T^* are less than T_3^* because the average amount of damage is greater than that in the continuous model (Sect. 7.2.1.3). However, if the damage level would be estimated at a lowest level and additional damage suffered from environment change and frequent disasters would be caused randomly, this would be applied to such models (Problem 7.8). On the other hand, if $Z(t) = \omega t - B_t$, i.e., the total damage changes negatively according to an exponential distribution, then

Table 7.4 Optimum T^* and $C_1(T^*)/c_T$ when $\omega = \sigma^2 = 1$

c_K/c_T	K							
	5		10		15		20	
	T^*	$C_1(T^*)/c_T$	T^*	$C_1(T^*)/c_T$	T^*	$C_1(T^*)/c_T$	T^*	$C_1(T^*)/c_T$
5	2.2	0.7619	4.8	0.2920	7.8	0.1696	10.9	0.1163
10	1.6	1.1029	3.8	0.3648	6.4	0.2045	9.2	0.1371
20	1.1	1.3361	3.1	0.4461	5.3	0.2424	7.8	0.1594
50	0.8	1.8131	2.3	0.5683	4.3	0.2983	6.5	0.1918

$$\begin{aligned} \Pr\{Z(t) \leq x\} &= \Pr\{B_t \geq \omega t - x\} = \exp[-(\omega t - x)/\sigma\sqrt{t}], \\ E\{Z(t)\} &= \omega t - \sigma\sqrt{t}, \quad V\{Z(t)\} = \sigma^2 t. \end{aligned} \tag{7.28}$$

Thus, the expected cost rate is (Problem 7.9)

$$C_2(T) = \frac{c_K - (c_K - c_T)\exp[-(\omega T - K)/\sigma\sqrt{T}]}{\int_0^T \exp[-(\omega t - K)/\sigma\sqrt{t}]\mathrm{d}t}. \tag{7.29}$$

Example 7.4 [6, p. 35] Consider the damage model with two kinds of damage in Sect. 6.3.4. One of them is caused by shock and is additive in Sect. 6.3.1, and the other increases proportionately with time t, i.e., the total damage is $Y(t) \equiv \alpha t + Z(t)$, where $Z(t)$ is given in (7.12). The unit fails whether the total damage exceeds a failure level with time or has exceeded in both cases, at some shock, and the failure is detected only at time of shocks.

Denote that $S_n \equiv X_1 + X_2 + \cdots + X_n, Z_n = W_1 + W_2 + \cdots + W_n (n = 1, 2, \ldots)$, where $\mu \equiv E\{X_n\}$. The distribution of time to detect a failure at some shock is, from (6.36),

$$\begin{aligned} &\sum_{n=0}^{\infty} \Pr\{Z_n + \alpha S_n < K \leq Z_{n+1} + \alpha S_{n+1}, S_{n+1} \leq t\} \\ &= \sum_{n=0}^{\infty} \int_0^t \left\{ \int_0^{t-u} \left[G^{(n)}(K - \alpha u) - G^{(n+1)}(K - \alpha(u + x)) \right] \mathrm{d}F(x) \right\} \mathrm{d}F^{(n)}(u). \end{aligned} \tag{7.30}$$

The mean time to detect a failure at time shock is, from (6.37),

$$\alpha \sum_{n=0}^{\infty} \int_0^{K/\alpha} G^{(n)}(K - \alpha t)\mathrm{d}F^{(n)}(t). \tag{7.31}$$

Next, suppose that the unit is replaced at time T or at failure, i.e., when the total damage has exceeded K at some shock, whichever occurs first [6, p. 35, 15]. Then, the probability that the unit is replaced because failure is detected at some shock is, from (7.30),

$$\sum_{n=0}^{\infty} \Pr\{Z_n + \alpha S_n \le K, Z_{n+1} + \alpha S_{n+1} > K, S_{n+1} \le T\}$$
$$= \sum_{n=0}^{\infty} \int_0^T \left[G^{(n)}(K - \alpha t) - G^{(n+1)}(K - \alpha t)\right] \mathrm{d}F^{(n+1)}(t)$$
$$+ \sum_{n=0}^{\infty} \int_0^T \left\{ \int_0^{T-t} \left[G^{(n)}(K - \alpha t) - G^{(n+1)}(K - \alpha(t + x))\right] \mathrm{d}F(x) \right\} \mathrm{d}F^{(n)}(t), \tag{7.32}$$

and the probability that it is replaced at time T before failure is

$$\sum_{n=0}^{\infty} \Pr\{Z_n + \alpha S_n \le K, S_n \le T \le S_{n+1}\}$$
$$= \sum_{n=0}^{\infty} \int_0^T \overline{F}(T - t) G^{(n)}(K - \alpha t) \mathrm{d}F^{(n)}(t), \tag{7.33}$$

where (7.32) + (7.33) = 1. The mean time to replacement is, from (7.32) and (7.33),

$$T \sum_{n=0}^{\infty} \int_0^T \overline{F}(T - t) G^{(n)}(K - \alpha t) \mathrm{d}F^{(n)}(t)$$
$$+ \sum_{n=0}^{\infty} \int_0^T t\left[G^{(n)}(K - \alpha t) - G^{(n+1)}(K - \alpha t)\right] \mathrm{d}F^{(n+1)}(t)$$
$$+ \sum_{n=0}^{\infty} \int_0^T \left\{ \int_0^{T-t} (t + x)\left[G^{(n)}(K - \alpha t) - G^{(n)}(K - \alpha(t + x))\right] \mathrm{d}F(x) \right\} \mathrm{d}F^{(n)}(t)$$
$$= \sum_{n=0}^{\infty} \int_0^T \left[\int_0^{T-t} \overline{F}(x) \mathrm{d}x \right] G^{(n)}(K - \alpha t) \mathrm{d}F^{(n)}(t). \tag{7.34}$$

Thus, the expected cost rate is, from (7.33) and (7.34),

$$C(T) = \frac{c_K - (c_K - c_T)\sum_{n=0}^{\infty} \int_0^T \overline{F}(T - t) G^{(n)}(K - \alpha t) \mathrm{d}F^{(n)}(t)}{\sum_{n=0}^{\infty} \int_0^T \left[\int_0^{T-t} \overline{F}(x) \mathrm{d}x\right] G^{(n)}(K - \alpha t) \mathrm{d}F^{(n)}(t)}, \tag{7.35}$$

where c_K is the replacement cost at failure, and c_T is the replacement cost at time T, where $c_K > c_T$.

Table 7.5 Optimum T^* when $1/\lambda = 1/\theta = 1.0$

α	$K = 100$				$K = 200$			
	c_K/c_T				c_K/c_T			
	2	5	10	100	2	5	10	100
0	81	72	68	59	169	157	151	138
0.2	69	62	58	51	142	133	128	118
0.4	59	54	51	45	123	115	112	103
0.6	52	48	45	40	108	102	99	92
0.8	47	43	41	37	97	92	89	83
1.0	42	39	37	33	87	83	81	76
1.2	39	36	34	31	80	76	74	69
1.4	36	33	32	29	74	70	68	64
1.6	33	31	29	27	68	65	64	60
1.8	31	29	28	25	63	61	59	56
2.0	29	27	26	24	59	57	56	53
2.5	25	23	22	21	51	49	48	46
3.0	22	21	20	18	45	43	43	40
4.0	18	17	16	15	36	35	34	33
5.0	15	14	14	13	30	29	29	28
6.0	13	12	12	11	26	25	25	24

We find an optimum T^* that minimizes $C(T)$ when shocks occur in a Poisson process, i.e., $F^{(n)}(t) = \sum_{k=n}^{\infty}[(\lambda t)^k/k!]\mathrm{e}^{-\lambda t}$ $(n = 0, 1, 2, \ldots)$. Then, differentiating $C(T)$ with respect to T and setting it equal to zero,

$$\int_0^T \mathrm{e}^{-\lambda t}\mathrm{d}B(t) - \frac{B(T)}{1 + \int_0^T B(t)\,\mathrm{d}t}\left[1 + \int_0^T \mathrm{e}^{-\lambda t}\,B(t)\,\mathrm{d}t\right] = \frac{\lambda c_T}{c_K - c_T}, \tag{7.36}$$

where

$$B(t) \equiv \lambda \sum_{n=0}^{\infty} \frac{(\lambda t)^n}{n!} G^{(n+1)}(K - \alpha t).$$

Next, it is assumed that $G(x) = 1 - \mathrm{e}^{-\theta x}$. Then, the average damage per unit of time is $\alpha + \lambda/\theta$. So that, the mean time until the total damage exceeds a failure level K is $K/(\alpha + \lambda/\theta)$. Table 7.5 presents the optimum T^* that satisfies (7.36) for $K = 100, 200, c_K/c_T = 2, 5, 10, 100$ and α changes from 0 to 6 when $1/\lambda = 1/\theta = 1.0$. For example, when $K = 200, c_K/c_T = 5, \alpha = 0.2, K/(\alpha + \lambda/\theta) = 167$ and $T^* = 133$, i.e., the unit with the meantime 167 to failure should be replaced at time 133. In other words, if the operating time is $(133/167) \times 100 \approx 80\%$ of the mean time to failure, the unit should be replaced before failure. This indicates that optimum T^* decreases when both c_K and α increase. This means that if α is large, the total damage increases mainly with time, and hence, we should adopt the scheduled replacement time T (Problem 7.10). Therefore, the rate α is an important factor for deciding the optimum policy (Problem 7.11).

7.4 Problems 7

7.1 When the density is

$$f(x,t) = \frac{1}{\sqrt{2\pi t}\sigma} e^{-x^2/(2\sigma^2 t)},$$

show that

$$\int_{-\infty}^{\infty} e^{-sx} f(x,t) dx = e^{s^2\sigma^2 t/2}.$$

7.2 In (1) of Example 7.1, compute the reliability $R(t)$ for $t = 5, 10$ when $\mu = \sigma = 1$ and $K = 10$, and the MTTF.

7.3 Prove in Example 7.2 that an optimum T^* $(0 < T^* < \infty)$ is a finite and unique solution of (7.10), and the resulting cost rate is given in (7.11).

7.4 In Example 7.2, $C(T)$ in (7.9) is approximately given by

$$\widetilde{C}(T) = c_1\left(\mu + \frac{\sigma^2}{2}\right)\left[1 + \frac{(\mu + \frac{\sigma^2}{2})T}{2}\right] + \frac{c_2}{T}.$$

Derive $\widetilde{C}(T)$ and an optimum $\widetilde{T}$ that minimizes $\widetilde{C}(T)$. Furthermore, show numerically that $\widetilde{T}$ becomes a good approximation to T^*.

7.5 Prove that if $Q_1(T)$ increases strictly with T and $Q_1(\infty)[1 + M_G(K)] > c_K / [\mu(c_K - c_T)]$, then there exists a finite and unique T_1^* $(0 < T_1^* < \infty)$ that satisfies (7.15).

7.6 Prove that if $Q_2(N)$ increases strictly with N and $Q_2(\infty)[1 + M_{G_T}(K)] > c_K / (c_K - c_T)$, then there exists a finite and unique minimum $N^*(1 \le N^* < \infty)$ that satisfies (7.18).

7.7 Prove that if $r_A(t)$ increases strictly to infinity, then there exists a finite and unique T_3^* that satisfies (7.21).

7.8 In Example 7.3, rewrite all results when B_t has a Weibull distribution $\Pr\{B_t \le x\} = 1 - \exp\left[-\left(x^\alpha / \sigma\sqrt{t}\right)\right]$ for $\alpha \ge 1$.

7.9 Derive an optimum T^* that minimizes $C_2(T)$ in (7.29).

7.10 Explain the physical meaning of why optimum T^* decreases with α.

7.11 In Example 7.4, consider the damage model where the unit is replaced at shock number N or at failure, and discuss an optimum N^* that minimizes the expected cost [15].

References

1. Ross SM (1983) Stochastic processes. Wiley, New York
2. Karlin S, Taylor HM (1975) A first course in stochastic processes. Academic Press, New York
3. Birnbaum ZW, Saunders SC (1969) A new family of life distributions. J Appl Probab 6:319–327

4. Owen WJ, Padgett WJ (2003) Accelerated test models with the Birnbaum–Saunders distribution. In: Pham H (ed) Handbook of reliability engineering. Springer, London, pp 429–439
5. Park C, Padgett WJ (2005) New cumulative damage models for failure using stochastic processes as initial damage. IEEE Trans Reliab 54:530–540
6. Nakagawa T (2007) Shock and damage models in reliability theory. Springer, London
7. Beichelt FE (2006) Stochastic processes in science, engineering and finance. Chapman & Hall, Boca Raton
8. Ito K, Nakagawa T (2008) Comparison of three cumulative damage models. In: Sheu SH, Dohi T (eds) Advanced reliability modeling III. McGraw-Hill, Taipei, pp 332–338
9. Prabhu NU (2007) Stochastic processes. World Scientific, Singapore
10. Bertoin J (1996) Lévy processes. Cambridge University Press, Cambridge
11. Sato K (2001) Basic results on Lévy processes. In: Bandorff-Nielsen O, Mikosch T, Resnick S (eds) Lévy processes: theory and applications. Birkhäuser, Boston
12. Zuckerman D (1978) Optimal replacement policy for the case where the damage process is a one-sided Lévy process. Stoch Process Appl 7:141–151
13. Abdel-Hameed M (2010) Degradation process: an overview. In: Nikulin MS, Limnios N, Balakrishnan N, Kahle W, Huber-Carol C (eds) Advances in degradation modeling. Birkhäuser, Boston, pp 17–25
14. Lehmann A (2010) Failure time models based on degradation processes. In: Nikulin MS, Limnios N, Balakrishnan N, Kahle W, Huber-Carol C (eds) Advances in degradation modeling. Birkhäuser, Boston, pp 209–233
15. Satow T, Nakagawa T (1997) Replacement policies for a shock model with two kinds of damages. In: Osaki S (ed) Stochastic modeling in innovative manufacturing. Springer Lecture notes in economics and mathematical systems, vol 445. Springer, Berlin, pp 188–195

Chapter 8
Redundant Systems

High system reliability can be achieved by providing redundancy and maintenance. It is not too much to say that learning reliability is analyzing redundant systems and deriving optimum maintenance policies. In the preceding chapters, we have already given many useful examples of reliability models to understand naturally stochastic processes, and conversely, to apply the theory of stochastic processes to actual reliability models. As final examples of reliability models, we take up standard redundant systems and show systematically how to use well the techniques of stochastic processes to analyze them and to obtain their reliability properties theoretically. This would be greatly helpful for understanding stochastic processes and learning reliability theory [1, p. 160].

In Sect. 8.1, we analyze one-unit systems with no repair, that is, the simplest and fundamental reliability system, using Poisson, nonhomogeneous Poisson, and renewal processes. This section aims not only to make differences among three processes, but also to apply renewal processes to standard replacement policies. Furthermore, we obtain several reliability measures of a one-unit system with repair, using an alternating renewal process.

In Sect. 8.2, we take up two-unit systems with repair. When both failure and repair times have exponential distributions, we obtain transition probabilities, using a continuous-time Markov chain. Further, when failure and repair times have general distributions, we analyze the system, using a Markov renewal process. This indicates that Markov renewal processes are the most powerful tool for analyzing reliability systems.

It has been presented until now that states are well defined in stochastic processes and their behaviors are derived analytically. In Sect. 8.3, it can be shown as an example of a standby system with spare units that the system is analyzed and its several reliability measures are derived by forming renewal equations directly, and by taking their Laplace-Stieltjes (LS) transforms and forming generating functions. This would be useful for analyzing simple non-complicated reliability models.

T. Nakagawa, *Stochastic Processes*, Springer Series in Reliability Engineering,
DOI: 10.1007/978-0-85729-274-2_8,

8.1 One-Unit System

8.1.1 Poisson Process

As the simplest redundant system, we consider a one-unit system where a unit begins to operate at time 0 and has a failure distribution $(1 - e^{-\lambda t})$. When the unit fails, it is repaired, becomes like new, and begins to operate again. The repair time is very small compared to its operating time and is assumed to be negligible. In other words, the unit is assumed to be replaced immediately with a new one at each failure and its replacement time is negligible.

Let X_k $(k = 1, 2, \ldots)$ be the failure times of the kth operating unit that are independent and identically distributed random variables with an exponential distribution $(1 - e^{-\lambda t})$, and $S_n \equiv X_1 + X_2 + \cdots + X_n$ $(n = 1, 2, \ldots)$. From Chap. 2, the distribution of S_n is a gamma distribution with order n. Thus, the number $N(t)$ of failures in $[0, t]$ is, from (2.12) and (2.16),

$$\begin{aligned}\Pr\{N(t) = n\} &= \Pr\{S_n \le t < S_{n+1}\} = \Pr\{S_n \le t\} - \Pr\{S_{n+1} \le t\} \\ &= \sum_{k=n}^{\infty} \frac{(\lambda t)^k}{k!} e^{-\lambda t} - \sum_{k=n+1}^{\infty} \frac{(\lambda t)^k}{k!} e^{-\lambda t} \\ &= \frac{(\lambda t)^n}{n!} e^{-\lambda t} \qquad (k = 0, 1, 2, \ldots), \end{aligned} \tag{8.1}$$

that is a Poisson distribution with mean λt. It is said that failures occur in a *Poisson process* with rate λ, i.e., the failures occur randomly independent of its age and operating time, and its residual life at any time t also has the same exponential distribution $(1 - e^{-\lambda t})$ because of its memoryless property.

8.1.2 Nonhomogeneous Poisson Process

Suppose that the unit undergoes minimal repair at failures and its failure rate $h(t)$ remains unchanged by minimal repair given in Definition 2.3 of Sect. 2.3. The unit begins to operate again after the completion of minimal repair, where the time for minimal repair is negligible. Then, the expected number $E\{N(t)\}$ of failures in $[0, t]$ is $H(t) \equiv \int_0^t h(u)\mathrm{d}u$, and the number $N(t)$ of failures in [0, t] is, from (2.42),

$$\Pr\{N(t) = n\} = \frac{[H(t)]^n}{n!} e^{-H(t)} \qquad (n = 0, 1, 2, \ldots), \tag{8.2}$$

that is a Poisson distribution with mean $H(t)$. It is said that failures occur in a *non-homogeneous Poisson process* with an intensity function $h(t)$ and a mean value function $H(t)$. If $F(t)$ is IFR, i.e., $h(t)$ increases with time t, failures are apt to fail with age. Some variations for minimal repair and preventive maintenance are well-known as the imperfect repair [2] and preventive maintenance, and were summarized [3, p. 96].

8.1.3 Renewal Process

Suppose that the unit has a general failure distribution $F(t)$ with finite mean μ. Then, letting $F^{(n)}(t)$ be the n-fold Stieltjes convolution of $F(t)$ with itself, from (3.3), (3.4), and (3.18),

$$\Pr\{S_n \le t\} = F^{(n)}(t), \quad \Pr\{N(t) = n\} = F^{(n)}(t) - F^{(n+1)}(t),$$
$$\lim_{t\to\infty} \frac{M(t)}{t} = \frac{1}{\mu}, \quad E\{N(t)\} = M(t) = \sum_{n=1}^{\infty} F^{(n)}(t). \tag{8.3}$$

It is said that failures occur in a *renewal process* with an interarrival distribution $F(t)$. Most reliability systems fail in accordance with renewal processes. The renewal function $M(t)$ plays an important role in analyzing reliability systems. However, in general, it is practically impossible to explicitly express $M(t)$ for a general distribution $F(t)$ because it is very difficult to compute the n-fold Stieltjes convolution $F^{(n)}(t)$ of $F(t)$ explicitly except exponential and gamma distributions given in Example 3.2. Thus, we derive the Laplace-Stieltjes (LS) transform of $M(t)$ from

$$M^*(s) = \int_0^\infty \mathrm{e}^{-st} \mathrm{d}M(t) = \frac{F^*(s)}{1 - F^*(s)}, \tag{8.4}$$

and compute $M(t)$ explicitly only by inverting $M^*(s)$ numerically, as shown in Sect. 3.2.3.

A unit is planned to be replaced at periodic times kT $(k = 1, 2, \ldots)$ and failures occur in a renewal process between replacements, i.e., each failed unit is replaced with a new one. Then, from (3.44), the expected cost rate is

$$C_1(T) = \frac{c_1 M(T) + c_2}{T}, \tag{8.5}$$

where c_1 = cost of replacement at each failure and c_2 = cost of planned replacement at time T. Similarly, suppose that failures occur in a nonhomogeneous Poisson process with a mean value function $H(t)$, i.e., each failed unit undergoes minimal repair. Then, from (3.46), the expected cost rate is

$$C_2(T) = \frac{c_1 H(T) + c_2}{T}, \tag{8.6}$$

where c_1 = cost of minimal repair at each failure and c_2 is given in (8.5).

Next, suppose that a unit is replaced at time T or at failure, whichever occurs first. Then, defining $Z_k \equiv \min\{X_k, T\}$, the process $\{Z_k\}$ has a renewal process with interarrival distribution

$$\Pr\{Z_k \leq t\} = \begin{cases} F(t) & t \leq T, \\ 1 & t > T, \end{cases}$$

and hence,

$$E\{Z_k\} = \int_0^T t\,\mathrm{d}F(t) + T\overline{F}(T) = \int_0^T \overline{F}(t)\mathrm{d}t,$$

where $\overline{\varphi}(t) \equiv 1 - \varphi(t)$ for any function $\varphi(t)$. Then, from (3.42), the expected cost rate is

$$C_3(T) = \frac{c_1 F(T) + c_2 \overline{F}(T)}{\int_0^T \overline{F}(t)dt}, \tag{8.7}$$

where c_1 = replacement cost at failure and c_2 = replacement cost at time T.

8.1.4 Alternating Renewal Process

It is assumed in Sect. 8.1.3 that each repair time is not negligible and has an identical distribution $G(t)$ with finite mean β. Let Y_k $(k = 1, 2, \ldots)$ be the repair times of the kth failed unit that are independent and identically distributed random variables with $G(t)$. We define the following states:

State 0: Unit is operating.
State 1: Unit is under repair.

The unit repeats operating and repair alternately in Figs. 3.5 and 4.2. It is said that the unit forms an *alternating renewal process* with States 0 and 1. Clearly, the process $\{X_k + Y_k\}$ has a renewal process with interarrival distribution $F(t) * G(t) = \int_0^t G(t-u)\mathrm{d}F(u)$ and mean $(\mu + \beta)$.

Suppose that the unit begins to operate at time 0. Let $N_j(t)$ be the number of visits to State j in $[0, t]$, where the first visit to State 0 at time 0 is not counted, and $M_j(t) \equiv E\{N_j(t)\}$ that is called the *renewal function*. Then, from (3.53),

$$\begin{aligned} M_0^*(s) &= \int_0^\infty \mathrm{e}^{-st}\mathrm{d}M_0(t) = \frac{F^*(s)G^*(s)}{1 - F^*(s)G^*(s)}, \\ M_1^*(s) &= \int_0^\infty \mathrm{e}^{-st}\mathrm{d}M_1(t) = \frac{F^*(s)}{1 - F^*(s)G^*(s)}. \end{aligned} \tag{8.8}$$

Let $Z(t)$ be the process at time t, and $P_j(t) \equiv \Pr\{Z(t) = j\}$ $(j = 0, 1)$. Then, from (3.55),

$$P_0^*(s) = \frac{1 - F^*(s)}{1 - F^*(s)G^*(s)}, \qquad P_1^*(s) = \frac{F^*(s)[1 - G^*(s)]}{1 - F^*(s)G^*(s)}. \tag{8.9}$$

Thus, the limiting probabilities $P_j \equiv \lim_{t\to\infty} P_j(t)$ are

$$P_0 = \lim_{s\to 0} P_0^*(s) = \frac{\mu}{\mu+\beta} = 1 - P_1, \tag{8.10}$$

that represents the steady-state availability of the unit.

Furthermore, the interval reliability is the probability that at a specified time, the unit is operating and will continue to operate for an interval of duration [3, pp. 48, 140]. Then, the interval reliability $R(x;t)$ for an interval of duration x starting at time t is, from (3.62) to (3.64),

$$R(x;t) = \overline{F}(t+x) + \int_0^t \overline{F}(t+x-u)\mathrm{d}M_0(u),$$

and its Laplace transform in Appendix A is

$$R^*(x;s) = \int_0^\infty \mathrm{e}^{-st} R(x;t)\mathrm{d}t = \frac{\mathrm{e}^{sx}\int_x^\infty \mathrm{e}^{-st}\overline{F}(t)\mathrm{d}t}{1 - F^*(s)G^*(s)}.$$

Thus, the limiting interval reliability is

$$R(x) \equiv \lim_{t\to\infty} R(x;t) = \lim_{s\to 0} sR^*(x;s) = \frac{\int_x^\infty \overline{F}(t)\mathrm{d}t}{\mu+\beta}. \tag{8.11}$$

Next, suppose that an operating unit undergoes repair at time T or at failure, whichever occurs first, from the beginning of its operation, and define $Z_k \equiv \min\{X_k, T\}$. Then, the process $\{Z_k, Y_k\}$ forms an alternating renewal process. On the other hand, when the repair is not completed until time T, it is replaced with a new one, that is called a *repair limit policy* in Example 3.7. Then, defining $Z_k \equiv \min\{Y_k, T\}$, the process $\{X_k, Z_k\}$ forms an alternating renewal process. In this case, because $E\{Z_k\} = \int_0^T \overline{G}(t)\mathrm{d}t$, the expected cost rate is, from (3.82),

$$C(T) = \frac{c_1\overline{G}(T) + \int_0^T \overline{G}(t)\mathrm{d}c_r(t)}{\mu + \int_0^T \overline{G}(t)\mathrm{d}t}, \tag{8.12}$$

where c_1 = replacement cost of a failed unit when its repair is not completed until time T and $c_r(t)$ = repair cost during $[0,t]$ (Problem 8.1).

8.2 Two-Unit Standby System

A two-unit standby redundant system with a single repair facility is one of the most fundamental and important redundant systems in reliability theory [4]. The system consists of two units, where one unit is used for operation and the other is in standby as an initial condition. If an operating unit fails, then it undergoes repair immediately and the other standby unit takes over its operation. Either of the two units is alternately operating. It is said that system failure occurs when both units are down simultaneously. The transition probabilities, the first-passage distributions, and the renewal functions are derived by using the techniques of Markov processes.

8.2.1 Exponential Failure and Repair Times

An operating unit has an exponential failure distribution $(1 - e^{-\lambda t})$, and a failed unit also has an exponential repair distribution $(1 - e^{-\theta t})$. If an operating unit fails and the other unit is in standby, the failed unit undergoes repair immediately, and the standby unit takes over its operation immediately. However, if an operating unit fails while the other unit is under repair, the failed unit has to wait until the repair facility become empty. This means that system failure has occurred. It is also assumed that a failed unit recovers its functioning upon repair completion and becomes as good as new. The unit in standby neither deteriorates nor fails in the standby interval. Two units are used alternately for its operation, as described above. Even if system failure occurs, the system can operate again upon repair completion. That is, the system repeats up and down alternately. All random variables considered here are independent of each other.

To analyze the above system, we define the following three system states that represent the states of the process:

State 0: One unit is operating and the other unit is in standby.
State 1: One unit is operating and the other unit is under repair.
State 2: One unit is under repair and the other unit is waiting for repair.

The system states defined above are regeneration points because the failure and repair times are exponential. Let $\{X(t) = j\}$ denote that the process is in State j $(j = 0, 1, 2)$ at time t. Then, the process $\{X(t)\}$ forms a *continuous-time Markov chain* with finite state set $\{0, 1, 2\}$. Let $P_{ij}(t) \equiv \Pr\{X(t) = j | X(0) = i\}$ be the transition probabilities from State i to State j at time t. By the similar method in Example 4.11,

$$
\begin{aligned}
P'_{00}(t) &= -\lambda P_{00}(t) + \theta P_{01}(t), \\
P'_{01}(t) &= \lambda P_{00}(t) - (\lambda + \theta) P_{01}(t) + \theta P_{02}(t), \\
P'_{02}(t) &= \lambda P_{01}(t) - \theta P_{02}(t).
\end{aligned}
$$

Taking the LS transformations of the above equations,

$$
\begin{aligned}
sP^*_{00}(s) - s &= -\lambda P^*_{00}(s) + \theta P^*_{01}(s), \\
sP^*_{01}(s) &= \lambda P^*_{00}(s) - (\lambda + \theta) P^*_{01}(s) + \theta P^*_{02}(s), \\
sP^*_{02}(s) &= \lambda P^*_{01}(s) - \theta P^*_{02}(s).
\end{aligned}
$$

Solving these equations,

$$P_{00}^{*}(s) = \frac{s^2 + (\lambda + 2\theta)s + \theta^2}{s^2 + 2(\lambda + \theta)s + \lambda^2 + \lambda\theta + \theta^2},$$

$$P_{01}^{*}(s) = \frac{\lambda(s + \theta)}{s^2 + 2(\lambda + \theta)s + \lambda^2 + \lambda\theta + \theta^2},$$

$$P_{02}^{*}(s) = \frac{\lambda^2}{s^2 + 2(\lambda + \theta)s + \lambda^2 + \lambda\theta + \theta^2}.$$

Taking the inverse LS transforms,

$$P_{00}(t) = \frac{\theta^2}{\lambda^2 + \lambda\theta + \theta^2} + \frac{\lambda}{\omega_1 - \omega_2}\left(\frac{\omega_1 - \lambda - \theta}{\omega_1}\mathrm{e}^{-\omega_1 t} - \frac{\omega_2 - \lambda - \theta}{\omega_2}\mathrm{e}^{-\omega_2 t}\right),$$

$$P_{01}(t) = \frac{\lambda\theta}{\lambda^2 + \lambda\theta + \theta^2} + \frac{\lambda}{\omega_1 - \omega_2}\left(\frac{-\omega_1 + \theta}{\omega_1}\mathrm{e}^{-\omega_1 t} - \frac{-\omega_2 + \theta}{\omega_2}\mathrm{e}^{-\omega_2 t}\right),$$

$$P_{02}(t) = \frac{\lambda^2}{\lambda^2 + \lambda\theta + \theta^2} + \frac{\lambda^2}{\omega_1 - \omega_2}\left(\frac{1}{\omega_1}\mathrm{e}^{-\omega_1 t} - \frac{1}{\omega_2}\mathrm{e}^{-\omega_2 t}\right), \qquad (8.13)$$

where $P_{00}(t) + P_{01}(t) + P_{02}(t) = 1$ (Problem 8.2), and

$$\omega_1 \equiv \lambda + \theta + \sqrt{\lambda\theta}, \qquad \omega_2 \equiv \lambda + \theta - \sqrt{\lambda\theta}.$$

8.2.2 General Failure and Repair Times

An operating unit has a general failure distribution $F(t)$ with finite mean μ, and a failed unit has a general repair distribution $G(t)$ with finite mean β. The other assumptions are the same ones in Sect. 8.2.1. However, State 0 is not a regeneration point except when an operating unit fails for the first time from the beginning of its operation at time 0, and the failure time is exponential. From this reason, we define an initial time instant:

State -1: One unit begins to operate and the other unit is in standby.

The other states are the same as those in Sect. 8.2.1. An epoch at which the process makes a transition into State 1 is a regeneration point. However, the epochs at which the system makes a transition into State j ($j = 0, 2$) are not regeneration points, except the failure and repair times are exponential (Fig. 5.3).

We analyze the system by using the techniques of a Markov renewal process in Sect. 5.3. Define a mass function $Q_{ij}(t)$ from State i ($i = -1, 1$) to State j ($j = 0, 1, 2$) by the probability that after making a transition into State i, the process next makes a transition into State j, in a smaller amount of time than time t in a Markov renewal process. Then, from Sect. 5.3.2, we have the following mass functions:

$$\begin{aligned} Q_{-11}(t) &= F(t), \\ Q_{10}(t) &= \int_0^t \overline{F}(u)\mathrm{d}G(u), \quad Q_{12}(t) = \int_0^t \overline{G}(u)\mathrm{d}F(u). \end{aligned} \tag{8.14}$$

It is impossible to define the mass function $Q_{01}(t)$ and $Q_{21}(t)$ because the epochs for States 0 and 2 are not regeneration points. Thus, define a new mass function $Q_{11}^{(0)}(t)$ $(Q_{11}^{(2)}(t))$ that is the recurrence time distribution for State 1 via State 0 (State 2). That is, $Q_{11}^{(0)}(t)$ $(Q_{11}^{(2)}(t))$ is the probability that after making a transition into State 1, the process next makes a transition into State 0 (State 2) and returns to State 1 in time t. Then, from Sect. 5.3.2,

$$Q_{11}^{(0)}(t) = \int_0^t G(u)\mathrm{d}F(u), \qquad Q_{11}^{(2)}(t) = \int_0^t F(u)\mathrm{d}G(u). \tag{8.15}$$

Therefore, the LS transforms of the mass functions are

$$\begin{aligned} Q_{-11}^*(s) &= F^*(s), \\ Q_{10}^*(s) &= \int_0^\infty \mathrm{e}^{-st}\overline{F}(t)\mathrm{d}G(t), \qquad Q_{12}^*(s) = \int_0^\infty \mathrm{e}^{-st}\overline{G}(t)\mathrm{d}F(t), \\ Q_{11}^{*(0)}(s) &= \int_0^\infty \mathrm{e}^{-st}G(t)\mathrm{d}F(t), \quad Q_{11}^{*(2)}(s) = \int_0^\infty \mathrm{e}^{-st}F(t)\mathrm{d}G(t). \end{aligned} \tag{8.16}$$

8.2.2.1 First-Passage Distribution

We obtain the first-passage distributions and their mean times by using the mass functions. Let $H_{ij}(t)$ denote the first-passage distributions from State i $(i = -1, 1)$ to State j $(j = 0, 1, 2)$. First, suppose that the process starts in the epoch for State 1. Then, from (8.15), the recurrence time distribution $H_{11}(t)$ for State 1 is

$$H_{11}(t) = Q_{11}^{(0)}(t) + Q_{11}^{(2)}(t) = F(t)G(t),$$

that is the probability that the process comes back to State 1 via State 0 or State 2, whichever occurs first, and its LS transform is, from (8.16),

$$H_{11}^*(s) = \int_0^\infty \mathrm{e}^{-st}\mathrm{d}H_{11}(t) = \int_0^\infty \mathrm{e}^{-st}F(t)\mathrm{d}G(t) + \int_0^\infty \mathrm{e}^{-st}G(t)\mathrm{d}F(t). \tag{8.17}$$

Thus, the mean recurrence time l_{11} for State 1 is

$$l_{11} = \int_0^\infty t\,\mathrm{d}H_{11}(t) = \mu + \beta - \gamma, \tag{8.18}$$

where $\gamma \equiv \int_0^\infty \overline{F}(t)\overline{G}(t)\mathrm{d}t$.

The first-passage distributions $H_{1j}(t)$ $(j = 0, 2)$ are given by solving the following renewal-type equation:

$$H_{10}(t) = Q_{10}(t) + Q_{11}^{(2)}(t) * H_{10}(t), \tag{8.19}$$

where the asterisk mark represents the Stieltjes convolution, i.e., $a(t) * b(t) \equiv \int_0^t b(t-u)\mathrm{d}a(u)$ for any functions $a(t)$ and $b(t)$. The first term at the right-hand side in (8.19) represents the probability that the process goes to State 0 directly, and the second term is the probability that the process goes to State 0 after returning to State 1 via State 2. Similarly,

$$H_{12}(t) = Q_{12}(t) + Q_{11}^{(0)}(t) * H_{12}(t). \tag{8.20}$$

If the process starts from State -1, then,

$$H_{-1j}(t) = Q_{-11}(t) * H_{1j}(t) \qquad (j = 0, 2). \tag{8.21}$$

Therefore, taking the LS transforms of (8.20)–(8.22) and solving them,

$$\begin{aligned} H_{-10}^*(s) &= \frac{Q_{-11}^*(s)Q_{10}^*(s)}{1 - Q_{11}^{*(2)}(s)} = \frac{F^*(s)\int_0^\infty \mathrm{e}^{-st}\overline{F}(t)\mathrm{d}G(t)}{1 - \int_0^\infty \mathrm{e}^{-st}F(t)\mathrm{d}G(t)}, \\ H_{-12}^*(s) &= \frac{Q_{-11}^*(s)Q_{12}^*(s)}{1 - Q_{11}^{*(0)}(s)} = \frac{F^*(s)\int_0^\infty \mathrm{e}^{-st}\overline{G}(t)\mathrm{d}F(t)}{1 - \int_0^\infty \mathrm{e}^{-st}G(t)\mathrm{d}F(t)}. \end{aligned} \tag{8.22}$$

The mean first-passage times l_{-1j} $(j = 0, 2)$, starting from State -1, are

$$\begin{aligned} l_{-10} &= \lim_{s\to 0}\frac{1 - H_{-10}^*(s)}{s} = \mu + \frac{\beta}{\int_0^\infty \overline{F}(t)\mathrm{d}G(t)}, \\ l_{-12} &= \lim_{s\to 0}\frac{1 - H_{-12}^*(s)}{s} = \mu\left[1 + \frac{1}{\int_0^\infty \overline{G}(t)\mathrm{d}F(t)}\right], \end{aligned} \tag{8.23}$$

that represent the mean times to system recovery and system failure, respectively. Furthermore, the mean time of the busy period of a repair facility is

$$l_{10} = \lim_{s\to 0}\frac{1 - H_{10}^*(s)}{s} = l_{-10} - \mu = \frac{\beta}{\int_0^\infty \overline{F}(t)\mathrm{d}G(t)}.$$

The mean downtime l_d after system failure is

$$\begin{aligned} l_d &= \int_0^\infty t\,\mathrm{d}Q_{11}^{(2)}(t) - \int_0^\infty t\,\mathrm{d}Q_{12}(t) \\ &= \int_0^\infty F(t)\overline{G}(t)\mathrm{d}t = \beta - \gamma. \end{aligned}$$

8.2.2.2 Expected Number of Visits to States

Consider the expected number of visits to State j ($j = 0, 1, 2$) during the interval $[0, t]$. Let the renewal function $M_{ij}(t)$ be the expected number of occurrences of State j in $[0, t]$, starting in the epoch for State i ($i = -1, 1$) at time 0, where the first visit to j is not counted if $i = j$.

First, when the process starts from State 1 at time 0, we have the following renewal-type equation:

$$\begin{aligned} M_{10}(t) &= Q_{10}(t) - Q_{11}^{(0)}(t) + Q_{11}^{(0)}(t) * [1 + M_{10}(t)] + Q_{11}^{(2)}(t) * M_{10}(t) \\ &= Q_{10}(t) + H_{11}(t) * M_{10}(t), \end{aligned} \tag{8.24}$$

where the first term is the probability that the process goes to State 0 directly and then remains in State 0, the second term is the expected number that the process returns to State 1 via State 0 and goes to State 0, and finally, the third term is the expected number that the process returns to State 1 via State 2 and then goes to State 0. Similarly,

$$\begin{aligned} M_{11}(t) &= Q_{11}^{(0)}(t) * [1 + M_{11}(t)] + Q_{11}^{(2)}(t) * [1 + M_{11}(t)] \\ &= H_{11}(t) * [1 + M_{11}(t)], \\ M_{12}(t) &= Q_{12}(t) - Q_{11}^{(2)}(t) + Q_{11}^{(0)}(t) * M_{12}(t) + Q_{11}^{(2)}(t) * [1 + M_{12}(t)] \\ &= Q_{12}(t) + H_{11}(t) * M_{12}(t). \end{aligned} \tag{8.25}$$

The expected numbers of visits to State j ($j = 0, 1, 2$) in $[0, t]$, starting from State -1, are

$$\begin{aligned} M_{-11}(t) &= Q_{-11}(t) * [1 + M_{11}(t)], \\ M_{-1j}(t) &= Q_{-11}(t) * M_{1j}(t) \qquad (j = 0, 2). \end{aligned} \tag{8.26}$$

Taking the LS transforms of (8.24)–(8.26) and solving them,

$$\begin{aligned} M_{-10}^*(s) &= \frac{Q_{-11}^*(s) Q_{10}^*(s)}{1 - H_{11}^*(s)} = \frac{F^*(s) \int_0^\infty \mathrm{e}^{-st} \overline{F}(t) \mathrm{d}G(t)}{1 - H_{11}^*(s)}, \\ M_{-11}^*(s) &= \frac{Q_{-11}^*(s)}{1 - H_{11}^*(s)} = \frac{F^*(s)}{1 - H_{11}^*(s)}, \\ M_{-12}^*(s) &= \frac{Q_{-11}^*(s) Q_{12}^*(s)}{1 - H_{11}^*(s)} = \frac{F^*(s) \int_0^\infty \mathrm{e}^{-st} \overline{G}(t) \mathrm{d}F(t)}{1 - H_{11}^*(s)}. \end{aligned} \tag{8.27}$$

In addition, if $F(t)$ and $G(t)$ are non-lattice and their mean times are finite, from Theorem 3.2, there exist the limits $M_j \equiv \lim_{t \to \infty} M_{ij}(t)/t$ ($j = 0, 1, 2$) of expected numbers of visits to State j per unit of time in the steady-state that are independent of an initial State i. Thus, by applying a Tauberian theorem in Appendix A.3 to (8.27), respectively,

$$M_0 = \lim_{s\to 0} sM^*_{-10}(s) = \frac{1}{l_{11}}\int_0^\infty \overline{F}(t)\mathrm{d}G(t),$$
$$M_2 = \lim_{s\to 0} sM^*_{-12}(s) = \frac{1}{l_{11}}\int_0^\infty \overline{G}(t)\mathrm{d}F(t),$$
$$M_1 = \lim_{s\to 0} sM^*_{-11}(s) = \frac{1}{l_{11}} = M_0 + M_2.$$

A quantity of some interest is the total number of units that have failed and have been repaired in $[0, t]$, respectively. Letting $M_{if}(t)$ and $M_{ir}(t)$ be the respected number of failed and repaired units in $[0, t]$, when the process starts from State i $(i = -1, 1)$,

$$\begin{aligned} M_{1f}(t) &= Q_{12}(t) - Q_{11}^{(2)}(t) + H_{11}(t) * [1 + M_{1f}(t)] \\ &= F(t)\overline{G}(t) + H_{11}(t) * [1 + M_{1f}(t)], \\ M_{1r}(t) &= Q_{10}(t) - Q_{11}^{(0)}(t) + H_{11}(t) * [1 + M_{1r}(t)] \\ &= \overline{F}(t)G(t) + H_{11}(t) * [1 + M_{1r}(t)]. \end{aligned} \tag{8.28}$$

If the process starts from State -1, then

$$\begin{aligned} M_{-1f}(t) &= Q_{-11}(t) * [1 + M_{1f}(t)], \\ M_{-1r}(t) &= Q_{-11}(t) * M_{1r}(t). \end{aligned} \tag{8.29}$$

Thus, combining (8.28) and (8.29),

$$\begin{aligned} M^*_{-1f}(s) &= \frac{Q^*_{-11}(s)\left[1 + Q^*_{12}(s) - Q_{11}^{*(2)}(s)\right]}{1 - H^*_{11}(s)}, \\ M^*_{-1r}(s) &= \frac{Q^*_{-11}(s)\left[Q^*_{10}(s) + Q_{11}^{*(2)}(s)\right]}{1 - H^*_{11}(s)}. \end{aligned} \tag{8.30}$$

Clearly,

$$M_f \equiv \lim_{t\to\infty} \frac{M_{-1f}(t)}{t} = M_r = \lim_{t\to\infty} \frac{M_{-1r}(t)}{t} = M_1.$$

8.2.2.3 Transition Probabilities

Denote that $P_{ij}(t)$ is the transition probability that the process is in State j $(j = 0, 1, 2)$ at time t, starting from State i $(i = -1, 1)$ at time 0. Then, we have the following renewal equations:

$$\begin{aligned}
P_{10}(t) &= Q_{10}(t) - Q_{11}^{(0)}(t) + H_{11}(t) * P_{10}(t) \\
&= \overline{F}(t)G(t) + H_{11}(t) * P_{10}(t), \\
P_{11}(t) &= 1 - Q_{10}(t) - Q_{12}(t) + H_{11}(t) * P_{11}(t) \\
&= \overline{F}(t)\overline{G}(t) + H_{11}(t) * P_{11}(t), \\
P_{12}(t) &= Q_{12}(t) - Q_{11}^{(2)}(t) + H_{11}(t) * P_{12}(t) \\
&= F(t)\overline{G}(t) + H_{11}(t) * P_{12}(t).
\end{aligned} \tag{8.31}$$

Furthermore, if the process starts from State -1, then

$$\begin{aligned}
P_{-10}(t) &= 1 - Q_{-11}(t) + Q_{-11}(t) * P_{10}(t) \\
&= \overline{F}(t) + Q_{-11}(t) * P_{10}(t), \\
P_{-1j}(t) &= Q_{-11}(t) * P_{1j}(t) \qquad (j = 0, 2).
\end{aligned} \tag{8.32}$$

Thus, the LS transforms of $P_{-1j}(t)$ $(j = 0, 1, 2)$ are

$$\begin{aligned}
P^*_{-10}(s) &= 1 - Q^*_{-11}(s) + \frac{Q^*_{-11}(s)\left[Q^*_{10}(s) - Q^{*(0)}_{11}(s)\right]}{1 - H^*_{11}(s)}, \\
P^*_{-11}(s) &= \frac{Q^*_{-11}(s)\left[1 - Q^*_{10}(s) - Q^*_{12}(s)\right]}{1 - H^*_{11}(s)}, \\
P^*_{-12}(s) &= \frac{Q^*_{-11}(s)\left[Q^*_{12}(s) - Q^{*(2)}_{11}(s)\right]}{1 - H^*_{11}(s)}.
\end{aligned} \tag{8.33}$$

It is evident that because $P^*_{-10}(s) + P^*_{-11}(s) + P^*_{-12}(s) = 1$, $P_{-10}(t) + P_{-11}(t) + P_{-12}(t) = 1$. Note that $P_{-11}(t) = M_{-11}(t) - M_{-10}(t) - M_{-12}(t)$.

If $F(t)$ and $G(t)$ are non-lattice and their means are finite, there exist the limiting probabilities $P_j \equiv \lim_{t\to\infty} P_{ij}(t)$ $(j = 0, 1, 2)$ that are independent of an initial State i. Thus, by applying a Tauberian theorem to (8.33),

$$\begin{aligned}
P_0 &= \lim_{s\to 0} P^*_{-10}(s) = 1 - \frac{\beta}{l_{11}}, \\
P_1 &= \lim_{s\to 0} P^*_{-11}(s) = -1 + \frac{\mu + \beta}{l_{11}}, \\
P_2 &= \lim_{s\to 0} P^*_{-12}(s) = 1 - \frac{\mu}{l_{11}}.
\end{aligned}$$

It is noted that $A(t) = P_{-10}(t) + P_{-11}(t)$ represents the pointwise availability of the system at time t, and $\overline{A}(t) = P_{-12}(t)$ represents the pointwise unavailability at time t, given that two units are initial good. It is also noted that $A = P_0 + P_1 = \mu/l_{11}$ represents the steady-state availability and $\overline{A} = P_2$ represents the steady-state unavailability (Problem 8.3). In other words, the so-called availability represents the ratio of the mean operating time to the total of mean operating and repair times [3, p. 9].

8.3 Standby System with Spare Units

Consider the system with a main unit and n spare units [5]. It is assumed that the spare units are statistically not identical to the main unit, but have the same functions as the main unit, if they take over the system operation.

The system operates in such a way that if the main unit fails then it undergoes repair immediately and one of the spare units replaces it. As soon as the repair of the main unit is completed, it begins to operate and the operating spare unit is available for further use. A failed spare unit is scrapped. The system can operate until the nth spare unit fails. In other words, the system fails when the last spare unit fails while the main unit is under repair.

The operating behaviors of this model require the following knowledge:

(1) Distribution of time and mean time to first system failure, given that n spare units are provided at time 0.
(2) Probability that the number of failed spare units is exactly equal to n and its expected number in $[0, t]$.

These quantities are derived from renewal equations and used for two optimization problems and to determine an initial number of spare units to stock.

8.3.1 First-Passage Time to System Failure

Assume that the failure time of the main unit has a general distribution $F(t)$ with mean μ and its repair time has a general distribution $G(t)$ with mean β. That is, the main unit forms an alternating renewal process that repeats operating and failed states. It is also assumed that the failure time of each operating spare unit has a general distribution $F_S(t)$ with mean μ_S, even if it has been used before, i.e., the life of spare units that have already been in service is not affected by the past. All random variables considered here are independent. Unless otherwise stated, all units are good at time 0, and any failures are instantaneously detected and repaired or scrapped. Each switchover is perfect and each switchover time is instantaneous.

Using the modified methods of an alternating renewal process in Sect. 3.4, we obtain the reliability properties of the system. First, suppose that n spare units are provided at time 0. Let $H_k(t)$ $(k = 1, 2, \ldots, n)$ denote the first-passage distribution to system failure when k spare units are provided at time 0. Then, we have the following renewal equation:

$$H_n(t) = F(t) * \left\{ \int_0^t \overline{G}(u) \mathrm{d}F_S^{(n)}(u) + \sum_{k=0}^{n-1} H_{n-k}(t) * \int_0^t \left[F_S^{(k)}(u) - F_S^{(k+1)}(u) \right] \mathrm{d}G(u) \right\} \quad (n = 1, 2, \ldots), \tag{8.34}$$

where the asterisk mark represents the convolution, and $F_S^{(k)}(t)$ $(k = 1, 2, \ldots)$ represents the k-fold Stieltjes convolution of $F_S(t)$ with itself and $F_S^{(0)}(t) \equiv 1$ for $t \geq 0$. The first term of the bracket on the right-hand side in (8.34) is the distribution of the time that all of n spare units fail before the first repair completion of the failed main unit. The second term is the distribution of the time that k $(k = 0, 1, \ldots, n-1)$ spare units fail exactly until the first repair completion of the main unit, and then, the system with $n - k$ spare units operates again.

The first-passage distribution $H_n(t)$ can be calculated recursively and uniquely determined from (8.34). Introduce the notation of the generating function of the LS transform:

$$H^*(z, s) \equiv \sum_{n=1}^{\infty} z^n \int_0^{\infty} \mathrm{e}^{-st} \mathrm{d}H_n(t)$$

for $|z| \leq 1$. Thus, taking the LS transform of (8.34) and forming the generating function $H^*(z, s)$,

$$H^*(z, s) = \frac{F^*(s) \sum_{n=1}^{\infty} z^n \int_0^{\infty} \mathrm{e}^{-st} \overline{G}(t) \mathrm{d}F_S^{(n)}(t)}{1 - F^*(s) \sum_{n=0}^{\infty} z^n \int_0^{\infty} \mathrm{e}^{-st} \left[F_S^{(n)}(t) - F_S^{(n+1)}(t) \right] \mathrm{d}G(t)}. \tag{8.35}$$

Moreover, let l_n denote the mean first-passage time to system failure. Then, the renewal equation for l_n is easily given, from (8.34),

$$l_n = \mu + \int_0^{\infty} \left[1 - F_S^{(n)}(t) \right] \overline{G}(t) \mathrm{d}t + \sum_{k=0}^{n-1} l_{n-k} \int_0^{\infty} \left[F_S^{(k)}(t) - F_S^{(k+1)}(t) \right] \mathrm{d}G(t) \quad (n = 1, 2, \ldots), \tag{8.36}$$

and hence, its generating function is

$$l^*(z) \equiv \sum_{n=1}^{\infty} z^n l_n = \frac{\mu z/(1 - z) + \sum_{n=1}^{\infty} z^n \int_0^{\infty} \left[1 - F_S^{(n)}(t) \right] \overline{G}(t) \mathrm{d}t}{1 - \sum_{n=0}^{\infty} z^n \int_0^{\infty} \left[F_S^{(n)}(t) - F_S^{(n+1)}(t) \right] \mathrm{d}G(t)}. \tag{8.37}$$

8.3.2 Expected Number of Failed Units

We derive the expected number of failed spare units in $[0,t]$. Let $M_n(t)$ $(n=0,1,\ldots)$ be the probability that the total number of failed spare units in $[0,t]$ are n, and $M(t) \equiv \sum_{n=1}^{\infty} nM_n(t)$. Then, in a similar way of obtaining (8.34),

$$M_0(t) = \overline{F}(t) + F(t) * \left[\overline{F}_S(t)\overline{G}(t) + M_0(t) * \int_0^t \overline{F}_S(u)\mathrm{d}G(u)\right],$$
$$M_n(t) = F(t) * \left\{\left[F_S^{(n)}(t) - F_S^{(n+1)}(t)\right]\overline{G}(t) + \sum_{k=0}^{n} M_{n-k}(t) * \int_0^t \left[F_S^{(k)}(u) - F_S^{(k+1)}(u)\right]\mathrm{d}G(u)\right\} \quad (n=1,2,\ldots). \tag{8.38}$$

Introduce the notation

$$M^*(z,s) \equiv \sum_{n=0}^{\infty} z^n \int_0^{\infty} \mathrm{e}^{-st}\mathrm{d}M_n(t).$$

Then, from (8.38),

$$M^*(z,s) = \frac{1 - F^*(s)\left(1 - \sum_{n=0}^{\infty} z^n \int_0^{\infty} \mathrm{e}^{-st}\mathrm{d}\left\{\left[F_S^{(n)}(t) - F_S^{(n+1)}(t)\right]\overline{G}(t)\right\}\right)}{1 - F^*(s)\sum_{n=0}^{\infty} z^n \int_0^{\infty} \mathrm{e}^{-st}\left[F_S^{(n)}(t) - F_S^{(n+1)}(t)\right]\mathrm{d}G(t)}. \tag{8.39}$$

It can be easily shown that $\lim_{z\to 1} M^*(z,s) = 1$. The LS transform of the expected number of failed spare units in $[0,t]$ is

$$M^*(s) \equiv \sum_{n=1}^{\infty} n \int_0^{\infty} \mathrm{e}^{-st}\mathrm{d}M_n(t) = \lim_{z\to 1}\frac{\mathrm{d}M^*(z,s)}{\mathrm{d}z}$$
$$= \frac{F^*(s)\int_0^{\infty} \mathrm{e}^{-st}\overline{G}(t)\mathrm{d}M_S(t)}{1 - F^*(s)G^*(s)}, \tag{8.40}$$

where note that $M_S(t) \equiv \sum_{n=1}^{\infty} F_S^{(n)}(t)$ is the renewal function for spare units.

Furthermore, the expected number of failed spare units per unit of time in the steady-state is

$$M \equiv \lim_{t\to\infty}\frac{M(t)}{t} = \frac{\int_0^{\infty} M_S(t)\mathrm{d}G(t)}{\mu + \beta}. \tag{8.41}$$

Equation (8.41) can be intuitively explained because the numerator represents the total expected number of failed spare units during the repair time of the main unit, and the denominator represents the mean time from the beginning of the operating main unit to its repair completion.

Further, we obtain the expected number m_n of failed main units until system failure for the system with n spare units. Because

$$\int_0^\infty \overline{G}(t)\mathrm{d}F_S^{(n)}(t) + \sum_{k=0}^{n-1}\int_0^\infty \left[F_S^{(k)}(t) - F_S^{(k+1)}(t)\right]\mathrm{d}G(t) = 1,$$

and by the similar method of obtaining (8.36),

$$m_n = 1 + \sum_{k=0}^{n-1} m_{n-k}\int_0^\infty \left[F_S^{(k)}(t) - F_S^{(k+1)}(t)\right]\mathrm{d}G(t).$$

Thus, the generating function of m_n is

$$m^*(z) \equiv \sum_{n=1}^{\infty} z^n m_n = \frac{z/(1-z)}{1 - \sum_{n=0}^{\infty} z^n \int_0^\infty \left[F_S^{(n)}(t) - F_S^{(n+1)}(t)\right]\mathrm{d}G(t)}. \tag{8.42}$$

Example 8.1 (Exponential repair time) Suppose that $G(t) = 1 - \mathrm{e}^{-\theta t}$. Then, from (8.37), the mean time to system failure, given that n pare units are provided at time 0, is (Problem 8.4)

$$l_n = \mu + n\left(\mu + \frac{1}{\theta}\right)\frac{1 - F_S^*(\theta)}{F_S^*(\theta)}.$$

Note that, by adding one spare unit to the system, the mean time to system increases by constant A, independent of the number of the former spare units, where $A \equiv (\mu + 1/\theta)[1 - F_S^*(\theta)]/F_S^*(\theta)$.

Furthermore, the LS transform of the expected number of failed spare units in $[0, t]$ is, from (8.40),

$$M^*(s) = \frac{F^*(s)F_S^*(s+\theta)}{[1 - F_S^*(s+\theta)]\{1 - [\theta/(s+\theta)]F^*(s)\}},$$

and its limit per unit of time is, from (8.41) (Problem 8.4),

$$M = \frac{1}{\mu + 1/\theta}\frac{F_S^*(\theta)}{1 - F_S^*(\theta)} = \frac{1}{A}.$$

In a similar way, the expected number of failed main units until system failure is, from (8.42) (Problem 8.4),

$$m_n = 1 + n\frac{1 - F_S^*(\theta)}{F_S^*(\theta)},$$

where note that

$$m_n = \frac{l_n + 1/\theta}{\mu + 1/\theta}.$$

8.3.3 Expected Cost and Optimization Problems

We derive the expected cost rate by introducing costs incurred for each failed main unit and failed spare units when infinite spare units are provided at time 0. This expected cost is easily deduced from the expected number of failed units. Then, we compare the expected cost rates for the system with both main unit and spare units, and for the system with only spare units, and determine which of them is more economical.

Cost c_1 is incurred for each failed main unit that is repaired, and cost c_S is incurred for each failed spare unit that is replaced and scrapped. Letting $C(t)$ be total expected cost incurred in $[0, t]$,

$$C(t) \equiv c_1 M_R(t) + c_S M(t),$$

where $M_R(t)$ represents the expected number of repairs for the failed main unit, and recall that $M(t)$ is the expected number of failed spare units in $[0, t]$. It is easily seen in Sect. 3.4 that

$$M_R \equiv \lim_{t\to\infty} \frac{M_R(t)}{t} = \frac{1}{\mu + \beta}.$$

Thus, the expected cost rate is, from (3.19) and (8.41),

$$C \equiv \lim_{t\to\infty} \frac{C(t)}{t} = c_1 M_R + c_S M$$
$$= \frac{c_1 + c_S \int_0^\infty M_S(t)\mathrm{d}G(t)}{\mu + \beta}, \tag{8.43}$$

that is also equal to the total expected cost of one cycle from the beginning of the operating main unit to its repair completion. On the other hand, if only a spare unit is allowed to operate, then the expected cost rate is

$$C_S = \frac{c_S}{\mu_s}. \tag{8.44}$$

Therefore, comparing (8.43) and (8.44), $C \leq (\geq) C_S$ if and only if

$$c_1 \leq (\geq) c_S \left[\frac{\mu + \beta}{\mu_s} - \int_0^\infty M_S(t)\mathrm{d}G(t) \right]. \tag{8.45}$$

In particular, when $G(t) = 1 - \mathrm{e}^{-\theta t}$, (8.45) becomes

$$c_1 \le (\ge) c_S \left[\frac{\mu + 1/\theta}{\mu_s} - \frac{F_S^*(\theta)}{1 - F_S^*(\theta)} \right]. \tag{8.46}$$

In this case, if the mean time l_n to system failure is required to be greater than D_T, then we can provide a minimum number of spare units that satisfies

$$n \ge \frac{D_T - \mu}{\mu + 1/\theta} \frac{1 - F_S^*(\theta)}{F_S^*(\theta)}.$$

If the expected cost until the mean time to system failure that is given by

$$c_S n + c_1 m_n = c_1 + n \left[c_S + c_1 \frac{1 - F_S^*(\theta)}{F_S^*(\theta)} \right]$$

is less than D_C, we can provide a maximum number of spare units that satisfies

$$n \le \frac{D_C - c_1}{c_S + c_1 \{ [1 - F_S^*(\theta)] / F_S^*(\theta) \}}$$

(Problem 8.5).

8.4 Problems 8

8.1 In (8.12), derive an optimum repair limit time T^* that minimizes $C(T)$ when $c_r(t) = at$ for $a > 0$ [3, p. 52].

8.2 Derive the transition probabilities of a two-unit standby system with standby failure, i.e., when the unit in standby has the failure rate λ_S, and the other assumptions are the same ones in Sect. 8.2.1 [4].

8.3 Consider the same two-unit standby system in Sect. 8.2.2 where each unit has a failure distribution $F(t)$ with mean μ and a failed unit has a repair distribution $G_1(t)$ with mean β_1. We adopt the following PM policy for an operating unit. When an operating unit works for a specified time T without failure, we stop its operation for PM. It is assumed that the time to PM completion has a general distribution $G_2(t)$ with mean β_2, and any unit becomes as good as new upon repair or PM completion. Furthermore, we make the following two assumptions.

(a) PM of an operating unit is done only if the other unit is in standby.

(b) An operating unit that has forfeited PM because of (1) undergoes PM just upon repair or PM completion of the other unit.

Under the above assumptions, obtain the availability and derive an optimum PM time that maximizes it [3, p. 144].

8.4 In Example 8.1, derive l_n, m_n and M.

8.5 Consider the PM policy in Sect. 8.3 where the operating main unit undergoes PM at time T $(0<T\leq\infty)$ after its installation or is repaired at failure, whichever occurs first. Spare units are infinite and work temporarily during the interval of the repair or PM time of the main unit. It is assumed that the PM time has the same distribution $G(t)$ as the repair time. The main unit is as good as new upon repair or PM and begins to operate immediately. The costs incurred for each failed main unit and each spare unit are c_1 and c_s, respectively. The PM cost c_2 with $c_2<c_1$ is incurred for each non-failed main unit. Then, obtain the expected cost rate and derive an optimum PM time T^* that minimizes it [5].

References

1. Birolini A (1999) Reliability engineering theory and practice. Springer, New York
2. Brown M, Proschan F (1983) Imperfect repair. J Appl Prob 20:851–859
3. Nakagawa T (2005) Maintenance theory of reliability. Springer, London
4. Nakagawa T (2002) Two-unit redundant models. In: Osaki S (ed) Stochastic models in reliability and maintenance. Springer, Berlin, pp 165–185
5. Nakagawa T, Osaki S (1976) Reliability analysis of a one-unit system with unrepairable spare units and it optimization applications. Oper Res Q 27:101–110

Appendix A: Laplace Transform

This appendix presents some results of Laplace and Laplace–Stieltjes (LS) transforms and give useful tools used in reliability theory [1–4].

A.1 Laplace and Laplace–Stieltjes Transforms

Let $f(t)$ be a function of t specified for $t \geq 0$ and s be a complex number. Then, the Laplace transform of $f(t)$ is defined by

$$L\{f(t)\} \equiv f^*(s) = \int_0^\infty e^{-st} f(t) dt, \tag{A.1}$$

that is a function of the complex number s. Although s is defined as the complex number, we do not need the knowledge of complex analysis and may use the Laplace transform as if it is a real number in reliability theory. If the above integral (A.1) converges for $Re(s) > s_0$, then it always exists for $Re(s) > s_0$. For example, suppose that $f(t)$ is a function of $t \geq 0$ such that $|f(t)| \leq M e^{\alpha t}$ for some M and α. Then, $f^*(s)$ exists for all $Re(s) > \alpha$.

The function $f(t)$ is called the inverse Laplace transform of $f^*(s)$ and is written

$$f(t) = \mathcal{L}^{-1}\{f^*(s)\}. \tag{A.2}$$

When $f(t)$ is a probability density of a nonnegative random variable X,

$$E\{e^{-sX}\} = \int_0^\infty e^{-st} f(t) dt = f^*(s). \tag{A.3}$$

Similarly, let $F(t)$ be a function of $t \geq 0$. The Laplace–Stieltjes (LS) transform of $F(t)$ is defined by

$$F^*(s) \equiv \int_0^\infty e^{-st} dF(t) \tag{A.4}$$

for $Re(s) > s_0$. When $F(t)$ is a probability distribution of a random variable X, the LS transform of $F(t)$ is

$$E\{e^{-sX}\} = \int_0^\infty e^{-st} dF(t) = F^*(s), \tag{A.5}$$

that agrees with the Laplace transform $f^*(s)$ in (A.3). From (A.5), the nth moment of a random variable X is

$$E\{X\} = -\frac{dF^*(s)}{ds}\bigg|_{s=0} = \lim_{s\to 0} \frac{1 - F^*(s)}{s},$$

$$E\{X^n\} = -\frac{d^n F^*(s)}{ds^n}\bigg|_{s=0} \quad (n = 1, 2, \ldots),$$

$$V\{X\} = E\{X^2\} - (E\{X\})^2.$$

There is no theoretical difference greatly between Laplace and LS transforms, because

$$F^*(s) = \int_0^\infty e^{-st} dF(t) = s \int_0^\infty e^{-st} F(t) dt. \tag{A.6}$$

In other words, the LS transform can be obtained by s times the Laplace transform. It would be more convenient to use LS transforms because we deal mainly with distribution forms more than density ones in reliability theory. However, we sometimes use the Laplace transform for some functions in p. 61, p. 77, p. 78, p.184 and p. 196.

A.2 Properties

We always use LS transforms, except for some special functions through this book and summarize their brief properties. LS transforms play significant, important roles in making integral and differential calculuses, and in calculating distributions of the sum of random variables.

Let $F^*(s)$ be the LS transform of $F(t)$ in (A.5).

1. The LS transform of $\int_0^t F(u)\mathrm{d}u$ is

$$\int_0^\infty \mathrm{e}^{-st}\mathrm{d}\left[\int_0^t F(u)\mathrm{d}u\right] = \int_0^\infty \mathrm{e}^{-st}F(t)\mathrm{d}t = \frac{1}{s}F^*(s). \tag{A.7}$$

2. The LS transform of $F'(t)$ is

$$\int_0^\infty \mathrm{e}^{-st}\mathrm{d}F'(t) = s\int_0^\infty \mathrm{e}^{-st}F'(t)\mathrm{d}t = s[F^*(s) - F(0)]. \tag{A.8}$$

3. If the function $H(t)$ is given by the convolution $H(t) = \int_0^t G(t-u)\mathrm{d}F(u)$, and the LS transforms of $G(t)$ and $F(t)$ are $G^*(s)$ and $F^*(s)$, respectively, the LS transform of $H(t)$ is

$$\begin{aligned} H^*(s) &= \int_0^\infty \mathrm{e}^{-st}\mathrm{d}\left[\int_0^t G(t-u)\mathrm{d}F(u)\right] \\ &= \int_0^\infty \mathrm{e}^{-st}\mathrm{d}G(t)\int_0^\infty \mathrm{e}^{-st}\mathrm{d}F(t) \\ &= G^*(s)F^*(s). \end{aligned} \tag{A.9}$$

   This is easily obtained by another method: If two random variables X and Y are independent and have distributions $G(t)$ and $F(t)$, respectively, the LS transforms of distribution of $Z \equiv X + Y$ is

$$\begin{aligned} E\{\mathrm{e}^{-sZ}\} &= E\{\mathrm{e}^{-s(X+Y)}\} = E\{\mathrm{e}^{-sX}\}E\{\mathrm{e}^{-sY}\} \\ &= G^*(s)F^*(s). \end{aligned}$$

Table A.1 presents some properties of LS transforms and Table A.2 presents LS transforms of some functions [1, 4]

A.3 Abelian and Tauberian Theorems

We show two theorems that are very useful for performing limiting operations with transforms:

Theorem A.1 (An Abelian Theorem) *If*

$$\lim_{t\to\infty}\frac{F(t)}{t^\alpha} = \frac{C}{\Gamma(\alpha+1)} \quad \text{for } \alpha \geq 0, \tag{A.10}$$

Table A.1 Properties of LS transforms

$F(t)$	$F^*(s)$
$F_1(t)+F_2(t)$	$F_1^*(s)+F_2^*(s)$
$aF(t)$	$aF^*(s)$
$F(t-a)(a>0)$	$\mathrm{e}^{-sa}F^*(s)$
$F(at)$	$F^*(s/a)$
$\mathrm{e}^{-at}F(t)\ (a>0)$	$[s/(s+a)]F^*(s+a)$
$F'(t)=\mathrm{d}F(t)/\mathrm{d}t$	$s[F^*(s)-F(0)]$
$tF'(t)$	$-s\mathrm{d}F^*(s)/\mathrm{d}s$
$\int_0^t F(u)\mathrm{d}u$	$F^*(s)/s$
$\int_0^t F_2(t-u)\mathrm{d}F_1(u)$	$F_1^*(s)F_2^*(s)$
$\lim_{t\to 0}F(t)$	$\lim_{s\to\infty}F^*(s)$
$\lim_{t\to\infty}F(t)$	$\lim_{s\to 0}F^*(s)$

Table A.2 LS transforms of functions

$F(t)$	$F^*(s)$
1	1
t	$1/s$
$t^n\ (n=1,2,\ldots)$	$n!/s^n$
$t^\alpha\ (\alpha>-1)$	$\Gamma(\alpha+1)/s^\alpha$
$\mathrm{e}^{-\alpha t}\ (\alpha>0)$	$s/(s+\alpha)$
$t\mathrm{e}^{-\alpha t}\ (\alpha>0)$	$s/(s+\alpha)^2$
$t^n\mathrm{e}^{-\alpha t}\ (\alpha>0)$	$sn!/(s+\alpha)^{n+1}$
$t^\beta\mathrm{e}^{-\alpha t}\ (\alpha>0,\beta>-1)$	$s\Gamma(\beta+1)/(s+\alpha)^{\beta+1}$
$\cos\alpha t$	$s^2/(s^2+\alpha^2)$
$\sin\alpha t$	$s\alpha/(s^2+\alpha^2)$
$\log t$	$-C-\log s^\dagger$

$^\dagger C$ Euler's constant and $C=0.577215\ldots$

then

$$\lim_{s\to 0} s^\alpha F^*(s)=C, \tag{A.11}$$

where $\Gamma(\alpha)\equiv\int_0^\infty t^{\alpha-1}\mathrm{e}^{-t}\mathrm{d}t$ *is a gamma function defined in* (2.13).

Theorem A.2 (A Tauberian Theorem) *If* $F(t)$ *increases and its LS transform* $F^*(s)=\int_0^\infty \mathrm{e}^{-st}\mathrm{d}F(t)$ *exists for Re*$(s)>0$, *and if*

$$\lim_{s\to 0} s^\alpha F^*(s)=C \quad \textit{for}\ \alpha\geq 0, \tag{A.12}$$

then

$$\lim_{t\to\infty}\frac{F(t)}{t^\alpha}=\frac{C}{\Gamma(\alpha+1)}. \tag{A.13}$$

The readers refer to Example 4.11, Problems 4.5, and Sect. 8.2.1 as applications to differential equations of LS transforms.

In general, it is very difficult to making the inversion of LS transforms. However, with rapid progress of personal computers and a significant development of Laplace inversion methods, it is not so difficult to compute numerically some probabilities and other quantities in reliability models. Tijms [2] summarized some useful methods of Laplace inversion (see Sect. 3.2).

A.4 Generating Function

LS transforms are used for continuous functions, while generating functions [2, p. 449, 5, p. 446] are used for discrete probability distributions and functions with variables n ($n = 0, 1, 2, \ldots$, or $n = 1, 2, \ldots$) as shown in Sects. 3.5.2, 4.1.2 and 8.3.

Let a_n $(n = 0, 1, 2, \ldots)$ be a sequence of real number. If

$$a^*(z) \equiv \sum_{n=0}^{\infty} a_n z^n \tag{A.14}$$

exists for $|z| \leq 1$, then $a^*(z)$ is called the *generating function* of the sequence $\{a_n\}_{n=0}^{\infty}$. The discrete function can be recovered from $a_0 = a^*(0)$ and

$$a_n = \frac{1}{n!} \frac{\mathrm{d}^n a^*(z)}{\mathrm{d}z^n}\bigg|_{z=0} \quad (n = 1, 2, \ldots). \tag{A.15}$$

When p_n $(n = 0, 1, 2)$ is a probability function of random variable X,

$$E\{X\} = \sum_{n=1}^{\infty} n p_n = \frac{\mathrm{d}p^*(z)}{\mathrm{d}z}\bigg|_{z=1},$$

$$E\{X(X-1)\} = \sum_{n=1}^{\infty} n(n-1) p_n = \frac{\mathrm{d}^2 p^*(z)}{\mathrm{d}z^2}\bigg|_{z=1},$$

$$V\{X\} = E\{X^2\} - (E\{X\})^2.$$

For example, when a probability distribution is geometric distribution, i.e., $p_n = pq^{n-1}$ $(n = 1, 2, \ldots)$, the generating function is

$$p^*(z) = \sum_{n=1}^{\infty} z^n p q^{n-1} = \frac{pz}{1-qz},$$

and

$$\frac{\mathrm{d}p^*(z)}{\mathrm{d}z}\bigg|_{z=1} = \frac{p}{(1-qz)^2}\bigg|_{z=1} = \frac{1}{p}, \quad \frac{\mathrm{d}^2 p^*(z)}{\mathrm{d}z^2}\bigg|_{z=1} = \frac{2pq}{(1-qz)^3}\bigg|_{z=1} = \frac{2q}{p^2}.$$

Thus, the mean and variance of a geometric distribution are

$$E\{X\} = \frac{1}{p}, \quad V\{X\} = \frac{2q}{p^2} + \frac{1}{p} - \frac{1}{p^2} = \frac{q}{p^2}.$$

The sequence $c_n = \sum_{k=0}^{n} a_k b_{n-k}$ $(n = 0, 1, 2, \ldots)$ where the generating functions of $\{a_n\}$ and $\{b_n\}$ are $a^*(z)$ and $b^*(z)$ for $|z| \leq 1$, respectively, the generating function of $\{c_n\}$ is

$$c^*(z) \equiv \sum_{n=0}^{\infty} z^n \sum_{k=0}^{n} a_k b_{n-k} = a^*(z) b^*(z)$$

for $|z| \leq 1$, that corresponds to the property of (A.9) in continuous case.

For example, when both probability functions are Poisson distributions, i.e., $a_n = [(\lambda_1 t)^n / n!] \mathrm{e}^{-\lambda_1 t}$ and $b_n = [(\lambda_2 t)^n / n!] \mathrm{e}^{-\lambda_2 t}$ $(n = 0, 1, 2, \ldots)$, their generating functions are

$$a^*(z) \equiv \mathrm{e}^{-\lambda_1 t(1-z)}, \quad b^*(z) \equiv \mathrm{e}^{-\lambda_2 t(1-z)},$$

and hence,

$$c^*(z) = \exp[-(\lambda_1 + \lambda_2) t (1 - z)],$$

that shows the properties of superposition and decomposition of a Poisson process as shown in Examples 2.8 and 2.9, respectively.

Abelian and Tauberian theorems are rewritten in discrete case [5 p. 447, 6, p. 418]:

Theorem A.3 (An Abelian Theorem) *If* $\lim_{n\to\infty} a_n = a$, *then*

$$\lim_{z\to 1}(1 - z) \sum_{n=0}^{\infty} a_n z^n = a.$$

Theorem A.4 (A Tauberian Theorem) *If* $\lim_{z\to 1}(1 - z) \sum_{n=0}^{\infty} a_n z^n = a$ *and* $\lim_{n\to\infty} n(a_n - a_{n-1}) = 0$, *then*

$$\lim_{n\to\infty} a_n = a.$$

Theorem A.5 *Let* $\{a_n\}_{n=0}^{\infty}$ *be a nonnegative sequence of real numbers with generating function in* (A.14). *For constant a and b,*

$$\sum_{n=0}^{\infty} a_n = a \quad \Leftrightarrow \quad \lim_{z\to 1} a^*(z) = a,$$

$$\lim_{m\to\infty} \left(\frac{1}{m} \sum_{n=0}^{m} a_n \right) = c \quad \Leftrightarrow \quad \lim_{z\to 1}(1 - z) a^*(z) = c.$$

Extending such a generating function, the universal generating function is introduced and is applied to reliability analysis of multi-state systems [7–9].

Appendix B: Answers to Selected Problems

Chapter 2

2.1 From S_n has a gamma distribution with order n in (2.12) [10, p. 84,],

$$E\{g(S_n)\} = \int_0^\infty g(t)\mathrm{d}\Pr\{S_n \le t\} = \int_0^\infty g(t)\frac{\lambda(\lambda t)^{n-1}}{(n-1)!}\mathrm{e}^{-\lambda t}\mathrm{d}t.$$

Hence,

$$E\left\{\sum_{n=1}^\infty g(S_n)\right\} = \sum_{n=1}^\infty \int_0^\infty g(t)\frac{\lambda(\lambda t)^{n-1}}{(n-1)!}\mathrm{e}^{-\lambda t}\mathrm{d}t = \lambda\int_0^\infty g(t)\mathrm{d}t.$$

2.2 See [11, p. 24].

2.3

$$E\{X_{(1)}\} = \int_0^\infty \mathrm{e}^{-n\lambda t}\mathrm{d}t = \frac{1}{n\lambda},$$

$$E\{X_{(n)}\} = \int_0^\infty \left[1-(1-\mathrm{e}^{-n\lambda t})\right]\mathrm{d}t = \frac{1}{\lambda}\sum_{k=1}^n \frac{1}{k}.$$

2.5 Compute numerically when $F(t) = 1 - \mathrm{e}^{-t^\alpha}$ [12],

$$\mathrm{MTTF} = \int_0^\infty \left[1-(1-\mathrm{e}^{-t^\alpha})^n\right]\mathrm{d}t,$$

and compare it with $\left(\sum_{k=1}^n 1/k\right)^{1/\alpha}$ and $(C+\log n)^{1/\alpha}$.

2.6 From [13, p. 9], an optimum number n^* $(1 \leq n^* < \infty)$ is given by a finite and unique minimum that satisfies

$$\sum_{k=1}^{n+1} \frac{n+1}{k+1} \geq \frac{c_2}{c_1} \quad (n = 1, 2, \ldots).$$

2.7 From [13, p. 12],

$$\text{MTTF} = \sum_{j=k}^{n} \binom{n}{j} \int_0^\infty \mathrm{e}^{-j\lambda t}(1 - \mathrm{e}^{-\lambda t})^{n-j} \mathrm{d}t = \frac{1}{\lambda} \sum_{j=k}^{n} \frac{1}{j}.$$

2.8 From [11, p. 286], let $I_k = 1$ if there is a record at time t, $I_k = 0$ otherwise. Then, because $\Pr\{I_k = 1\} = (k-1)!/k! = 1/k$,

$$N_n = \sum_{k=1}^{n} I_k, \quad E\{N_n\} = \sum_{k=1}^{n} E\{I_k\} = \sum_{k=1}^{n} \frac{1}{k},$$

$$\text{Var}\{N_n\} = \sum_{k=1}^{n} \text{Var}\{I_k\} = \sum_{k=1}^{n} \frac{1}{k}\left(1 - \frac{1}{k}\right).$$

When $n = 10, 100, 1000$, respectively,

$$\sum_{k=1}^{n} \frac{1}{k} = 2.92897, \quad 5.18738, \quad 7.48547.$$

2.9 See [1, p. 40].

2.11 The number of spare units is given a minimum number that satisfies

$$\sum_{k=0}^{n} \frac{(\lambda_1 + \lambda_2)^k}{k!} \mathrm{e}^{-(\lambda_1 + \lambda_2)} \geq \alpha.$$

2.12 Compute the probabilities, for $p = 0.90, 0.10$ and $\lambda t = 1, 2, 3$,

$$\frac{(p\lambda t)^k}{k!} \mathrm{e}^{-p\lambda t} \quad (k = 0, 1, 2, 3).$$

2.13 From the inequality $C(N+1) - C(N) \geq 0$,

$$N(N+1) \geq \frac{B}{A}, \quad \text{i.e.,} \quad \sum_{k=1}^{N} k \geq \frac{B}{2A}.$$

2.14 Consider the multiple backward times T_k $(k = 1, 2, \ldots, N)$, where $S_k \equiv \sum_{j=1}^{k} T_j$, $S_0 \equiv 0$ and $S_N \equiv t$ [14]. If we cannot find the actual failure time until time S_{k-1} from time t by the backward operation, we execute again the next backward time with T_k. If we can find the failure time, i.e., the system fails in $(t - S_k, t - S_{k-1}]$, then the process ends. Finally, if we cannot find the failure time until S_{N-1}, we make the final backward with $T_N = t - S_{N-1}$, and can find it certainly.

It is assumed that $c_1 + c_2 T_k$ is the cost for the kth backward. When the failure time is detected at the kth backward, the expected cost is

$$\frac{1}{F(t)}(c_1 k + c_2 S_k)[F(t - S_{k-1}) - F(t - S_k)],$$

and hence, the total expected cost is

$$C(T_1, T_2, \ldots, T_N) = \frac{1}{F(t)} \sum_{k=1}^{N-1} (c_1 k + c_2 S_k)[F(t - S_{k-1}) - F(t - S_k)].$$

We find optimum times T_k $(k = 1, 2, \ldots, N)$ that minimize $C(T_1, T_2, \ldots, T_N)$ for a given $t > 0$. Differentiating $C(T_1, T_2, \ldots, T_N)$ with respect to S_k and setting it equal to zero,

$$S_{k+1} - S_k = \frac{F(t - S_{k-1}) - F(t - S_k)}{f(t - S_k)} - \frac{c_1}{c_2} \quad (k = 1, 2, \ldots, N-1).$$

Setting that $x_k \equiv t - S_k$,

$$x_k - x_{k+1} = \frac{F(x_{k-1}) - F(x_k)}{f(x_k)} - \frac{c_1}{c_2} \quad (k = 1, 2, \ldots, N-1), \tag{B.1}$$

where $x_N \equiv 0$ and $x_0 \equiv t$. Then, the total expected cost is

$$C(x_1, x_2, \ldots, x_N) = \frac{1}{F(t)} \sum_{k=1}^{N} [c_1 k + c_2(t - x_k)][F(x_{k-1}) - F(x_k)]. \tag{B.2}$$

Therefore, we solve the simultaneous equations in (B.1) for given $N \geq 1$ and compute the total expected cost in (B.2). Finally, comparing the total expected cost for $N \geq 1$, we can get an optimum number N^* and backward times $S_k^* = t - x_k^*$ $(k = 1, 2, \ldots, N^*)$.

2.15

$$E\{S_k | N(t) = n\} = \sum_{j=0}^{k-1} \binom{n}{j} \int_0^t \left(\frac{u}{t}\right)^j \left(1 - \frac{u}{t}\right)^{n-j} \mathrm{d}u = t \sum_{j=0}^{k-1} \binom{n}{j} \frac{j!(n-j)!}{(n+1)!} = \frac{kt}{n+1},$$

$$E\{S_k - S_{k-1} | N(t) = n\} = \frac{kt}{n+1} - \frac{(k-1)t}{n+1} = \frac{t}{n+1}.$$

Compute when $k = 0, 1, 2,$

$$\Pr\{N(10) = k | N(100) = 10\} = \binom{10}{k} \left(\frac{10}{100}\right)^k \left(1 - \frac{10}{100}\right)^{10-k},$$

$$\Pr\{S_3 \leq 50 | N(100) = 10\} = \sum_{k=3}^{10} \binom{10}{k} \left(\frac{50}{100}\right)^k \left(1 - \frac{50}{100}\right)^{10-k} = 0.945.$$

2.16 Differentiate $[H(t)]^n/n!$ with t and integrate it from 0 to t, where note that $H(0) \equiv 0$.

2.17 Noting that $\mu = \int_0^\infty P_0(t)\mathrm{d}t$,

$$\frac{\sum_{k=1}^{N-1}\int_0^\infty P_k(t)\mathrm{d}t}{\int_0^\infty P_N(t)\mathrm{d}t} - (N-1) = \frac{1}{\int_0^\infty P_N(t)\mathrm{d}t}\sum_{k=1}^{N-1}\int_0^\infty [P_k(t) - P_N(t)]\mathrm{d}t > 0,$$

because when $h(t)$ increases strictly, $\int_0^\infty P_k(t)\mathrm{d}t$ decreases strictly from (2.44) for $N \geq 2$.

2.19 Setting that $K(t) \equiv 1 - H(t)/a$, solve the differential equation by using the separation of variables method.

2.20 From (2.57),

$$E\{Z(t)\} = -\frac{\mathrm{d}\exp[-\lambda ts/(s+\theta)]}{\mathrm{d}s}\bigg|_{s=0} = \frac{\lambda t}{\theta},$$

$$V\{Z(t)\} = \frac{\mathrm{d}\exp[-\lambda ts/(s+\theta)]}{\mathrm{d}s^2}\bigg|_{s=0} - \left(\frac{\lambda t}{\theta}\right)^2 = \frac{2\lambda t}{\theta^2},$$

and from (2.60),

$$E\{Y\} = \frac{1}{\lambda}(1+\theta K),$$

$$V\{Y\} = \frac{2}{\lambda^2}\sum_{k=0}^{\infty}(k+1)\sum_{j=k}^{\infty}\frac{(\theta K)^j}{j!}\mathrm{e}^{-\theta K} - \left[\frac{1}{\lambda}(1+\theta K)\right]^2 = \frac{2\theta K + 1}{\lambda^2}.$$

Chapter 3

3.2 Considering that the inverse LS transform of $1/[(s+a)^2 + b^2]$ is given by

$$\frac{1}{a^2+b^2}\left[1 - \frac{\sqrt{a^2+b^2}}{b}\mathrm{e}^{-at}\sin\left(bt + \tan^{-1}\frac{b}{a}\right)\right],$$

derive when $k = 3, 4$, respectively,

$$M^*(s) = \frac{\lambda^3}{s[(s+3\lambda/2)^2 + 3\lambda^2/4]} = \frac{\lambda}{3s} - \frac{\lambda s + 3\lambda^2}{3[(s+3\lambda/2)^2 + 3\lambda^2/4]},$$

$$M^*(s) = \frac{\lambda^4}{s(s+2\lambda)[(s+\lambda)^2 + \lambda^2]} = \frac{\lambda}{4s} - \frac{\lambda}{4(s+2\lambda)} - \frac{\lambda^2}{2[(s+\lambda)^2 + \lambda^2]}.$$

3.3 Use the LS transform of a gamma distribution in (2.14).

Table B.1 $M(t)$ and $t/E\{X\}$ when $F(t) = 1 - e^{-(\lambda t)^\alpha}$ and $\lambda = 1$

t	α				
	1.0	1.5	2.0	2.5	3.0
1	1.00	0.84	0.75	0.70	0.67
2	2.00	1.95	1.89	1.84	1.80
3	3.00	3.05	3.02	2.97	2.93
4	4.00	4.16	4.15	4.10	4.05
5	5.00	5.27	5.28	5.23	5.17
6	6.00	6.38	6.41	6.35	6.29
7	7.00	7.49	7.54	7.48	7.41
8	8.00	8.59	8.86	8.61	8.53
9	9.00	9.70	9.79	9.74	9.65
10	10.00	10.81	10.92	10.86	10.77
$10/E\{X\}$	10.00	11.08	11.28	11.27	11.20

3.4 Table B.1 is computed.

3.5 From Example 2.4, $F(t) = (1 - e^{-\lambda t})^2$, and hence,

$$F^*(s) = \int_0^\infty e^{-st} dF(t) = \frac{2\lambda^2}{(s+\lambda)(s+2\lambda)},$$

$$M^*(s) = \frac{F^*(s)}{1 - F^*(s)} = \frac{2\lambda^2}{s^2 + 3\lambda s}.$$

Use the inverse LS transform in Table A.2.

3.7

$$n \approx 10 + 1.64\sqrt{2} = 12.32 \quad \text{for } \alpha = 0.95,$$
$$n \approx 10 + 2.33\sqrt{2} = 13.29 \quad \text{for } \alpha = 0.99.$$

3.8 From (3.25),

$$\int_t^\infty \left[\overline{F}(x) + \int_0^t \overline{F}(x-u) dM(u) \right] dx$$
$$= \int_0^\infty \overline{F}(x) dx - \int_0^t \overline{F}(x) dx + \int_t^\infty \left[\int_0^t \overline{F}(x-u) dM(u) \right] dx$$
$$= \mu - t + \int_0^t F(x) dx + \int_0^t \left[\int_0^\infty \overline{F}(x) dx - \int_0^{t-u} \overline{F}(x) dx \right] dM(u)$$
$$= \mu - t + \mu M(t) + \int_0^t F(x) dx - \int_0^t M(t-x) \overline{F}(x) dx$$
$$= \mu[1 + M(t)] - t.$$

From (3.26),

$$\frac{1}{\mu}\int_0^\infty t\overline{F}(t)\mathrm{d}t = \frac{1}{2\mu}\int_0^\infty \overline{F}(t)\mathrm{d}t^2 = \frac{1}{2\mu}\int_0^\infty t^2\mathrm{d}F(t) = \frac{\mu^2+\sigma^2}{2\mu}.$$

3.9 From [3, p. 128] and Fig. B.1,

$$\Pr\{\beta(t)\le x\} = \Pr\{X_{N(t)+1}\le x\} = \int_{t-x}^{t} [F(x)-F(t-u)]\mathrm{d}M(u).$$

Thus, by Theorem 3.6,

$$\lim_{t\to\infty}\Pr\{\beta(t)\le x\} = \frac{1}{\mu}\int_0^x [F(x)-F(u)]\mathrm{d}u = \frac{1}{\mu}\int_0^x u\mathrm{d}F(u).$$

3.10 Let $N_A(t)$ and $N_B(t)$ be the number of replacements of units for the age and block replacements with planned replacement time T in $[0,t]$ of Sect. 3.3.2. Then, from [16, p. 67],

$$E\{N_B(t)\}\ge E\{N_A(t)\},$$

$$\lim_{t\to\infty}\frac{E\{N_B(t)\}}{t} = \frac{M(T)+1}{T} \ge \lim_{t\to\infty}\frac{E\{N_A(t)\}}{t} = \frac{1}{\int_0^T \overline{F}(t)\mathrm{d}t}.$$

Thus,

$$M(t)\ge\frac{t}{\int_0^t \overline{F}(u)du} - 1.$$

Next, let $\widetilde{N}_A(t)$ and $\widetilde{N}_B(t)$ be the number of failures of units for the age and block replacements in $[0,t]$. Then, from [16, p. 69], when $F(t)$ is IFR,

$$E\{\widetilde{N}_B(t)\}\le E\{\widetilde{N}_A(t)\},$$

$$\lim_{t\to\infty}\frac{E\{\widetilde{N}_B(t)\}}{t} = \frac{M(T)}{T} \le \lim_{t\to\infty}\frac{E\{\widetilde{N}_A(t)\}}{t} = \frac{F(T)}{\int_0^T \overline{F}(t)\mathrm{d}t}.$$

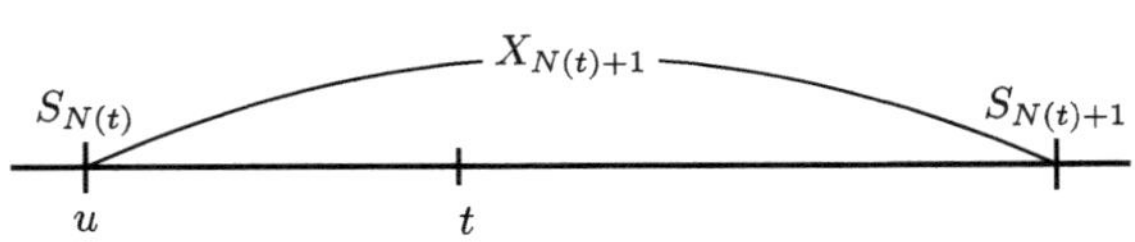

Fig. B.1 Process of the life in a renewal process

Thus,

$$M(t) \leq \frac{tF(t)}{\int_0^t \overline{F}(u)du}.$$

3.12

$$\lim_{T\to 0}\frac{1}{F(T)}\int_0^T \overline{F}(t)\mathrm{d}t = \frac{1}{h(0)}, \quad \lim_{T\to\infty}\frac{1}{F(T)}\int_0^T \overline{F}(t)\mathrm{d}t = \mu,$$

where if $h(t) \to 0$ as $t \to 0$ then $1/h(0)$ is assumed to be infinity. In addition, when $F(t)$ is IFR,

$$\left(\frac{1}{F(T)}\int_0^T \overline{F}(t)\mathrm{d}t\right)' = \frac{1}{\overline{F}(T)[F(T)]^2 \int_0^T \overline{F}(t)\mathrm{d}t}\left[\frac{F(T)}{\int_0^T \overline{F}(t)\mathrm{d}t} - h(T)\right] \leq 0.$$

Similarly,

$$\lim_{T\to 0}\frac{1}{\overline{F}(T)}\int_0^T \overline{F}(t)\mathrm{d}t = 0, \quad \lim_{T\to\infty} = \frac{1}{\overline{F}(T)}\int_0^T \overline{F}(t)\mathrm{d}t = \infty.$$

Clearly, $\int_0^T \overline{F}(t)\mathrm{d}t/\overline{F}(T)$ increases even when $F(t)$ is not IFR.

3.13

$$\begin{aligned}
A(T) &+ \int_0^T B(T-t)\mathrm{d}F(t) \\
&= c_1F(T) + c_2\overline{F}(T) + \int_0^T [c_1M(T-t) + c_2]\mathrm{d}F(t) \\
&= c_1M(T) + c_2, \\
A(T) &+ \int_0^T A(T-t)\mathrm{d}M(t) \\
&= c_1F(T) + c_2\overline{F}(T) + \int_0^T [c_1F(T-t) + c_2\overline{F}(T-t)]\mathrm{d}M(t) \\
&= c_1M(T) + c_2.
\end{aligned}$$

3.14 When T_0 has an exponential distribution $(1 - \mathrm{e}^{-\alpha t})$, from (3.67),

$$R(T;x,\alpha)=\frac{\alpha e^{\alpha x}\int_x^{T+x}e^{-\alpha t}\overline{F}(t)dt}{1-G^*(\alpha)+\alpha G^*(\alpha)\int_0^T e^{-\alpha t}\overline{F}(t)dt}.$$

Clearly,

$$R(0;x,\alpha)=0,\quad R(\infty;x,\alpha)=\frac{\alpha e^{\alpha x}\int_x^{\infty}e^{-\alpha t}\overline{F}(t)dt}{1-G^*(\alpha)F^*(\alpha)}.$$

Differentiating $R(T;x,\alpha)$ with respect to T and setting it equal to zero,

$$\lambda(T;x)\left[1-G^*(\alpha)+\alpha G^*(\alpha)\int_0^T e^{-\alpha t}\overline{F}(t)dt\right]$$
$$-\alpha G^*(\alpha)\int_0^T e^{-\alpha t}[\overline{F}(t)-\overline{F}(t+x)]dt=1-G^*(\alpha),\tag{B.3}$$

where

$$\lambda(t;x)\equiv\frac{F(t+x)-F(t)}{\overline{F}(t)}.$$

Note that $\lambda(t;x)$ and $h(t)$ have the same property. Letting $L(T)$ be the left-hand side of (B.3),

$$L(0)=F(x)[1-G^*(\alpha)]\le 1-G^*(\alpha),$$
$$L(\infty)=[\lambda(\infty;x)-1][1-G^*(\alpha)F^*(\alpha)]$$
$$+1-G^*(\alpha)+\alpha G^*(\alpha)e^{\alpha x}\int_x^{\infty}e^{-\alpha t}\overline{F}(t)dt.$$

If the failure rate increase strictly, then

$$L'(T)=\lambda'(T;x)\left[1-G^*(\alpha)+\alpha G^*(\alpha)\int_0^T e^{-\alpha t}\overline{F}(t)dt\right]>0.$$

Thus, we have the following optimum policy:

(i) If $L(\infty)>1-G^*(\alpha)$ then there exists a finite and unique $T^*(0<T^*<\infty)$ that satisfies (B.3), and the interval reliability is

$$R(T^*;x,\alpha)=\frac{\overline{F}(T^*+x)}{G^*(\alpha)\overline{F}(T^*)}.$$

(ii) If $L(\infty)\le 1-G^*(\alpha)$ then $T^*=\infty$.

3.16 From Example 3.5, when $\alpha = 0.95$ and 0.99, respectively,

$$\delta \approx 909.09 - 102.56 \times 1.64 = 740.89,$$
$$\delta \approx 909.09 - 102.56 \times 2.33 = 670.13.$$

3.17 When $G(t) = 1 - \mathrm{e}^{-\theta\sqrt{t}}$, (3.84) is

$$\frac{a\mu}{c_1}\sqrt{T} - \frac{1}{\theta}\left(1 - \mathrm{e}^{-\theta\sqrt{T}}\right) = \frac{\theta\mu}{2}, \tag{B.4}$$

and when $c_r(t) = at^2$ and $G(t) = 1 - \mathrm{e}^{-\theta t}$, from (3.82),

$$T - \frac{1 - \mathrm{e}^{-\theta T}}{\theta(\mu\theta + 1)} = \frac{c_1\theta}{2a}. \tag{B.5}$$

In both cases, there exist optimum T^* that satisfies (B.4) and (B.5).

3.18 From [17, p. 156], the LS transform of an exponential distribution $F_k(t) = (1 - \mathrm{e}^{-\lambda_k t})$ is $\lambda_k/(s + \lambda_k)$, and hence, the LS transform of the nth convolution $F^{(n)}(t)$ of $F_k(t)$ is

$$F^{(n)*}(s) = \int_0^\infty \mathrm{e}^{-st}\mathrm{d}F^{(n)}(t) = \prod_{k=1}^n \frac{\lambda_k}{s + \lambda_k}.$$

Inverting $F^{(n)*}(s)$,

$$F^{(n)}(t) = 1 - \sum_{k=1}^n \left(\prod_{j=1, j\neq k}^n \frac{\lambda_j}{\lambda_j - \lambda_k}\right)\mathrm{e}^{-\lambda_k t}.$$

In particular, when $\lambda_k = a^{k-1}\lambda$,

$$F^{(n)}(t) = 1 - \sum_{k=1}^n \left(\prod_{j=1, j\neq k}^n \frac{1}{1 - a^{k-j}}\right)\exp(-a^{k-1}\lambda t).$$

3.19 The mean time to replacement is

$$\sum_{k=1}^N \frac{\mu}{a^{k-1}} = \frac{\mu(a - a^{-N+1})}{a - 1}.$$

3.20 The mean operating and repair times until the unit is replaced at the Nth failure are, respectively,

$$\sum_{k=1}^N \frac{\mu}{a^{k-1}}, \quad \sum_{k=1}^{N-1} \frac{\beta}{b^{k-1}}.$$

3.21

$$\frac{pz}{(1-z)[1-(\beta-p)z]} = \frac{p}{p+\alpha}\left[\frac{1}{1-z} - \frac{1}{1-(\beta-p)z}\right]$$
$$= \frac{p}{p+\alpha}\sum_{n=0}^{\infty}[1-(\beta-p)^n]z^n,$$

$$\frac{pz}{1-z(1-z)[1-(\beta-p)z]}\frac{1-\beta z}{} = \frac{p}{p+\alpha}\left[\frac{\alpha}{1-z} + \frac{p}{1-(\beta-p)z}\right]\frac{z}{1-z}$$
$$= \frac{p}{p+\alpha}\sum_{n=0}^{\infty}[\alpha + p(\beta-p)^n]z^n\sum_{k=1}^{\infty}z^k$$
$$= \frac{p}{p+\alpha}\sum_{n=0}^{\infty}\left\{n\alpha + \frac{p[1-(\beta-p)^n]}{p+\alpha}\right\}z^n.$$

3.22

$$\Pr\{X_{n+1} \le x | S_n = t\} = \frac{F(x+at) - F(at)}{1 - F(at)}.$$

Chapter 4

4.1

$$\boldsymbol{P}\cdot\boldsymbol{P}^n = \begin{pmatrix} 1-a & a \\ b & 1-b \end{pmatrix}\left[\frac{1}{a+b}\begin{pmatrix} b & a \\ b & a \end{pmatrix} + \frac{(1-a-b)^n}{a+b}\begin{pmatrix} a & -a \\ -b & b \end{pmatrix}\right]$$
$$= \boldsymbol{P}^{n+1}.$$
$$\boldsymbol{\pi}(n)\cdot\boldsymbol{P} = \left(\frac{b+a(1-a-b)^n}{a+b}, \frac{a[1-(1-a-b)^n]}{a+b}\right)\begin{pmatrix} 1-a & a \\ b & 1-b \end{pmatrix}$$
$$= \boldsymbol{\pi}(n+1).$$

4.2

$$\boldsymbol{P} = \begin{array}{c} \\ 1 \\ 2 \\ \vdots \\ N-1 \\ N \end{array} \begin{array}{c} \begin{array}{cccccc} 1 & 2 & 3 & \dots & N-1 & N \end{array} \\ \begin{pmatrix} r_1 & 1-r_1 & 0 & \dots & 0 & 0 \\ r_2 & 0 & 1-r_2 & \dots & 0 & 0 \\ \vdots & \vdots & \vdots & & \vdots & \vdots \\ r_{N-1} & 0 & 0 & \dots & 0 & 1-r_{N-1} \\ 1 & 0 & 0 & \dots & 0 & 0 \end{pmatrix} \end{array}.$$

From $\boldsymbol{\pi} = \boldsymbol{\pi P}$,

$$\pi_1 = \sum_{j=1}^{N} \pi_j r_j,$$

$$\pi_{j+1} = \pi_j(1 - r_j) \quad (j = 1, 2, \ldots, N - 1).$$

Thus,

$$\pi_j = \frac{\Pi_{i=1}^{j-1}(1 - r_i)}{\sum_{j=1}^{N}\left[\Pi_{i=1}^{j-1}(1 - r_i)\right]},$$

where $r_N \equiv 1$ and $\Pi_{i=1}^{0} \equiv 1$,

4.4 Let $F(t)$ be the identical failure probability of each unit at time t. Then, the probability that k units have failed and $n - k$ units have not failed independently at time t is

$$\binom{n}{k}[F(t)]^k[1 - F(t)]^{n-k} \quad (k = 0, 1, 2, \ldots, n).$$

4.5 For one repair person, the LS transforms of the differential equations are

$$(s + \lambda)P_{00}^*(s) = s + \theta P_{01}^*(s),$$
$$(s + \lambda + \theta)P_{01}^*(s) = \lambda P_{00}^*(s) + \theta P_{02}^*(s),$$
$$(s + \theta)P_{02}^*(s) = \lambda P_{01}^*(s).$$

Solving these equations for $P_{0i}^*(s)$ $(i = 0, 1, 2)$ and inverting them,

$$P_{00}(t) = \frac{\theta^2}{\lambda^2 + \lambda\theta + \theta^2} + \frac{\lambda}{\omega_1 - \omega_2}\left(\frac{\omega_1 - \lambda - \theta}{\omega_1}e^{-\omega_1 t} - \frac{\omega_2 - \lambda - \theta}{\omega_2}e^{-\omega_2 t}\right),$$

$$P_{01}(t) = \frac{\lambda\theta}{\lambda^2 + \lambda\theta + \theta^2} + \frac{\lambda}{\omega_1 - \omega_2}\left(\frac{-\omega_1 + \theta}{\omega_1}e^{-\omega_1 t} - \frac{-\omega_2 + \theta}{\omega_2}e^{-\omega_2 t}\right),$$

$$P_{02}(t) = \frac{\lambda^2}{\lambda^2 + \lambda\theta + \theta^2} + \frac{\lambda^2}{\omega_1 - \omega_2}\left(\frac{1}{\omega_1}e^{-\omega_1 t} - \frac{1}{\omega_2}e^{-\omega_2 t}\right),$$

where

$$\omega_1 \equiv \lambda + \theta + \sqrt{\lambda\theta}, \quad \omega_2 \equiv \lambda + \theta - \sqrt{\lambda\theta}.$$

For two repair persons,

$$P_{00}(t) = \frac{2\theta^2}{\lambda^2 + 2\lambda\theta + 2\theta^2} + \frac{\lambda}{\omega_1 - \omega_2}\left(\frac{\omega_1 - \lambda - 2\theta}{\omega_1}e^{-\omega_1 t} - \frac{\omega_2 - \lambda - 2\theta}{\omega_2}e^{-\omega_2 t}\right),$$

$$P_{01}(t) = \frac{2\lambda\theta}{\lambda^2 + 2\lambda\theta + \theta^2} + \frac{\lambda}{\omega_1 - \omega_2}\left(\frac{-\omega_1 + 2\theta}{\omega_1}e^{-\omega_1 t} - \frac{-\omega_2 + 2\theta}{\omega_2}e^{-\omega_2 t}\right),$$

$$P_{02}(t) = \frac{\lambda^2}{\lambda^2 + 2\lambda\theta + 2\theta^2} + \frac{\lambda^2}{\omega_1 - \omega_2}\left(\frac{1}{\omega_1}e^{-\omega_1 t} - \frac{1}{\omega_2}e^{-\omega_2 t}\right),$$

where

$$\omega_1 \equiv \frac{1}{2}\left(2\lambda + 3\theta + \sqrt{4\lambda\theta + \theta^2}\right),$$
$$\omega_2 \equiv \frac{1}{2}\left(2\lambda + 3\theta - \sqrt{4\lambda\theta + \theta^2}\right).$$

Chapter 5

5.1

$$F_0^*(s) = F^*(s)[\alpha G_1^*(s) + (1-\alpha)G_2^*(s)],$$
$$F_1^*(s) = \frac{\alpha F^*(s)}{1-(1-\alpha)F^*(s)G_2^*(s)}, \quad F_2^*(s) = \frac{(1-\alpha)F^*(s)}{1-\alpha F^*(s)G_1^*(s)},$$
$$P_0^*(s) = \frac{1-F^*(s)}{1-F_0^*(s)}, \quad P_j^*(s) = \frac{F_j^*(s)[1-G_j^*(s)]}{1-F_j^*(s)G_j^*(s)} \quad (j=1,2),$$
$$M_0^*(s) = \frac{F_0^*(s)}{1-F_0^*(s)}, \quad M_j^*(s) = \frac{F_j^*(s)}{1-F_j^*(s)G_j^*(s)} \quad (j=1,2).$$

5.2 Let Q_{ij} be the probability that when the system is in State $i\,(i=0,1)$, the data transmission has failed at j times $(j=1,2,\ldots,N)$ consecutively. Then, using (5.15), we have the following equations related to Q_{ij} [13, p. 116]:

$$Q_{0j} = P_{00}(T_1)p_0Q_{0j-1} + P_{01}(T_1)p_1Q_{1j-1} + P_{02}(T_1)p_2^j,$$
$$Q_{1j} = P_{10}(T_1)p_0Q_{0j-1} + P_{11}(T_1)p_1Q_{1j-1} + P_{12}(T_1)p_2^j.$$

Forming the generating functions of Q_{0j} and Q_{1j}, solving them, and inverting their generating functions,

$$\begin{aligned}Q_{0j} = &\frac{1}{\omega_1-\omega_2}(\{\omega_1 - p_1[P_{11}(T_1) - P_{01}(T_1)]\}\omega_1^j \\ &- \{\omega_2 - p_1[P_{11}(T_1) - P_{01}(T_1)]\}\omega_2^j) \\ &+ \frac{p_2}{\omega_1-\omega_2}\Bigg(\{\omega_1 P_{02}(T_1) - p_1[P_{11}(T_1)P_{02}(T_1) - P_{01}(T_1)P_{12}(T_1)]\} \\ &\times \frac{\omega_1^j - p_2^j}{\omega_1 - p_2} - \{\omega_2 P_{02}(T_1) - p_1[P_{11}(T_1)P_{02}(T_1) - P_{01}(T_1)P_{12}(T_1)]\} \\ &\times \frac{\omega_2^j - p_2^j}{\omega_2 - p_2}\Bigg) \quad (j=0,1,2,\ldots,N),\end{aligned}$$

where

$$\omega_1 \equiv \frac{1}{2}\Big\{p_0P_{00}(T_1) + p_1P_{11}(T_1) + \sqrt{[p_0P_{00}(T_1) - p_1P_{11}(T_1)]^2 + 4p_0p_1P_{01}(T_1)P_{10}(T_1)}\Big\},$$

$$\omega_2 \equiv \frac{1}{2}\Big\{p_0P_{00}(T_1) + p_1P_{11}(T_1) - \sqrt{[p_0P_{00}(T_1) - p_1P_{11}(T_1)]^2 + 4p_0p_1P_{01}(T_1)P_{10}(T_1)}\Big\}.$$

Thus, the mean time $l_3(N)$ is

$$l_3(N) = \frac{T_1\sum_{j=0}^{N-1} Q_{0j} + T_2Q_{0N}}{1 - Q_{0N}} \quad (N = 1, 2, \ldots).$$

From $l_3(N+1) - l_3(N) \geq 0$,

$$\frac{Q_{0N}}{Q_{0N} - Q_{0N+1}}(1 - Q_{0N}) - \sum_{j=0}^{N-1} Q_{0j} \geq \frac{T_2}{T_1}. \tag{B.6}$$

Let $L(N)$ be the left-hand side of (B.6). If Q_{0j+1}/Q_{0j} increases strictly, then $L(N)$ increases strictly. Therefore, if $L(\infty) > T_2/T_1$, then there exists a finite and unique minimum N^* $(1 \leq N^* < \infty)$ that satisfies (B.6).

5.3 $f_{i;ij}$ is the probability of first-passage from State i to State j without returning to State i, and $f_{j;ii}$ is the probability of returning from State i to State i without passing through State j, and hence, $f_{i;ij} + f_{j;ii} = 1$.

5.4 From (5.44) and (5.45),

$$\mu C(n) = \frac{(n+1)(c_0 - c_1) + c_2}{\sum_{j=1}^{n+1}\binom{n+1}{j}\Big/C_{j-1}} + c_1.$$

Thus, because

$$\frac{1}{n+1}\sum_{j=1}^{n+1}\binom{n+1}{j}\frac{1}{C_{j-1}} - \frac{1}{n}\sum_{j=1}^{n}\sum_{j=1}^{n}\binom{n}{j}\frac{1}{C_{j-1}} \geq 0,$$

$$\lim_{n\to\infty} C(n) = \frac{c_0}{\mu},$$

$C(n)$ decreases to c_0/μ.

5.6

$$C(T) = c_1M_4 + c_2M_1 + c_3(P_3 + P_4) + c_4P_2 = \frac{c_1F(T) + c_2\overline{F}(T) + c_3\int_T^{T+L} F(t)\mathrm{d}t + c_4\int_{T+L}^{\infty}\overline{F}(t)\mathrm{d}t}{\mu + \int_T^{T+L} F(t)\mathrm{d}t}.$$

Clearly,

$$C(0) = \frac{c_2 + c_3 \int_0^L F(t)\mathrm{d}t + c_4 \int_L^\infty \overline{F}(t)\mathrm{d}t}{\mu + \int_0^L F(t)\mathrm{d}t}, \quad C(\infty) = \frac{c_1 + c_3 L}{\mu + L}.$$

Differentiating $C(T)$ with respect to T and setting it equal to zero,

$$L(T) \equiv \lambda(T;L)\left\{c_3\mu + c_4\left[L + \int_0^T \overline{F}(t)\mathrm{d}t\right] - c_1 F(T) - c_2\overline{F}(T)\right\}$$
$$+ [(c_1 - c_2)h(T) - c_4]\left[\mu + \int_T^{T+L} F(t)\mathrm{d}t\right] = 0, \tag{B.7}$$

where $h(t) \equiv f(t)/\overline{F}(t)$ and $\lambda(t;L) \equiv [F(t+L) - F(t)]/\overline{F}(t)$. If $h(t)$ increases strictly, then $\lambda(t;L)$ increases strictly with t, and hence, $L(T)$ increases strictly with T from $L(0)$ to $L(\infty)$, where

$$L(0) = F(L)(c_3\mu + c_4 L - c_2) + \left[\mu + \int_0^L F(t)\mathrm{d}t\right][(c_1 - c_2)h(0) - c_4],$$
$$L(\infty) = \lambda(\infty;L)(c_3\mu - c_1) + (\mu + L)\{(c_1 - c_2)h(\infty) - c_4[1 - \lambda(\infty;L)]\}.$$

Therefore, we have the following optimum policy:

(i) If $L(0) \geq 0$ then $T^* = 0$, i.e., the order is made at the same time as the beginning of operation.
(ii) If $L(0) < 0$ and $L(\infty) > 0$ then there exists a finite and unique T^* $(0 < T^* < \infty)$ that satisfies (B.7).
(iii) If $L(\infty) \leq 0$ then $T^* = \infty$.

Chapter 6

6.3 From Problem 3.7 and 3.16, when $\alpha = 0.95, 0.99$, respectively,

$$z = 15 - 5.12 \times 1.64 = 6.60,$$
$$z = 15 - 5.12 \times 2.33 = 3.07,$$

$$\Pr\{Z(30) > 15\} = 0.500, \quad \Pr\{Z(30) > 20\} = 0.164.$$

6.6

$$\frac{(\theta K)^{n+1}}{(n+1)!}\sum_{k=n}^{\infty}\frac{(\theta K)^k}{k!}-\frac{(\theta K)^n}{n!}\sum_{k=n+1}^{\infty}\frac{(\theta K)^k}{k!}$$
$$=\frac{(\theta K)^n}{(n+1)!}\left[\sum_{k=n}^{\infty}\frac{(\theta K)^{k+1}}{k!}-(n+1)\sum_{k=n+1}^{\infty}\frac{(\theta K)^k}{k!}\right]>0.$$

6.7 Differentiating (6.25) with respect to T,

$$[(c_K-c_T)r'(T)+c_M h'(T)]\sum_{n=0}^{\infty}G^{(n)}(K)\int_0^T[F^{(n)}(t)-F^{(n+1)}(t)]\mathrm{d}t.$$

6.8 From [17, p. 62], the probability that the system is replaced after failure is

$$P_n=\sum_{k=0}^{N}\binom{n}{k}(-1)^k p^{n-k}\sum_{i=0}^{k}\binom{k}{i}(-1)^i\frac{1}{1-q^{n-i}},$$

the probability that it is replaced before failure, i.e., the total number of failed units is $N+1,N+2,\ldots,n-1$, is

$$P_N=\sum_{k=N+1}^{n-1}\binom{n}{k}(-1)^k p^{n-k}\sum_{i=0}^{k}\binom{k}{i}(-1)^i\frac{1}{1-q^{n-i}},$$

and the mean time to replacement is

$$l_{N+1}=\mu\sum_{k=0}^{N}\binom{n}{k}(-1)^k\sum_{i=0}^{N-k}\binom{n-k}{i}\frac{1}{1-q^{n-i}}\quad(N=0,1,2,\ldots,n-1).$$

Thus, the expected cost rate is

$$C(N)\equiv\frac{c_nP_n+c_NP_N}{l_{N+1}}$$
$$=\frac{c_N+(c_n-c_N)\sum_{k=0}^{N}\binom{n}{k}(-1)^k p^{n-k}\sum_{i=0}^{k}\binom{k}{i}(-1)^i[1/(1-q^{n-i})]}{\mu\sum_{k=0}^{N}\binom{n}{k}(-1)^k\sum_{i=0}^{N-k}\binom{n-k}{i}[1/(1-q^{n-i})]}$$
$$(N=0,1,2,\ldots,n-1),$$

where c_n = replacement cost of a failed system and c_N = replacement cost before system failure with $c_N<c_n$. When the number n of units is given, we can obtain an optimum number n^* that minimizes $C(N)$.

6.10 Differentiating the left-hand side of (6.44) with respect to T,

$$r'(T)\sum_{n=0}^{\infty}G^{(n)}(K)\int_0^T[F^{(n)}(t)-F^{(n+1)}(t)]\mathrm{d}t.$$

From (6.48),

$$r_{N+2}\sum_{n=0}^{N}G^{(n)}(K)-[1-G^{(N+1)}(K)]-r_{N+1}\sum_{n=0}^{N-1}G^{(n)}(K)+[1-G^{(N)}(K)]$$
$$=(r_{N+2}-r_{N+1})\sum_{n=0}^{N}G^{(n)}(K).$$

6.11 When $c_T=c_Z$ and shocks occur in a Poisson process, i.e., $F^{(n)}(t)=\sum_{j=n}^{\infty}[(\lambda t)^j/j!]\mathrm{e}^{-\lambda t}$, from (6.40),

$$C_5(T,Z)\equiv\lim_{N\to\infty}C(T,N,Z)$$
$$=\frac{c_T+(c_K-c_T)\sum_{n=0}^{\infty}F^{(n+1)}(T)\int_0^Z\overline{G}(K-x)\mathrm{d}G^{(n)}(x)}{(1/\lambda)\sum_{n=0}^{\infty}F^{(n+1)}(T)G^{(n)}(Z)}.$$

Differentiating $C_5(T,Z)$ with respect to T and setting it equal to zero,

$$Q(T,Z)\sum_{n=0}^{\infty}F^{(n+1)}(T)G^{(n)}(Z)$$
$$-\sum_{n=0}^{\infty}F^{(n+1)}(T)\int_0^Z\overline{G}(K-x)\mathrm{d}G^{(n)}(x)=\frac{c_T}{c_K-c_T}, \tag{B.8}$$

and differentiating $C_5(T,Z)$ with respect to Z and setting it equal to zero,

$$\overline{G}(K-Z)\sum_{n=0}^{\infty}F^{(n+1)}(T)G^{(n)}(Z)$$
$$-\sum_{n=0}^{\infty}F^{(n+1)}(T)\int_0^Z\overline{G}(K-x)\mathrm{d}G^{(n)}(x)=\frac{c_T}{c_K-c_T}, \tag{B.9}$$

where

$$Q(T,Z)\equiv\frac{\sum_{n=0}^{\infty}[F^{(n)}(T)-F^{(n+1)}(T)]\int_0^Z\overline{G}(K-x)\mathrm{d}G^{(n)}(x)}{\sum_{n=0}^{\infty}[F^{(n)}(T)-F^{(n+1)}(T)]G^{(n)}(Z)}.$$

Thus, there does not exist both T^* $(0<T^*<\infty)$ and Z^* $(0<Z^*<K)$ that satisfy (B.8) and (B.9) simultaneously, because $Q(T,Z)<\overline{G}(K-Z)$. As further studies, when $c_Z>c_T$, obtain both optimum T^* and Z^* that minimize

$$C_5(T,Z) = \frac{\begin{array}{l} c_K - (c_K - c_T)\sum_{n=0}^{\infty}[F^{(n)}(T) - F^{(n+1)}(T)]G^{(n)}(Z) \\ - (c_K - c_Z)\sum_{n=0}^{\infty} F^{(n+1)}(T) \\ \times \int_0^Z [G(K-x) - G(Z-x)]\mathrm{d}G^{(n)}(x) \end{array}}{\sum_{n=0}^{\infty} G^{(n)}(Z) \int_0^T [F^{(n)}(t) - F^{(n+1)}(t)]\mathrm{d}t}.$$

6.13 Suppose that the unit is replaced at time T or at shock number N, whichever occurs first [17, p. 90]. The probability that the unit is replaced at time T is $\sum_{n=0}^{N-1} F^{(n)}(T)$, the probability that it is replaced at shock number N is $\sum_{n=N}^{\infty} F^{(n)}(T)$, and the expected number of minimal maintenance is

$$\begin{aligned} &\sum_{n=1}^{N-1}[F^{(n)}(T) - F^{(n+1)}(T)]\sum_{i=1}^{n}[1 - G^{(i)}(K)] \\ &\quad + \sum_{n=N}^{\infty}[F^{(n)}(T) - F^{(n+1)}(T)]\sum_{i=1}^{N-1}[1 - G^{(i)}(K)] \\ &= \sum_{n=1}^{N-1}[1 - G^{(n)}(K)]F^{(n)}(T). \end{aligned}$$

The mean time to replacement is

$$T[1 - F^{(N)}(T)] + \int_0^T t\mathrm{d}F^{(N)}(t) = \int_0^T [1 - F^{(N)}(t)]\mathrm{d}t.$$

Thus, the expected cost rate is

$$C(T,N) = \frac{c_N + (c_T - c_N)\sum_{n=1}^{N-1} F^{(n)}(T) + c_M \sum_{n=1}^{N-1}[1 - G^{(n)}(K)]F^{(n)}(T)}{\int_0^T [1 - F^{(N)}(t)]\mathrm{d}t}.$$

Chapter 7

7.1

$$\int_{-\infty}^{\infty} f(x;t)\mathrm{d}x = \mathrm{e}^{s^2\sigma^2 t/2} \int_{-\infty}^{\infty} \frac{1}{\sqrt{2\pi t}\sigma}\exp\left[-\frac{(x + s\sigma^2 t)^2}{2\sigma^2 t}\right]\mathrm{d}x = \mathrm{e}^{s^2\sigma^2 t/2}.$$

7.2 When $t = 5$, $R(5) = \Phi((10-5)/\sqrt{5}) = 0.987$, and

$$\mathrm{MTTF} = \int_0^\infty \Phi\left(\frac{10-t}{\sqrt{t}}\right)\mathrm{d}t = \int_0^\infty \left[\frac{1}{\sqrt{2\pi t}}\int_0^t \mathrm{e}^{-(10-x)^2/2\sqrt{x}}\mathrm{d}x\right]\mathrm{d}t \approx 10.5.$$

When $t = 10$, $R(10) = \Phi(0) = 0.500$.

7.3 Letting $L(T)$ be the left-hand side of (7.10), where $A \equiv \mathrm{e}^{\mu+\sigma^2/2}$, $L(0) = 0$, $L(\infty) = \infty$ and

$$L'(T) = A^2 T \mathrm{e}^{AT} > 0, \quad C(T^*) = c_1 A \mathrm{e}^{AT^*}.$$

7.4 Using $\mathrm{e}^a \approx 1 + a + a^2/2$, from (7.9),

$$\widetilde{C}(T) = c_1\left(\mu + \frac{\sigma^2}{2}\right)\left[1 + \frac{(\mu+\sigma^2/2)T}{2}\right] + \frac{c_2}{T}.$$

Clearly, $\widetilde{C}(0) = \widetilde{C}(\infty) = \infty$. Differentiating $\widetilde{C}(T)$ with respect to T and setting it equal to zero,

$$\widetilde{T} = \frac{1}{\mu + \sigma^2/2}\sqrt{\frac{2c_2}{c_1}}$$

Compute $\widetilde{T}$ and T^* from (7.10), and compare them, and assure that $T^* < \widetilde{T}$, because for $T > 0$ and $\mu + \sigma^2/2 > 1$,

$$AT\mathrm{e}^{AT} - (\mathrm{e}^{AT} - 1) > \frac{(AT)^2}{2}.$$

7.5 Letting $L(T)$ be the left-hand side of (7.15),

$$L(0) = 0, \quad L(\infty) = \mu Q_1(\infty)[1 + M_G(K)] - 1,$$

$$L'(T) = Q_1'(T)\sum_{k=0}^{\infty} G^{(k)}(K)\int_0^T [F^{(k)}(t) - F^{(k+1)}(t)]\mathrm{d}t.$$

7.6 Letting $L(N)$ be the left-hand side of (7.18),

$$L(\infty) = Q_2(\infty)[1 + M_{G_T}(K)] - 1,$$

$$L(N+1) - L(N) = [Q_2(N+2) - Q_2(N+1)]\sum_{n=0}^{N} G_T^{(n)}(K) > 0.$$

7.7 Letting $L(T)$ be the left-hand side of (7.21),

$$L(0) = 0, \quad L(\infty) = \infty,$$

$$L'(T) = r_A'(K/T)\int_0^T L_A(K/t)\mathrm{d}t.$$

7.8 (7.26) and (7.27) are rewritten as, respectively,

$$C(T) = \frac{c_K - (c_K - c_T)\{1 - \exp[-(K - \omega T)^\alpha/\sigma\sqrt{T}]\}}{\int_0^T \{1 - \exp[-(K - \omega t)^\alpha/\sigma\sqrt{t}]\}\mathrm{d}t},$$

$$\frac{(K - \omega T)^{\alpha-1}[K + (2\alpha - 1)\omega T]}{2\sigma T\sqrt{T}} \frac{\exp[-(K - \omega T)^\alpha/\sigma\sqrt{T}]}{1 - \exp[-(K - \omega T)^\alpha/\sigma\sqrt{T}]}$$

$$\times \int_0^T \{1 - \exp[-(K - \omega t)^\alpha/\sigma\sqrt{t}]\}\mathrm{d}t - \exp[-(K - \omega T)^\alpha/\sigma\sqrt{T}]$$

$$= \frac{1}{c_K/c_T - 1}.$$

7.9 From (7.29), $C_2(0) = \infty$ and

$$C_2(\infty) = \frac{c_K}{\int_0^\infty \exp[-(\omega t - K)/\sigma\sqrt{t}]\mathrm{d}t}.$$

Differentiating $C_2(T)$ with respect to T and setting it equal to zero,

$$\frac{K + \omega T}{2\sigma T\sqrt{T}} \int_0^T \exp[-(\omega t - K)/\sigma\sqrt{t}]\mathrm{d}t + \exp[-(\omega T - K)/\sigma\sqrt{T}] = \frac{c_K}{c_K - c_T},$$

whose left-hand decreases strictly to 0. Hence, $T^* = \infty$.

Chapter 8

8.1 When $c_r(t) = at$, (8.12) is [18, p. 53]

$$C(T) = \frac{c_1\overline{G}(T) + a\int_0^T \overline{G}(t)\mathrm{d}t}{\mu + \int_0^T \overline{G}(t)\mathrm{d}t}.$$

If $\int_0^\infty \overline{G}(t)\mathrm{d}t \equiv \beta$ then $C(\infty) = a\beta/(\mu + \beta)$ and $C(0)=c_1/\mu$. Let $r(t) \equiv g(t)/\overline{G}(t)$, where $g(t)$ is a density function of $G(t)$. Differentiating $C(T)$ with respect to T and setting it equal to zero,

$$r(T)\left[\mu + \int_0^T \overline{G}(t)\mathrm{d}t\right] + \overline{G}(T) = \frac{a\mu}{c_1}. \tag{B.10}$$

If $r(t)$ decreases strictly then the left-hand side of (B.10) decreases from $\mu r(0)+1$ to $(\mu+\beta)r(\infty)$. Therefore, we have the following optimum policy where

$$k \equiv \frac{a\mu - c_1}{c_1\mu}, \quad K \equiv \frac{a\mu}{c_1(\mu+\beta)}:$$

(i) If $r(0) \le k$ then $T^* = 0$.
(ii) If $r(0) > k$ and $r(\infty) < K$ then there exists a finite and unique $T^*(0 < T^* < \infty)$ that satisfies (B.10).
(iii) If $r(\infty) \ge K$ then $T^* = \infty$.

8.2

$$P'_{00}(t) = -(\lambda + \lambda s)P_{00}(t) + \theta P_{01}(t),$$
$$P'_{01}(t) = (\lambda + \lambda s)P_{00}(t) - (\lambda + \theta)P_{01}(t) + \theta P_{02}(t),$$
$$P'_{02}(t) = \lambda P_{01}(t) - \theta P_{02}(t).$$

Taking the LS transforms of the above equations, solving and inverting them,

$$P_{00}(t) = \frac{\theta^2}{(\lambda+\theta)(\lambda+\lambda s)+\theta^2} + \frac{\lambda+\lambda_s}{\omega_1-\omega_2}\left[\frac{\omega_1-\lambda-\theta}{\omega_1}e^{-\omega_1 t} - \frac{\omega_2-\lambda-\theta}{\omega_2}e^{-\omega_2 t}\right],$$

$$P_{01}(t) = \frac{\theta(\lambda+\lambda s)}{(\lambda+\theta)(\lambda+\lambda s)+\theta^2} + \frac{\lambda+\lambda s}{\omega_1-\omega_2}\left[\frac{-\omega_1+\theta}{\omega_1}e^{-\omega_1 t} - \frac{-\omega_2+\theta}{\omega_2}e^{-\omega_2 t}\right],$$

$$P_{02}(t) = \frac{\lambda(\lambda+\lambda s)}{(\lambda+\theta)(\lambda+\lambda s)+\theta^2} + \frac{\lambda(\lambda+\lambda s)}{\omega_1-\omega_2}\left[\frac{1}{\omega_1}e^{-\omega_1 t} - \frac{1}{\omega_2}e^{-\omega_2 t}\right],$$

where

$$\omega_1 \equiv \frac{1}{2}[2\lambda + 2\theta + \lambda s + \sqrt{4\lambda\theta + \lambda_s^2}],$$
$$\omega_2 \equiv \frac{1}{2}[2\lambda + 2\theta + \lambda s - \sqrt{4\lambda\theta + \lambda_s^2}].$$

8.3

State 0: One unit is operating and the other is on standby.
State 1: One unit is operating and the other is under repair.
State 2: One unit is operating and the other is under PM.
State 3: One unit is under repair or PM and the other waits for repair.

From [18, p. 149], the limiting probabilities P_j $(j = 0, 1, 2, 3)$ of transition probabilities are

$$P_0 = \frac{1}{\gamma_{11}} \int_0^T G_1(t)\overline{F}(t)\mathrm{d}t + \frac{1}{\gamma_{22}} \int_0^T G_2(t)\overline{F}(t)\mathrm{d}t,$$

$$P_j = \frac{1}{\gamma_{jj}} \int_0^T \overline{G}_j(t)\overline{F}(t)\mathrm{d}t \quad (j = 1, 2),$$

$$P_3 = \frac{1}{\gamma_{11}} \int_0^T \overline{G}_1(t)F(t)\mathrm{d}t + \frac{1}{\gamma_{22}} \int_0^T \overline{G}_2(t)F(t)\mathrm{d}t,$$

where

$$\gamma_{11} \equiv \beta_1 + \int_0^T G_1(t)\overline{F}(t)\mathrm{d}t + \left[\beta_2 + \int_0^T G_2(t)\overline{F}(t)\mathrm{d}t\right] \frac{\int_T^\infty G_1(t)\mathrm{d}F(t)}{1 - \int_T^\infty G_2(t)\mathrm{d}F(t)},$$

$$\gamma_{22} \equiv \beta_2 + \int_0^T G_2(t)\overline{F}(t)\mathrm{d}t + \left[\beta_1 + \int_0^T G_1(t)\overline{F}(t)\mathrm{d}t\right] \frac{1 - \int_T^\infty G_2(t)\mathrm{d}F(t)}{\int_T^\infty G_1(t)\mathrm{d}F(t)}.$$

The availability is

$$A(T) \equiv P_0 + P_1 + P_2$$
$$= \frac{\left[\gamma_1 + \int_0^T G_1(t)\overline{F}(t)\mathrm{d}t\right]\left[1 - \int_T^\infty G_2(t)\mathrm{d}F(t)\right] + \left[\gamma_2 + \int_0^T G_2(t)\overline{F}(t)\mathrm{d}t\right]\int_T^\infty G_1(t)\mathrm{d}F(t)}{\left[\beta_1 + \int_0^T G_1(t)\overline{F}(t)\mathrm{d}t\right]\left[1 - \int_T^\infty G_2(t)\mathrm{d}F(t)\right] + \left[\beta_2 + \int_0^T G_2(t)\overline{F}(t)\mathrm{d}t\right]\int_T^\infty G_1(t)\mathrm{d}F(t)},$$

where

$$\gamma_i \equiv \int_0^\infty \overline{G}_i(t)\overline{F}(t)\mathrm{d}t, \quad q_i \equiv \int_0^\infty \overline{G}_i(t)\mathrm{d}F(t) \quad (i = 1, 2).$$

Let

$$Q(t) \equiv \frac{q_1 G_2(t) - q_2 G_1(t)}{q_1 - q_2}, \quad k \equiv \frac{q_2}{q_1\gamma_2 - q_2\gamma_1}, \quad K \equiv \frac{q_1}{\mu(q_1 - q_2)}.$$

Suppose that $G_1(t) < G_2(t)$ for $0 < t < \infty$ and the failure rate increases strictly.

(i) If $h(\infty) > K, q_1\gamma_2 > q_2\gamma_1$, and $h(0) < k$ or $h(\infty) > K$ and $q_1\gamma_2 \le q_2\gamma_1$, then there exists a finite and unique $T^*(0 < T^* < \infty)$ that satisfies

$$h(T)\left[\int_0^T Q(t)\overline{F}(t)\mathrm{d}t + \int_0^\infty \overline{Q}(t)\overline{F}(t)\mathrm{d}t\right] - \int_0^T Q(t)\mathrm{d}F(t)$$
$$= \frac{q_1\int_0^\infty \overline{Q}(t)\mathrm{d}F(t) + q_2\int_0^T Q(t)\mathrm{d}F(t)}{q_1 - q_2}.$$

(ii) If $h(\infty) < K$ then $T^* = \infty$.

(iii) If $q_1\gamma_2 > q_2\gamma_1$ and $h(0) > k$, then $T^* = 0$.

8.4 When $G(t) = 1 - \mathrm{e}^{-\theta t}$,

$$l^*(z) = \frac{\mu z/(1-z) + (1/\theta)\sum_{n=1}^\infty z^n[1 - F_S^*(\theta)^j]}{1 - [1 - F_S^*(\theta)]/[1 - zF_S^*(\theta)]}$$
$$= \left(\mu + \frac{1}{\theta}\right)\frac{1 - F_S^*(\theta)}{F_S^*(\theta)}\sum_{n=1}^\infty nz^n + \mu\sum_{n=1}^\infty z^n,$$
$$m^*(z) = \frac{z/(1-z)}{1 - [1 - F_S^*(\theta)]/[1 - zF_S^*(\theta)]}$$
$$= \frac{1 - F_S^*(\theta)}{F_S^*(\theta)}\sum_{n=1}^\infty nz^n + \sum_{n=1}^\infty z^n,$$
$$M = \frac{1}{\mu + 1/\theta}M_S^*(\theta) = \frac{1}{\mu + 1/\theta}\frac{F_S^*(\theta)}{1 - F_S^*(\theta)}.$$

8.5 From [19], the total expected cost of one cycle from the beginning of the operation of the main unit to repair or PM completion is

$$\overline{F}(T)\left[c_2 + c_S\int_0^\infty M_S(t)\mathrm{d}G(t)\right] + F(T)\left[c_1 + c_S\int_0^\infty M_S(t)\mathrm{d}G(t)\right],$$

and the mean time of one cycle is

$$\overline{F}(T)(\mu + T) + F(T)\mu + \int_0^T t\mathrm{d}F(t) = \mu + \int_0^T \overline{F}(t)\mathrm{d}t,$$

where $M_S(t) \equiv \sum_{n=1}^\infty F_S^{(n)}(t)$. The expected cost rate is

$$C(T) = \frac{\widetilde{c}_1 F(T) + \widetilde{c}_2\overline{F}(T)}{\mu + \int_0^T \overline{F}(t)\mathrm{d}t},$$

where

$$\widetilde{c}_1 \equiv c_1 + c_S \int_0^\infty M_S(t)\mathrm{d}G(t), \quad \widetilde{c}_2 \equiv c_2 + c_S \int_0^\infty M_S(t)\mathrm{d}G(t).$$

Using the results in Sect. 3.3.2.1, we can obtain an optimum policy.

References

1. Osaki S (1992) Applied stochastic system modeling. Springer, Berlin
2. Tijms HC (2003) A first course in stochastic models. Wiley, Chichester
3. Beichelt FE, Fatti LP (2002) Stochastic process and their applications. CRC Press, Boca Raton
4. Rausand M, Høyland A (2004) System reliability theory. Wiley, Hoboken
5. Bhat UN, Miller GK (2002) Elements of applied stochastic processes. Wiley, Hoboken
6. Feller W (1968) An introduction to probability theory and its application, vol I, 3rd edn. Wiley, New York
7. Gnedenko B, Ushakov I (1995) Probabilistic reliability engineering. Wiley, New York
8. Levitin L, Lisnianski A (2003) Multi-state systems reliability analysis and optimization. In: Pham H (ed) Handbook of reliability engineering. Springer, London
9. Levitin G (2005) The universal generating function in reliability analysis and optimization, Springer, London
10. Çinlar E (1975) Introduction to stochastic processes. Prentice-Hall, Englewood Cliffs
11. Ross SM (1983) Stochastic processes. Wiley, New York
12. Yun WY, Nakagawa T (2010) Note on MTTF of a parallel system. In: 16th ISSAT International conference on reliability and quality in design, pp 235–237
13. Nakagawa T (2008) Advanced reliability models and maintenance policies. Springer, London
14. Chen M, Mizutani S, Nakagawa T (2010) Optimal backward and backup policies in reliability theory. J Oper Res Soc Jpn 53:101–118
15. Prabhu NU (2007) Stochastic processes. World Scientific, Singapore
16. Barlow RE, Proschan F (1965) Mathematical theory of reliability. Wiley, New York
17. Nakagawa T (2007) Shock and damage models in reliability theory. Springer, London
18. Nakagawa T (2005) Maintenance theory of reliability. Springer, London
19. Nakagawa T, Osaki S (1976) Reliability analysis of a one-unit system with unrepairable spare units and it optimization applications. Oper Res Q 27:1019–110

Index

MIX
Papier aus verantwortungsvollen Quellen
Paper from responsible sources
FSC® C105338

If you have any concerns about our products,
you can contact us on
ProductSafety@springernature.com

In case Publisher is established outside the EU,
the EU authorized representative is:
Springer Nature Customer Service Center GmbH
Europaplatz 3, 69115 Heidelberg, Germany

Printed by Libri Plureos GmbH
in Hamburg, Germany